Cemetery and Related Records

Smartsville

Cemeteries

Yuba County, California

1852-2001

Compiled by
Renie Riccobuano

HERITAGE BOOKS
2006

HERITAGE BOOKS
AN IMPRINT OF HERITAGE BOOKS, INC.

Books, CDs, and more—Worldwide

For our listing of thousands of titles see our website
at
www.HeritageBooks.com

Published 2006 by
HERITAGE BOOKS, INC.
Publishing Division
65 East Main Street
Westminster, Maryland 21157-5026

Copyright © 2005 Renie Riccobuano

All rights reserved. No part of this book may be reproduced or transmitted in any form or by any means, electronic or mechanical, including photocopying, recording or by any information storage and retrieval system without written permission from the author, except for the inclusion of brief quotations in a review.

International Standard Book Number: 978-1-58549-919-6

Table of Contents

ACKNOWLEDGMENTS ... III
CEMETERY MAP ... V
INTRODUCTION ... VI
MASONIC CEMETERY LEGENDS ... 1
CATHOLIC CEMETERY LEGENDS ... 15
TIMBUCTOO CEMETERY LEGENDS ... 25
PIONEER CEMETERY LEGENDS AKA DAVIES CEMETERY ... 29
NEWSPAPER INFORMATION ... 31
BIRTH NOTICES ... 177
MARRIAGE NOTICES ... 179
DEATH NOTICES ... 185
APPENDIX ... 211
INDEX ... 213
COMBINED INFORMATION ANALYSES ... 249

Acknowledgments

I wish to thank my sweetheart and husband of twenty plus years for actively supporting me in this venture. He has helped me find cemeteries that have been forgotten and overgrown, without complaint.

I would like to thank Laura Nicholson, Editor of the *Appeal-Democrat Newspaper,* for giving me permission to use information from the *Marysville Weekly Appeal, Marysville Daily Appeal, Marysville Appeal, Marysville Democrat and The Appeal-Democrat.* The *Appeal-Democrat's* contribution to this effort has been of great value.

I would also like to thank the Yuba County Library and their staff for putting up with the hassel of shuffling me back and forth to the reference room and the microfilm readers.

Cemetery Map

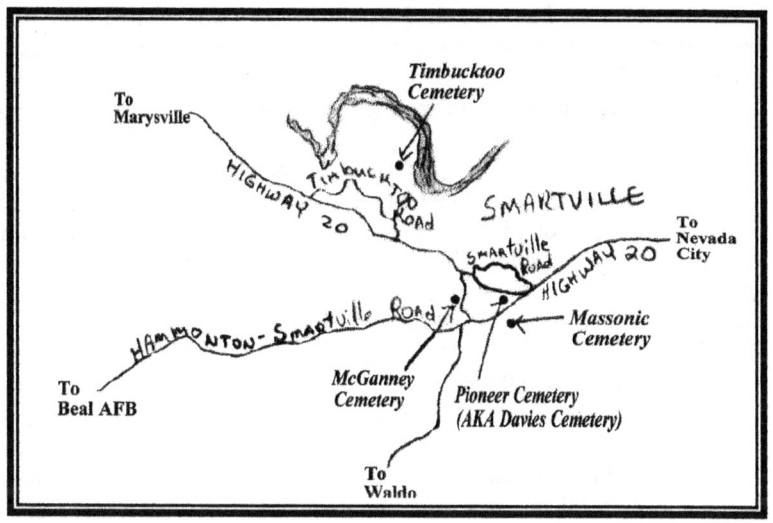

Smartsville Cemeteries

Masonic Cemetery (AKA Fraternal Cemetery)
McGanney Cemetery (AKA Catholic Cemetery)
Pioneer Cemetery (AKA Davies Cemetery)
Timbuctoo Cemetery

Introduction

Miners looking for gold settled this area of Yuba County and most people think that Jonas Spect first discovered gold about eighteen miles from Marysville on Jun 2, 1848. Small township's were settled on or near sand bars on the river and this first township was later known as Rose Bar, named after John Rose who started the first store at the bar. With this discovery of gold an influx of men willing to work hard came into Rose Bar and other nearby areas.

A man by the name of William Smart founded the town of Smartsville, when he built a hotel on the upper bank, overlooking the Yuba River in the spring of 1856. A few miners also occupied some cabins in this area near Smartsville and along the Yuba River, although Timbuctoo and Sucker Flat were the only other large settlements at this time. Soon there were several small mining camps along the Yuba River such as Rose Bar, Saw-Mill Bar, Landers Bar, Kennebec Bar, Sand Hill, Money Flat, Sucker Flat and Empire Ranch.

Hydraulic mining proved to be more profitable than panning for gold and to mine this way you needed a large amount of water, which was brought to the mines through a series of ditches. The water then was used to wash away a hillside to expose the gold and the debris washed down into the river ravine. As a result many of these small mining towns and bars were buried under the debis from the Smartsville and Timbuctoo Hydraulic Mining Co. The resulting flood of water and debris made it necessary for the miners to move up from the "Bars" at the riverbeds into the surrounding communities of Timbuctoo and Smartsville. All of the graves in these small river "bars" and what records that were kept, were buried under this massive amount of debris.

The Smartsville name stayed with the town until 1909, when the post office renamed it without the middle "s". For further history see the newspaper article in this work dated 23 Jan 1960.

The cemeteries in the Smartsville area are located off Highway 20, approximately 18 miles from Marysville, going northeast toward Grass Valley. There are a total of four cemeteries ranging from 14 headstones in the oldest to over 300 in the ones that are still active, with a total of approximately 985 individual graves in this book. In doing the research for this book I found and included over 200 names of individuals that were indicated as being buried in these cemeteries, yet have no headstones.

The Davies Cemetery was first used in 1852, for the burial of a man from Oregon. This was followed by the burial of several men who had cholera. In 1876, a mine caved in at Sucker Flat and killed seven men who were all buried the same day.

Up the road from the Davies Cemetery is the Masonic Cemetery, laid out by the Masons, Odd Fellows, and Good Templars, in 1875. It was originally called the Union, Smartsville, and Empire Cemetery, then the Fraternal Cemetery and now is called the Masonic Cemetery.

The Catholics in the towns surrounding Smartsville were taken to the Smartsville Catholic Cemetery until a cemetery was laid out in Marysville, in about 1879. It is said that Daniel McGanney donated the land for the Smartsville Catholic Cemetery and today it is also know as the McGanney Cemetery. There are a large number of young children and young women buried in this cemetery and the old timers say that it was because of the unsanitary conditions in the area.

Across Highway 20 and down the road is the Timbuctoo Cemetery. To get into this cemetery you have to hike or have a four-wheel drive vehicle. This was the first cemetery in this area and was started in 1855, when three men, who had been murdered by Jim Webster, were buried in this cemetery in Timbuctoo.

This book was finished in 2002 and as two of these cemeteries are active, any headstones added since this date will not be included in this project. Throughout this book I have typed all of the surnames in **bold** for ease in finding individuals. Wherever possible I used genealogical abreviations such as yrs for years, mos for months, and dys for days. I also used Jan for January, etc.

In this project I kept the cemeteries grouped in their separate sections so that you would know where each individual is located. Within these cemetery groups, the names and information on a headstone and in a family plot are grouped in paragraphs together. When information is at the bottom of a paragraph it is for everyone on the headstone or in the family plot. When the information is written beside the name or directly under just that name, it is for that individual only. In doing this research I found names of several individuals that were buried in these cemeteries, but did not have headstones. These names are entered under, "Research produced these individuals buried here".

The *Marysville Appeal - Democrat* is the newspaper that I used primarily in this work, as it was the newspaper most often used in Yuba County and has information on most of the individuals. When there was more information in another newspaper and I used that quote, I indicated so in the line next to the date. There were some instances where I used quotes from two sources, as it gave you more information about an individual. The information from the newspapers includes obituaries, marriages, and family histories. Some of the individuals in this cemetery were pioneers and their family histories were printed in the newspaper at a later date than their death notices. So that I did not miss any information that you might be looking for, I included all information about the individual. If you would like some history of this area, look in the newspaper section under 23 Jan 1960; this was the newspaper's Bicentenial Edition and gives information about this area.

The old newspapers are on microfilm at the county library, and for a fee you can have the library send you a copy of the information. (Address is in appendix) The *Marysville Appeal - Democrat Newspaper* started out as several small newspapers in this area including the *Marysville Weekly Appeal*, *The Marysville Daily Appeal*, *The Marysville Appeal* and *The Marysville Democrat*. It was combined into one paper and is now called the *Marysville Appeal - Democrat Newspaper*.

I researched information from the courthouses in Yuba County and Sutter Coounty. These searches were for birth, death, and marriage information. Some of the newer records give you more information than I have included in this book. Complete copies of these records can be obtained from the courthouses and their addresses are in the appendix. In the Death Notice section, I listed the state if the person did not die in California.

The appendix has the addresses, phone numbers and hours of operation of the places where my research was done. You can obtain copies of the originals by using these addresses as of 2003. Pictures of the headstones can also be obtained, address to write for these is in the appendix.

The Index is an all inclusive index listing every name in this book. The only names that I left out of this index are Reverends, Ministers, Justice of the Peace and Mortuaries mentioned in the text of records that I researched. I only used the first name or initial of an individual in the Index, to keep the index as short and complete as possible. So if you are looking for Ellen R. Moody, you will find her in the index under Ellen Moody, but in the text the actual whole name will be written.

In the back of the book is a "Combined Information Analyses" section that shows family connections and condensed information. This is a quick reference for every deceased individual in this book. The information that was used to make these conclusions is in this book, and can be looked up by using the index. The P beside the name means that you can get a picture of a headstone. (See appendix). The abbreviation in the Relationship section means: W (Wife of), H(Husband of), D(Daughter of), S(Son of), G(Grandparent of). I placed this section at the back of the book for ease of searching your individual family members.

Renie Riccobuano

Masonic Cemetery Legends

Welch, James   1861 – 1923   Father N.S.G.W.
Welch, Lillie E.   1868 – 1938   Mother at Rest Rainbow
Welch, James C.   1866 – 1953

Welch, James C.   9 Dec 1886 – 24 Jan 1953   CA PFC Med. Dept. WWI

Welch, Chester A.   14 Apr 1936 – 20 Jan 1937   Son of N. & Katherine **Welch**

Welch, Winifred H.   9 Apr 1914 - 25 Jun 1964
Welch, Clarence R. Sr.   13 Jan 1896 – 2 Jun 1970
      Together Forever

Welch, James R.   1932 – 1974   PFC US Army

Welch, Clarence R. Jr.   16 Mar 1934 – 22 Dec 1967

Taylor, Edward Roy   1942 – 1975
Welch, Georgina Louise   1924 – 1985
      Loving Mother and Son

Brunckhorst, Elmer W.   1912 – 1979   MOMMI US COAST GUARD

Brunckhorst, Fred James   2 Oct 1879 – 24 Nov 1972
Brunckhorst, Oscar   31 Oct 1914 – 26 May 1956

Puff, Ethel W.   6 Jul 1907
Puff, Edmond J.   18 Sep 1905 – 8 Nov 1969

Bitner, Captain Cyrus C.   1836 – 1917
Bitner, Mary   1839 – 1920
Austin, Ella Bitner   21 Jul 1858 – 4 Mar 1943
Austin, Bert C.   Age 79
Bitner, John   2 Sep 1831 – 2 Jul 1871

Anderson, August   1863 – 1942
Anderson, Emma   1865 – 1931

Gann, Wayne J.   18 Dec 1916 – 6 Nov 2000
Gann, Gladys D.   19 Apr 1917 – 23 Aug 1994
    Beloved Wife, Mother, & "Nana"   He saw through all of us, She saw the beauty in all of us

Campbell, Alan McKenzie   16 Jun 1947 – 25 Apr 1991
      You were the wind beneath our wings

Fisher, Grace E.   *Died* 13 Nov 1894   *Aged* 11 yrs
      A precious one from us has gone; a voice we loved is stilled
      A place is vacant in our home, which never can be filled

Brown, Elizabeth A.   *Died* 10 Sep 1908   *Aged* 87 yrs 4 mos   Our Mother

MASONIC CEMETERY LEGENDS

Fisher, Samuel   1839 – 1915
Fisher, Rose   1856 – 1945
Daniel, Elizabeth   *Died* 9 Aug 1868   *Aged* 95 yrs   Died at Mooney Flat
Otis, Walter B.   *Died* 17 Dec 1866   *Aged* 19 yrs   Died at Mooney Flat
Otis, Mrs. T. P.   *Died* 18 May 1867   *Aged* 34 yrs   Died at Mooney Flat
Otis, Isaac B.   *Died* 6 Jan 1853   *Aged* 40 yrs   Died at Mooney Flat
Otis, Elizabeth   *Age* 3 yrs
Otis, Eugene W.   *Age* 8 yrs
    Children of Isaac B. & ? **Otis**
Otis, M.   A footstone only

Chamberlin, W. W.   No dates given

Reese, Kyle Lawrence   1 Sep 1992 – 12 Sep 1992   God Bless

Carney, Cora May   1890   Loving Wife
Carney, William R.   1886 – 1981   Loving Husband
    We who believe shall have everlasting life

Carney, Richard William   19 Sep 1947 – 28 Nov 1970 CA FTG# US Navy Vietnam

Fiene, Henry   25 Mar 1825 – 17 Jul 1908   Native of Germany
Fiene, Annie Margaret   12 Jul 1832 – 22 Aug 1912

Dooley, Louisa A.   *Died* 11 Nov 1910   *Aged* 32 yrs 4 mos   Wife of James **Dooley**

Manford, Dorothy E.   17 Nov 1911 – 23 Sep 1982
Manford, James B.   2 Feb 1904 – 15 Jan 1985
    Together Forever

Poor, Mark E.   1862 – 1940
Poor, Emma   1872 – 1936
Poor, Mark Lewellyn   1892 – 1923 Mark, wounded Sep 28, 1918, N. W. of Vekaan, France
Havey, Mary Emma   1803 – 1828

Lessley, Betty   1942 – 1983   Mom

Greever, Jack W.   22 Nov 1943 – 22 Jul 1998   Beloved Husband & Dad

Greever, Aldon Lee   15 Feb 1919 – 4 Mar 1998
Greever, Maxine Minnie 11 Jan 1920 – 19 Apr 1998
    Married 58 yrs, in life, together forever in Heaven

Greever, Kyle A.   21 Aug 1893 – 26 Sep 1981
Greever, Hazelle C.   23 Jun 1895 – 17 Mar 1974

Singleton, Pete   No dates given
Singleton, Billie P.   1913 – 1978

Fitzhugh, James M.   5 Jan 1880 – 4 Jul 1967
Fitzhugh, Grace O.   18 Sep 1889 – 31 Oct 1969
Fitzhugh, Olive   No dates given

MASONIC CEMETERY LEGENDS

Davis, Adeline C.   30 Jun 1883 – 7 Jun 1978
Davis, Bert Benjamin   19 May 1890 – 25 Nov 1968   Montana
     PVT FORS REPL BN 20 ENGRS WWI

Winegar, Marvin E.   21 Jul 1939 – 1 Jun 1993
Winegar, Petra M. M.   11 Jan 1953
     Together Forever   Acts 2:21

Winegar, Madge A.   1919 – 2001

Winegar, Patricia M.   29 Feb 1944 – 23 Aug 1994

Whitmore, Alice E.   1909 – 1950   Wife

Whitmore, Jesse F.   1891 – 1958   Husband

Hinkle, Noble F.   1 Jun 1911 – 13 Feb 1965   California
     PVT INF ROPLING CENTER WW II

Hinkle, Archie W.   1879 – 1944

Parkison, John   1884 – 1981
Parkison, Chloe Mae   1888 – 1950

Wheaton, Allan W.   1900 – 1966
Wheaton, Allen G.   No dates given
Wheaton, Evelyn   No dates given
Wheaton, Florence Esther   12 Jan 1903 – 14 Sep 1903
Wheaton, Lucille   No dates given

Fippin, Sidney Wallis   1934 – 1992

Spencer, Carl Allen   1917 – 1935

Vargas, M. Ginger   1918 – 1981
Webb, Sandy M.   1963 – 1980
     Loving Friends
Vargas, Albert Dee   24 Mar 1917 – 21 Mar 1987

Weaver, Marie Cora   23 Oct 1898 – 31 May 1978   Our song Stardust

Weaver, Warren W.   26 Apr 1898 – 30 Dec 1978   PVT US Army WWI

Tifft, Ray W.   *Died* 25 Feb 1893   *Aged* 61 yrs 4 mos
Tifft, Julia A.   22 Sep 1835 – 23 Jul 1918   Wife of R. W. **Tifft**
Tifft, Julia A.   1 Dec 1870 – 23 Jan 1909   Wife of S. A. **Davis**
Tifft, George W.   *Died* 13 May 1857   *Aged* 1 yr 2 dys
Tifft, Alice M.   *Died* 25 Dec 1887   *Aged* 14 yrs 3 mos

Toland, John   1872 – 1943
Toland, Robert   1903 – 1938
Toland, Elizabeth   1871 – 1939
Webb, Alice Toland   1910 – 1987

MASONIC CEMETERY LEGENDS

Toland, Joseph   3 Feb 1878 – 16 Sep 1880
Toland, Martha   22 Aug 1832 – 13 Jul 1916
Toland, Robert   1832 – 26 Apr 1905
Toland, William Joseph   18 Dec 1869 – 26 Jan 1918

Gilbert, Absalom A.   5 Jan 1890
Gilbert, Pauline Ella   17 Aug 1889 – 5 Aug 1975   Beloved Wife and Mother

Hermann, Carl   31 Mar 1851 – 14 May 1921
Hermann, Caroline   23 Apr 1852 – 14 Mar 1946

Hermann, Charles H.   22 Sep 1886 – 1 Nov 1974

Bennion, Richard B.   5 Jan 1913 – 2 Jun 1996   CAPT US Army WWII

Bennion, Margaret A. Walsh   10 Aug 1917 – 9 Aug 1997

Gilliam, Wm. Glenn   24 Sep 1913 – 22 May 1989   He brought sunshine to everyone
Gilliam, Dorothy   *Died* 2 May 1998

Bennion, Robert W.   25 Mar 1943 – 27 Sep 1998   Free at last

Beatty, Frances E.   1861 – 1896
Beatty, Robert   1855 – 1926
Beatty, Infant Son   of R. & F. E. **Beatty**
Beatty, Richard   1822 – 1904
Yankey, Roberta B. "Bobbie"   29 Jul 1932 – 24 Jun 1995   Wife: John S. **Yankey**, III
     Daughter: R. R. & Grace V. Manasco **Beatty**   Mother: Paul Daniel **Yankey**
Beatty, Richard Ray   1891 – 1963
Beatty, Grace V.   1883 – 1954

Ehmann, Paul   1864 – 1929   Father
Ehmann, Rosa   1867 – 1940   Mother
Binet, LaTosca   1901 – 1989   Daughter
Ehmann, George   1904 – 1992   Son

Bach, Mary A.   1855 – 1929
Bach, Katie   No dates given
Bach, Ethel M.   1879 – 1936
Bach, John E.   1869 – 1931
Bach, John J.   1 May 1914 – 16 Nov 1991   S.SGT US Army WW II
Bach, Alice Baby   No dates given
Bach, John   1830 – 28 Jan 1898
Bach, Mary Ann   18 Jun 1837 – 24 Dec 1918

Heyne, Bernard W.   17 May 1905 – 17 Jun 1982
Heyne, Ethel K.   24 Nov 1917 – 29 Aug 2001   Sunny

Saunders, Karen J.   30 Apr 1941 – 3 Oct 1987

Ray, Charles Vance   1889 – 1958
Ray, Lavinia Hapgood   1897 – 1996
LeBourveau, Ernest   1877 – 1972

MASONIC CEMETERY LEGENDS

Sanford, N. B.   1858 – 1943
Sanford, Emily T.   27 Jan 1870 – 21 Jan 1953
Sanford, Alfred B.   29 Nov 1869 – 16 Nov 1956
Jackson, M. H.   Died 15 Aug 1882   Aged 69 yrs
Jackson, William H.   3 Jun 1820 – 4 Sep 1901

Bowman, Alexander   28 Jun 1813 – 19 Feb 1883
    Native of Aberdeen Shire Scotland   How lonely it is without thee

Loomer, Maggie   10 Mar 1852 – 1 Feb 1881   Beloved wife of N. H. **Loomer**
Loomer, Henry   No dates given
Loomer, Sarah I.   23 Feb 1824 – 24 Sep 1881   Beloved wife of J. W. **Loomer**
Loomer, J. W.   1820 – 23 Nov 1884   A native of Nova Scotia   Aged 66 yrs

Walsh, John E.   1864 – 1934
Walsh, Annie Quick   1860 – 1923
    Our Father & Mother
Walsh, Walter Lucien   27 Jul 1890 – 24 Aug 1975   Son
Walsh, Sadie Kane   18 Aug 1892 – 27 May 1984   Wife

Woehler, Otto A. Sr.   1859 – 1924
Woehler, Otto A. Jr.   1896 – 1922

Murdock, Mary   1848 – 4 Jul 1942   94 yrs 8 mos 9 dys

Fenton, Lois   8 Oct 1858 – 5 May 1945

Wallace, Michael   1854 – 1921   Native of West Gore, Nova Scotia

Trevena, Nicholas   Died 24 Jun 1877   Aged 57 yrs 2 mos 11 dys   Fare Well
Trevena, Mrs.   No dates given   Wife of Nicholas Trevena, Sr.
Trevena, Nicholas Jr.   No dates given

Woodroffe, John S.   1838 – 1903
Vineyard, Sarah E. Woodroffe   1848 – 1928
Woodroffe, Pearl   1886 – 1886

Woodroffe, Bruce D.   1884 – 1935

Woodroffe, Earl E.   23 Aug 1893 – 17 Jun 1962

Bebout, Charles W.   13 May 1921 – 7 Apr 1984   Lt Col USAF WWII Korea Vietnam

Fippin, Emma A.   1893 – 1981
Fippin, Asa D.   1885 – 1968

Fippin, John A.   1915 – 1922   Son

Roberts, Rev. C. D.   Died 12 Oct 1875   Aged 37 yrs 8 mos
    Blessed are the dead, which die in the Lord

Sanford, Sadie E.   1899 – 1919
Sanford, Wallace James   1864 – 1956

MASONIC CEMETERY LEGENDS

Sanford, Eva C.	1868 – 1915	
Sanford, Benjamin	1832 – 1931	
Sanford, Euphemia	1830 – 1926	
Sanford, Sarah	1866 – 1897	
Sanford, Thomas	1862 – 1888	
Wallace, Ellen Harreld	1803 – 1891	
Barnes, Glenn F.	27 Jun 1892 – 10 Jan 1897	
Sanford, Wallace L.	1893 – 1895	
Black, Eva	1855 – 1890	
Black, Robert	1853 – 1935	
Black, Effie A.	1880 – 1978	
Dykes, Hugh T.	1884 – 1954	
Dykes, Grace L.	1884 – 1973	
Davis, Edward H.	1881 – 1957	
Davis, Mae M.	1882 – 1962	
Verdier, Ernest E.	6 Apr 1904 – 24 May 1997	
Verdier, Lucille M.	21 May 1905 – 14 Mar 1986	
Hacker, Cora E.	1880 – 1962	
Hacker, William F.	1878 – 1952	
Saunders, Amelia	*Died* 30 Nov 1904 *Aged* 82 yrs Wife of John **Saunders**, Jr	
Saunders, John G., Jr.	1 Sep 1831 – 4 Mar 1911	
Foust, Louis Perry	4 Oct 1826 – 3 Nov 1900	
Foust, Nelson Lewis	29 Nov 1876 – 18 Nov 1902	
Warne, J. H.	1851 – 1930	**McKeel** Family Plot
Warne, Mary	*Died* 7 Sep 1909	*Aged* 72 yrs
McKeel, John	No dates given	
McKeel, Thomas	No dates given	
Woodroffe, Sherman E.	1880 – 1923	
Cooley, Janie E.	1881 – 1972	
Wilson, Linda	1937	Daughter
Boatwright, Alice	1917 – 1988	Mother
Wilson, Roy Wayne III	6 Mar 1961 – 17 Dec 1983	PVT US Army
Harriman, Elsie	1871 – 1906	
Harriman, Susan Ann	*Died* 31 Oct 1904	
Cramp, Wm.	No dates listed	Co. F. 27 Ill. Inf.
Congdon, George F.	1854 – 1937	
Congdon, Lucretia	1830 – 1892	
Congdon, Frances D.	1829 – 1916	
Keith, Granville C.	1812 – 1905	
Congdon, Fredrick H.	1869 – 1949	
Call, Harriett A.	1864 – 1950	

MASONIC CEMETERY LEGENDS

Call, S. M.   1861 – 1932

Widener, Russell   No dates given   Father
Widener, William   No dates given   Son
Widener, Alpha Frances   7 Jul 1852– 18 Jun 1921   Mother
Vineyard, Mary E.   7 Feb 1842 – 7 Jul 1908
Vineyard, Allen E. T.   8 Nov 1875 – 8 Oct 1881
Vineyard, Almora G.   24 Jun 1877 – 12 Mar 1877
Vineyard, John T.   *Died* 12 Oct 1920
Vineyard, Alice Elizabeth *Died* 9 Jul 1891   *Aged* 37 yrs 2 mos 22 dys

Gleason, Harold James   28 Feb 1903 – 26 Feb 1989 Beloved Husband and Father
             He trusted in Him who is able to save to the uttermost

Hall, J. C.   No dates given
Carter, Ida Hall   1869 – 1923
Jones, J. Y.   23 May 1912 – 16 Oct 1912
Jones, David   31 Jan 1834 – 12 Jul 1905
Jones, Elizabeth   1838 – 7 May 1902

Rose, Oscar J.   2 Mar 1872 – 3 Jan 1964
Rose, John   1817 – 1899
Emery, Mary Rose   1867 – 1901
Rose, Ann V.   1839 – 1910
Rose, James E.   1862 – 1902

Poole, Francis   1832 – 1922
Poole, Mary Ann   1840 – 1924
             Natives of England

Poole, Frank D.   1868 – 1948
Poole, Daisie E.   1878 – 1956

Bushby, Nora   1872 – 1951
Bushby, J. M.   1883 – 1924
Bushby, Edward   1868 – 9 Jan 1902
Bushby, Frank   29 May 1876 – 2 Aug 1901
Bushby, George Sr.   20 Sep 1830 – 18 Jul 1925
Bushby, Katherine   10 Sep 1838 – 2 Dec 1915
Bushby, Sarah   4 Sep 1874 – 8 Dec 1955

Bushby, Fannie   1873 – 1960
Bushby, Wm   1865 – 1955
Bushby, Baby Irma Jewel   *Died* 19 Jul 1911   *Aged* 2 yrs 4 mos 22 dys
Foreman, Baby Violet L.   *Died* 1 Feb 1918   *Aged* 5 mos 29 dys

Allinio, Wilberta Adkins 3 Dec 1904 – 21 Dec 2000
Adkins Wm. B.   *Died* 20 Jun 1907   *Aged* 49 yrs

Compton, Mary Jane   1876 – 1957   Mother
Compton, Brady Farrell   1872 – 1940   Father
Bartlett, Nellie G.   6 Jan 1903 – 15 Apr 1976

MASONIC CEMETERY LEGENDS

Perkins, Mrs. M. J.   *Died* 22 May 1828   *Aged* 62 yrs 4 mos 10 dys
    Oh, let us think of all she said, and all the kind advice she gave,
    And let us do it now, she is dead and sleeping on her lonely grave.

Frazier, Carrie   *Died* 1 Sep 1886   *Aged* 69 yrs 6 mos 9 dys   Wife of Benjamin **Frazier**
Frazier, Benjamin   24 Mar 1822 – 7 Aug 1902
Clayton, Esther   10 Nov 1904 – 8 Nov 1991
Clayton, Charles V.   18 Jul 1908 – 11 Jan 1991

Jewell, Henry C.   *Died* 28 Mar 1823   *Aged* 15 yrs 10 mos
    Son of E. H. & Elizabeth **Jewell**   Our little Eddie
    A light is from our household gone, a voice we loved is stilled.
    A place is vacant at our hearth, which never can be filled.

Hickeson, Katie J.   1870 – 1917

Adkins, Oliver	No dates given	Father
Adkins, Frances M.	No dates given	Mother
Adkins, Owen "Jeff"	No dates given	Son
Adkins, Oliver P.	No dates given	Son
Adkins, Benjamin F.	No dates given	Husband
Adkins, Clara	No dates given	Wife
Murch, Frances Marie	No dates given	Infant

Adkins, Oliver   No dates given   Company A 3rd Ind. Infantry Mexican War

McKeague   No names or dates given
Toland   No names or dates given

Dunham, Lucinda H.   *Died* 13 Sep 1880   *Aged* 51 yrs 9 mos 21 dys
    Wife of D. J. **Dunham**

Mooney, Thomas	1823 – 1884	Father
Mooney, Mary J.	1840 – 1875	Mother
Mooney, Thomas Jr.	1872 – 1948	
Mooney, Adelaide	1864 – 1953	
Mooney, Nellie	1874 – 1896	
Mooney, Arthur	1874 – 1950	
Mooney, Mary E.	1860 – 1941	
Mooney, Jessie	1870 – 1927	
Mooney, Clara L.	1858 – 1927	
Mooney, Lucy J.	1868 – 1923	
Mooney, Ruby	*Died* 24 Sep 1896	

Westerfield, W. P.   May 1820 – Oct 1888 Born in New York City Died at Mooney Flatt

Foreman, Ida   No dates given
Foreman, Jacob Washington   14 Dec 1848 – 14 may 1912
Foreman, Jermie   No dates given
Foreman, Frank Gilbert   1895 – 1963

Clark, Levi Blaisdell   16 Feb 1821 – 15 Jan 1886   To rest

MASONIC CEMETERY LEGENDS

Clark, Jane Haworth   16 Feb 1838 – 3 Jul 1927
Clark, Abigale Ann   *Died* 1859   *Aged* 3 mos
Clark, Mary Jane   *Died* 1861   *Aged* 12 yrs
Clark, Sherman Grant   *Died* 1868   *Aged* 4 yrs
    Children of L.B. & J. **Clark**
    Sleep on fair children and take thy rest. God called thee home, he thought it best.
Clark, Henry   1 Aug 1857 – 24 Mar 1941
Haworth, Ann   *Died* 1860   *Aged* 54 yrs
Haworth, Abraham   *Died* 1880   *Aged* 54 yrs
Haworth, Thomas   *Died* 1873   *Aged* 32 yrs
    Gone but not forgotten

Bridge, Betty Oldham   *Died* 28 Feb 1879 *Aged* 65 yrs. Wife of Thomas **Bridge**
    We miss thee at home Husband & Children I must leave you, yes leave you all alone, But my blessed Savior calls me, calls me to a Heavenly Home.

Fraser, Daniel   1843 – 1916
Fraser, Elizabeth   1847 – 1883

Fraser, James H.   16 Feb 1872 – 2 Dec 1945

McNutt, John F.   *Died* 5 Mar 1899   *Aged* 84 yrs
    Member of Yuba Lodge No. 39, Washington R. A. chapter No. 13
    Peace to his ashes, rest to his soul. Native of Tennessee

McNutt, Alice   *Died* 4 Aug 1896   *Aged* 75 yrs

Davies, Samuel   *Died* 2 Jun 1867   *Aged* 65 yrs
Davies, Elizabeth   1809 – 1889

Presley, Infant   *Died* 10 Jun 1877   Daughter of Allen & Rachel E. **Presley**
    One singing with voice so sweet, crown of gold, Dawn thee Savior sleep.
Presley, Allen   *Died* 6 Nov 1895
Presley, Gertie   No dates given
Presley, Mary   *Died* 25 May 1890   *Aged* 22 yrs 9 mos 2 dys

Hite, Herbert Leslie   24 Nov 1878 – 11 Nov 1960
    California CPL Co B 6 Regt Calif. Inf. Spanish American War
Hite, Ida M.   16 Nov 1882 – 20 Oct 1977
Hite, David C.   *Died* 7 Feb 1909   *Aged* 1 yr 4 mos   Son of H. L. & I. M. **Hite**

Monasco, George C.   1887 – 1943
Manasco, Clayton W.   1830 – 1915
Manasco, Veronica A.   1856 – 1922

Williams, James O.   *Died* 20 Sep 1879   *Aged* 29 yrs 6 mos 29 dys   My Husband
    Native of Madison Co., Missouri
Hicks, Joseph   *Died* 18 Mar 1878   *Aged* 31 yrs   May he rest in peace
    Farewell   Native of Cornwall, England

MASONIC CEMETERY LEGENDS

Reid, Samuel 1827 – 8 Nov 1881 Erected by his beloved wife Mary **Reid**
A native of County Dorun, Ireland

Gassaway, Bertha Colling 1865 – 1938
Colling, William Henry 1854 – 1899
Gassaway, Robert Samuel 1856 – 1915

Jeffery, Harry Clinton *Died* 15 Nov 1890 *Aged* 13 yrs 10 mos Son of G & M L **Jeffery**
Jeffery, Ella Clinton *Died* 7 May 1912 *Aged* 49 yrs 1 mos 16 dys
Jeffery, Ella *Died* 7 May 1912 *Aged* 49 yrs 1 mos 16 dys
Jeffery, Maria Louise *Died* 15 Jan 1902 *Aged* 61 yrs 16 dys Wife of George **Jeffery**
Jeffery, George *Died* 31 Jul 1907 *Aged* 77 yrs 7 mos 10 dys Native of Nova Scotia

Murdock, Henry *Died* 12 Dec 1876 *Aged* 69 yrs A native of Nova Scotia
Weep not, he is at rest

Hyatt, Jacob *Died* 13 Sep 1878 *Aged* 72 yrs 6 mos 20 dys

Revell, Martha *Died* 20 Dec 1870 *Aged* 75 yrs 16 dys
Revell, Joseph *Died* 1 Mar 1869 *Aged* 70 yrs 8 mos 13 dys
Hartley, Mary A. 2 Nov 1822 – 3 Feb 1877 Native of England
Hartley, W. M. 6 Apr 1816 – 27 Oct 1884

Wheaton, Robert 1894 – 1929

McConnell, James , M.D. *Died* 3 Oct 1824 *Aged* 19 yrs 8 mos 12 dys

McConnell, James *Aged* 18 mos Son of J. & Martha
McConnell, John Taylor 15 May 1855 – 2 Nov 1887
McConnell, Mary Jetta 17 Mar 1872 – 1 Mar 1879

Peardon, Ada E. 1887 – 1917
Newbert, Minnie E. 1850 – 1902
Newbert, Thomas A. 1830 – 1910
Hogarth, Rose Ann *Died* 22 Jan 1886 *Aged* 59 yrs 9 mos 12 dys Wife of M. **Hogarth**
A Native of Ireland Rest in Peace

McCrea, Arthur Wheaton 1913 – 1923 Son
Eiden, Mable McCrea 1896 – 1968 Mother
McCrea, Albert Ernest 1885 – 1931 Father

Peardon, F. G. 1888 – 1973 Brownie
Peardon, Frank R. 1888 – 1973 Brother
Peardon, John, Sr. 1843 – 1907 Father
Peardon, Elizabeth 1849 – 1906 Mother
Peardon, John, K. 1872 – 1910 Brother Here rests a woodman of the world
Smith, Bessie 1873 – 1905
Smith, Catherine A. 11 Dec 1909 – 10 Feb 1993
Smith, Frank Earl 8 Dec 1912 – 13 Nov 1981 S F II U. S. Navy WW II
Peardon, Gladys Elizabeth 25 Mar 1909 – 28 Dec 1911
Peardon, John H., Jr. 14 Jun 18722 – 27 May 1910

MASONIC CEMETERY LEGENDS

Jeffery, Ellen H.	1877 – 1943	
Jeffery, William B.	1868 – 1953	
Newbert, William L.	1873 – 1942	
Newbert, Nellie W.	1887 – 1963	
Mitchell, Elvira B.	1879 – 1914	
Newbert, Ella	11 Aug 1881 – 24 Jul 1900	
Newbert, William	17 Mar 1907 – 17 Mar 1907	
Mitchell, Catherine	No dates given	
Mitchell, John	No dates given	
Filcher, Eugenia Hapgood	1869 – 1957	
Hapgood, Annie Lily	1865 – 1881	
Hapgood, Elizabeth	1842 – 1905	
Hapgood, Eugene D.	1838 – 1913	
Smith, W. E.	Died 30 Jul 1892	Husband to Eugenia Hapgood
Hays, Isaac N.	No dates given	Co H S.LA Mil. Inf. Mexican War
Monk, James	Died 12 Apr 1877	Aged 59 yrs 1 mos 10 dys
Native of England	May his soul rest in Peace	
Monk, Betsy	Died 31 Jan 1894	Aged 76 yrs 1 mos 5 dys
Wife of James **Monk**	May she rest in Peace	
Magonigal, W.	Died 27 Feb 1891	Aged 72 yrs
Magonigal, William B.	Died 1933	
Magonigal, Nancy	Aged 71 yrs 10 mos	Wife of W. **Magonigal** Our Parents
Magonigal, Ellen J.	Died 7 Jan 1896	Aged 30 yrs Wife of W. B.
Magonigal		
Magonigal, S.	Died 19 Jul 1897	Aged 46 yrs
Magonigal, Lizzie	Died 22 Mar 1926	Aged 71 yrs
Murphy, Ida E.	1890 – 1986	
Murphy, John J.	1886 – 1963	
Murphy, Gladys L.	1911 – 1913	
Hord, Mary Wright	1862 – 1937	
Wright, Aden	1844 – 1921	
Murphy, Stephen	Died 30 Jun 1891	Aged 75 yrs
Murphy, Rosana	Died 18 Jan 1891	
Peckham, Thomas W.	1856 – 1935	
Peckham, Sophronia E.	1861 – 1957	
Hacker, Erma Pearl	1911 – 1913	Darling
Peckham, Miriam E.	1822 – 1901	
Taylor, Frank L.	Died 20 Jan 1916	Aged 32 yrs
Bowman, William H.	1880 – 1920	
Bowman, Harry	1822 – 1903	
Bowman, George	1851 – 1925	
Bowman, Emma	21 Nov 1860 – 29 Jun 1933	
Bowman, Mother	1825 – 1899	
Bowman, Lizzie	1863 – 1903	

MASONIC CEMETERY LEGENDS

Bowman, Gordon   1853 – 1908
Bowman, Louisa   1859 – 1934   (Double plot with James **Bowman**)
Bowman, James   1851 – 1913   (Double plot with Louisa **Bowman**)

Monk, James   *Died* 11 Dec 1895   *Aged* 55 yrs 11 mos 14 dys
Monk, Elizabeth   *Died* 11 Jan 1879   *Aged* 34 yrs 3 mos 4 dys

Magonigal, T. G.   7 Aug 1866 – 8 Jul 1945
Magonigal, Hannah   17 Jan 1872 – 7 Oct 1970
Magonigal, Frankie   No dates given
Magonigal, Clarence B. 1909 – 1938
Poole, Hazel Mae   29 Dec 1907 – 17 May 1994   Mother
Poole, Clarence Francis 19 Jul 1905 – 29 Dec 1995   Father

Magonigal, John L.   16 Nov 1900 – 11 Nov 1980   U. S. Army
Magonigal, Vera Lee   1906 – 1982
Magonigal, William Daniel 1898 – 1963
Magonigal, John   1863 – 1927   Father
Magonigal, Emma E.   1870 – 1961   Mother

Pisani, Leslie   1904 – 1959
Pisani, Bertha E.   1903 – 1959

Denney, Dorothy A.   1922 – 1966   In loving memory   Rest in Peace

Scogland, Grace D.   27 Jan 1900 – 10 Dec 1972
Scogland, Thomas K.   4 Dec 1884 – 27 Jun 1972

Cann, Winona F.   1918 – 1985   Beloved Mother & Grandmother
Cann, Philip   1911 – 2002   Beloved Husband
Cann, Edna M.   1906 – 1984   Beloved Wife & Mother
Cann, Bert   1916 – 1949   Brother
Cann, Maud   3 Apr 1884 – 29 Oct 1969   Mother
Cann, Thomas J.   24 Jan 1886 – 28 Mar 1945   Father

Trembly, Keith D.   1916 – 1982   Together Forever
Trembly, Kay M.   1923 – 2000

Shields, Lloyd Vernon   1904 – 1970

Shields, Mary Ellen   1903 – 1957

Collier, John A.   25 Sep 1923 – 17 Oct 1981   U. S. Army
Collier, Patricia E.   20 Oct 1926 – 19 Mar 1981

Kay, Samuel   *Died* 10 Jul 1899   *Aged* 68 yrs 7 mos
Hudson, Laverne J.   1909 – 1979
Hudson, Ralph R.   1905 – 1993
Clark, Annie   *Died* 19 Nov 1877   *Aged* 50 yrs   Asleep in Jesus

Clark, Wright   *Died* 18 Dec 1875   *Aged* 47 yrs 4 mos 4 dys
Clark, Mary   *Died* 3 Nov 1875   *Aged* 8 yrs 9 mos 8 dys   Daughter of W. & M. **Clark**

MASONIC CEMETERY LEGENDS

Magonigal, Henry E.   2 Nov 1905 – 15 Jun 1986   Together Forever
Magonigal, Alice R.   20 Nov 1912   No death date given
Magonigal, Gerald E.   7 May 1944 – 29 Aug 1994   Beloved son

Lane, Linda   1948 – 1993
Lane, Esther   1927 – 1998   Mother
Lane, Larry   1918 – 1999   Father
Lane, Louisa   *Died* 7 Jun 1876   *Aged* 2 yrs 7 mos

Wirth, Mable R.   1895 – 1981
Wirth, Fredrick C.   30 May 1891 – 13 May 1968   CA CGM US Navy WWI & II

Brett, Rose Mary   1900 – 1981
Brett, Nelson   1891 – 1971

Hawkins, Eldon John   1921 – 1993
Hawkins, Gloria Cecilia   1927 – 1987

Mills, Homer C.   1924 – 1985

Harris, Lee   8 Dec 1952 – 23 May 1986   Beloved Son & Brother At rest with love

Harris, Angel M.   31 Oct 1969 – 9 Feb 1987   Our Angel waits with God

Wiseman, Sara C.   1983 – 1986   **Poncherelli**

DuShane, L. June Bishop   13 June 1924 – 31 Dec 1998

Pott, Darrel A., Jr.   28 Feb 1959 – 10 Mar 1997   Our beloved Son & Father

Likens, R. C.   No dates given   Statues of hearts with doves
Likens, A. L.   No dates given   Big Angel statue
Likens, Alma Lou   1913 – 1986   Hooper & Weaver Funeral Home
Likens, H. B.   No dates given   Statue of Choirboy & Ceramic Bible with Hazel **Flowers** on the back of the Ceramic Bible

Miller, George C.   9 May 1930 – 19 Apr 1987   A II C   U. S. Air force Korea

Welch, William Gale   24 Aug 1851 – 23 Apr 1922
           Co. I 8 Regt. Calif. Inf   Spanish American War
Welch, Joseph B.   1867 – 1936
Welch, Thomas H.   1847 – 1903
Welch, Serena J.   1830 – 1893
Welch, Radford E.   1826 – 1892
Welch, Benjamin F.   *Died* 18 Dec 1876   *Aged* 11 yrs 3mos 25 dys
           Son of R. E. & S. J. **Welch**
           No kisses drop upon my cheek, Those lips are sealed to me,
           Dear Lord how could I give him up to any, but to thee

MASONIC CEMETERY LEGENDS

Research indicates that these individuals are buried in this cemetery, without headstones.

Allen, Edward	19 Feb 1854 – 24 Mar 1876
Bennett, Grace	6 May 1886 – 18 Jul 193?
Bone, Thomas	Unreadable – 30 Jan 1878
Broom, Bertha	Unreadable
Clifford, Keith Manville	1832 – 31 Oct 1905
Crocker, Fredona Damon	11 Jul 1830 – 7 Aug 1917
Crossett, J. G.	*Died* 10 Apr 1878
Crouch, H. W.	*Died* 16 Aug 1895
Crouch, H. W., Jr.	*Died* 1878
Curry, Artie B.	11 Jan 1880 – 26 May 1880
Curry, Bertram Earl	7 Jan 1889 – 30 Mar 1889
Curry, Flora J.	20 Mar 1873 – 28 Jan 1876
Curry, Wilbur R.	15 Sep 1875 – 20 May 1876
Curtis, Thomas	*Died* 9 Jul 1890
Dougherty, Ezekiel	1800 – Sep 1882
Davey, Carrie Bell	15 May 1879 – 5 Jul 1879
Davey, William H.	No dates given
Gratiot, Eliza	*Died* 2 Nov 1890
Grose, Harry	*Died* 1894
Holcomb, David	*Died* 12 Apr 1878
House, Anthony	No dates given
Hurling, Charles	4 Jul 1845 – 22 Feb 1930
Hurling, George W.	8 Nov 1872 – 23 May 1894
Hurling, Lucille	9 Jan 1854 – 2 Feb 1891
Hutchinson, David	*Aged* 24 yrs
Ingram, Mary	Nov 1801 – 2 Dec 1877
James, William Ray	28 Oct 1888 – 16 Dec 1974
McLean	*Died* 1864
Meredith, W. J.	No dates given
Pate, Derrel Austin, Jr.	28 Feb 1959 – 10 Mar 1997
Pendlebury, Mrs.	*Age* 65 yrs
Powers, Infant	No dates given
Powers, Infant	No dates given
Powers, Amos Mathew	12 Jul 1832 – 12 Apr 1919
Powers, Mrs.	No dates given
Reigh, Alexander	*Died* 1877
Rickey, John	18 May 1829 – 27 Sep 1903
Rickey, Peter S.	29 Sep 1804 – 16 May 1879
Robbins, Nathaniel	No dates given
Shilling, Charles	No dates given
Snelt, Mrs.	*Died* 1877
Spargo, T. J.	*Died* 1876
Teske, Edna	No dates given
Thrush, George W.	*Died* 29 May 1888
Thrush, Sarah Arvilla	*Died* 5 Jul 1885
Walker, John	*Died* 1899
Williamson, Linda	27 Sep 1948 – 10 Oct 1993
Williamson, Robert Bush	8 May 1913 – 22 Dec 1966
Woods, W. H.	1835 – 1 Nov 1882
Young, James	Jul 1836 – 13 Dec 1908

Catholic Cemetery Legends

LeBoeuf, Cecilia Ethelyn   22 Nov 1906 – 17 Dec 1967
Collins, Bertha LeBoeuf   14 Apr 1911 – 30 May 1995

Niles, Theresa J.   1883 – 1957   Dublin Ireland
Niles, George W.   1876 – 1964   So. Dakota
Niles, Anthony Joseph   20 Apr 1914 – 25 Dec 1985

Allen, Josephine A.   1920 – 1997   Age 76
Allen, Geo. B.   No dates given
Allen, Laura D.   No dates given

Martein, Charles   No dates given
Martein, Elizabeth   1869 – 1961

Zeisloft, Christopher Scott   10 Apr 1971 – 2 Jan 1993   Beloved Husband and Loving Daddy

Naglee, Joseph C.   17 May 1917 – 6 Nov 1970   New York   Lieutenant USNR WWII

Gruber, William Henry   1900 – 1991 "Wild Bill"   Truly Loved Deeply Missed

Grace, John Thomas   25 Oct 1894 – 19 Feb 1984   PVT US Army WWI
Grace, Rose   Died 1999

Compton, Clarence Albert   1898 – 31 Jul 1955

Heggerty, Maurice   1830 – 1873
Heggerty, Mary   1832 – 1875
Heggerty, William   1827 – 1891   Natives of Ireland
    Erected by Charles J. Heggerty son of Maurice and Mary Heggerty

McGovern, Mary Agnes   1875 – 1957
McGovern, Thomas James   20 Jan 1865 – 8 Jun 1926
McGovern, John   Died 9 Jul 1891   Aged 64 yrs
McGovern, Mrs. John   Died 23 Feb 1893   Aged 60 yrs   Native of Ireland

Linehan   Family plot no names or dates given in the plot
Linehan, Daniel   Died 11 May 1918
Linehan, Edward Thomas   6 Jun 1870 – 18 Jan 1904
Linehan, Ellen   Died 8 Jan 1912
Linehan, John   No dates given
Linehan, Katherine   No dates given
Linehan, Mary   No dates given
Linehan, Peter   No dates given
Linehan, Timothy   No dates given

Flanagan   Family plot no names or dates given in the plot
Flanagan, John Henry   20 Jan 1868 – 7 Dec 1907
Flanagan, Mary   Died 4 Dec 1926
Flanagan, Peter J.   Died 11 May 1904
Flanagan, Thomas J.   9 Jul 1880 – 12 Oct 1935

CATHOLIC CEMETERY LEGENDS

Flanagan, Michael			*Died* 20 Feb 1895

Smethurst, Elizabeth		1853 – 1913		Wife of G. W. **Smethurst**
McQuaid, Mary Margaret	31 May 1907 – 21 Aug 1990
Magonigal, Eleanor Germaine	5 Apr 1903 – 8 May 1987
    Granddaughters of John H. **McQuaid** – Homesteader 1853
McQuaid, Frank			*Died* 18 Aug 1911			*Aged* 49 yrs

Holbrook, Elise D.		1895 – 1963
Holbrook, James E.		1896 – 1975

Rigby, Joseph			1906 – 1962
Rigby, Ida				1871 – 1951
Rigby, George A.		18 Jul 1909 – 25 Jan 1992 T. SGT U. S. Army WWII
Rigby, Albert			26 Sep 1867 – 16 Dec 1945
Rigby, Frances M.		2 Oct 1917 – 30 May 1994		We who love you

Johnson, Wm. F.			14 Sep 1864 – 3 Jul 1961
Johnson, Carrie R.		30 May 1865 – 31 Dec 1945
Miles, E. Adele			30 May 1865 – 31 Dec 1945
Johnson, Wm. Allen		3 Mar 1930
Johnson, Francis V.		18 Jul 1895 – 8 Feb 1925

Driscoll				Family plot no dates or names given

Herboth, Elizabeth E.	1890 – 1972
Herboth, George			1889 – 1964

Quinkert, John			1857 – 1941
Quinkert, Clara C.		1865 – 1937		Clara
Quinkert, John			1824 – 1899		Father
Quinkert, Elizabeth		1826 – 1896		Mother
Conrath, Gustav			1857 – 1930		Brother		Rest in Peace
Conrath, Louis			1854 – 1904		Father		Rest in Peace
Conrath, Elizabeth		1860 – 1937		Mother		Rest in Peace

Beatty, Ann				1819 – 1899
Beatty, Alice			*Died* 30 Jan 1860			*Aged* 4 yrs 10 mos 3 dys

McQuaid, Mary S. Milhare	*Died* 20 Jun 1899		*Aged* 63 yrs
    Native of Ireland	Beloved wife of John **McQuaid**
McQuaid, John			*Died* 5 Nov 1913 *Aged* 75 yrs Native of Ireland
Allread, Margaret E.	1876 – 1923
McQuaid, Mary			1866 – 1925
McQuaid, John H.		1870 – 1932
McQuaid, Katherine R.	1875 – 1962

O'Brien, Mary Kirby		15 Apr 1832 – 13 Nov 1894	Native of Ireland
    Beloved wife of James **O'Brien**   R.I.P.
Holbrook, Kathleen Adele	*Died* 27 Dec 1934
O'Brien, Josephine		*Died* 8 Jun 1935
O'Brien, Helen M.		*Died* 30 Nov 1944
O'Brien, Agnes			*Died* 13 Dec 1954

CATHOLIC CEMETERY LEGENDS

O'Brien, James	28 May 1830 – 29 Aug 1915	Native of Ireland
O'Brien, James K.	Died 23 May 1945	
O'Brien, Philip	Died 6 Jan 1892	Aged 80 yrs
Kirby, John	1882 – 1934	
Walsh, Anna Kirby	1838 – 1883	
Walsh, Richard	1829 – 1890	
Conlin, Thomas	Died 9 Feb 1898 Aged 65 yrs 9 mos	Native of Ireland
Conlin, Margaret	Died 13 Nov 1908	Aged 72 yrs 7 mos
Conlin, Sarah L.	Died 20 Jan 1885	Aged 14 yrs
Conlin, Thomas B.	Died 6 Oct 1887	Aged 22 yrs 8 mos
Conlin, John H.	Died 28 Sep 1912	Aged 45 yrs
Cooney, William	1865 – 1928	
Cooney, Mary	1868 – 1945	
Walsh, John	Died 9 Jun 1902 Aged 74 yrs	His wife Mary Walsh
Walsh, Mary	Died 15 Feb 1901	Aged 72 yrs
Walsh, Philip	Died 30 Nov 1899	Aged 33 yrs
Walsh, Charles	Died 13 Jan 1874	Aged 1 yrs
Walsh, Edward	1864 – 1904	
Walsh, Thomas	Died 1 Jul 1889	Aged 55 yrs Native of Ireland
Walsh, John	Died 20 Apr 1911	Aged 87 yrs Native of Ireland
Walsh, Anna	Died 14 Nov 1912	Aged 67 yrs
Walsh, Richard	Died 6 Dec 1878	Aged 32 yrs
Dempsey, Ann Breslin	13 Sep 1899	
Dempsey, John	1834 – 1909	
Dempsey, Catherine Loretta	Died 7 Aug 1913	Aged 31 yrs A Native of California
Dempsey, Joseph William	No dates given	
Dempsey, Rose	13 Dec 1903 – 13 Nov 1904	
McCarty, Andrew	1832 – 1909	
McCarty, Susan	1839 – 1928	
McCarty, James	1872 – 1902	
McCarty, John T.	Died 4 Feb 1860	Aged 31 yrs 5 mos 11 dys
Galligan, Hugh P.	1858 – 1931	
Galligan, Mary E.	1864 – 1957	
McCarty, Matthew H.	1869 – 1951	
McCarty, Robert E.	1880 – 1953	

Lyons, Jeremiah   Died 27 Sep 1880   Aged 52 yrs 5 mos
           Native of Buhopourny Co. Cork, Ireland   May his soul rest in peace
Altschul, Harold Chas.   7 Aug 1905 – 9 Jun 1910   Aged 4 yrs   Native of San Francisco, CA

Byrne, Mary R.	1882 – 1947	Wife of R. E. Byrne	
Ryan, Hannah M.	1845 – 1907	At Rest	Mother
Ryan, Patrick	1831 – 1910	At Rest	Father
Ryan, Thomas	15 Feb 1878 – 29 Nov 1900		
Ryan, Nellie	17 Aug 1879 – 29 Nov 1882		
Ryan, Ellen	17 Aug 1862 – 29 Nov 1883		

Murphy, Morris   1846 – 1915   Father

CATHOLIC CEMETERY LEGENDS

Murphy, Catherine   1852 – 1898   Mother
Murphy, Charles E.   *Died* 3 Aug 1888   Son of Morris **Murphy**
Doubt, Elizabeth   *Died* 5 Nov 1895   *Aged* 75 yrs
     Native of Co. Lonegall, Ireland   Wife of J. **Doubt**

Driscoll, John   1827 – 1907
Driscoll, Catherine   1837 – 1913
Driscoll, Mary E.   1871 – 1891
Driscoll, Theodore   1892 – 1893

Driscoll, Sarah Delilah   9 Sep 1903 – 18 May 1910

Fortune, Martin   *Died* 24 Mar 1891
Fortune, Mary   *Died* 2 Jul 1898   Of your charity   Pray for the soul of

Beasley   No names or dates given
Beasley, Richard   *Died* 20 Apr 1922   *Aged* 67 yrs
Beasley, Elizabeth E.   *Died* 17 Apr 1900   *Aged* 39 yrs

Byrne, Robert E.   24 Feb 1879 – 8 Apr 1954   Loving Husband of Mary Ryan **Byrne**

Byrne, Selma F. E.   1885 – 1977

Byrne, James D.   1873 – 1943   U. S. Marines Vet WWI & Spanish Am.

Byrne, Rose B.   1878 – 1943
Byrne, John B.   1875 – 1924
Byrne, James   1843 – 24 Dec 1916
Byrne, Rose Anna   *Died* 30 Sep 1902   *Aged* 55 yrs   Native of New Jersey
Byrne, John   1841 – 1911
Byrne, Mary Roseann   *Died* 20 Sep 1902
Byrne, Margaret   No dates given
Byrne, Sarah M.   *Died* 13 Aug 1908
Byrne, Thomas P.   20 Feb 1881 – 30 Apr 1893

Havey, Pat Jr.   1838 – 1884   Uncle

Barrett, Joseph Jr.   1925 – 1989   CPL   U. S. Army   WWII

Daly, Jane   No dates given   In the Cramsie Family Plot

Horan, Michael   1858 – 1897

Meade Family Plot   No names or dates given
Meade, William Joseph   *Died* 30 Nov 1921   *Aged* 41 yrs 10 mos

Meade, John   No dates given
Meade, Leo   No dates given
Meade, Mary   No dates given

Barrett, John   1852 – 1900
Barrett, Catherine   1864 – 1924

CATHOLIC CEMETERY LEGENDS

Schmidt, Veronica   *Died* 9 Oct 1907   *Aged* 81 yrs 16 dys   Native of Germany
Schmidt, George   *Died* 8 Aug 1900   *Aged* 87 yrs 2 dys   Native of Germany
Schmidt, George, Jr.   *Died* 24 Jun 1852   (Also a Footstone W. H.)
Schmidt, William Henry   *Died* 5 Jan 1855

Newbert, Horace C.   1870 – 1953

Newbert, Mary J.   3 Mar 1942
McNally, Thomas   1836 – 1919   Father
McNally, Mary   1845 – 1900   Mother
McNally, Edwin   *Aged* 6 mos   Baby

Daly, M.   *Died* 14 Jun 1876   *Aged* 1 yrs
Daly, T.   *Died* 2 Jul 1879   *Aged* 2 yrs
Daly, Lawrence   *Died* 6 Oct 1890   *Aged* 58 yrs
Daly, Ann   *Died* 23 May 1906   *Aged* 68 yrs   Wife of Lawrence **Daly**
Daly, John J.   *Died* 9 Nov 1898   *Aged* 30 yrs
Daly, Cecelia   *Died* 18 Apr 1879   *Aged* 13 yrs
Daly, Kate   *Died* 23 May 1906
Daly, Michael   *Died* 15 Apr 1859   *Aged* 31 yrs   Native of Ireland
Daly, Infant son of Patrick Daly   *Died* 28 Dec 1860   *Aged* 8 mos
Daly, James   *Died* 30 Aug 1860   *Aged* 64 yrs

Casey, Peter   1824 – 1904
Casey, Winifred   1836 – 1906
Casey, William   1864 – 1917
Casey, Mary J.   1867 – 1882
Casey, Margaret   1871 – 1896
Casey, John J.   8 Sep 1853 – 18 Jan 1916

Twomey, Andrew, Rev.   *Died* 8 Mar 1902   *Aged* 36 yrs
    Native of Co. Cork, Ireland   R.I.P.   Erected by his loving Flock & Friends

Shea, Simon   *Died* 28 Feb, 1878   *Aged* 35 yrs   Native of Co. Cork, Ireland
Shea, M. J.   No dates given

Bristow, Geo W.   *Died* 10 Oct 1871   *Aged* 39 yrs.   This monument is erected by his wife   To the truth of a good and earnest man, to the sincerity of a faithful husband
Bristow, Virginia   No dates given
Bristow, Elmer   No dates given

Rooney, Mary E.   *Died* Dec 1899   *Aged* 43 yrs 6 mos   Native of New Jersey
Rooney, Eugene F.   *Died* 24 Jul 1909   *Aged* 28 yrs   Native of California
Sanders, Annie B.   *Died* 1 Feb 1913   *Aged* 60 yrs   Native of New Jersey
Sanders, Catherine   15 Aug 1822 – 12 Jan 1917
Sanders, Daniel   30 Apr 1825 – 7 Jul 1915
Sanders, John Henry   *Died* 9 Oct 1938   *Aged* 80 yrs   Son of Daniel & Catherine **Sanders**

Beirne, Patrick   *Died* 8 Oct 1884   *Aged* 49 yrs   Native of Co. Leitrim, Ireland
Beirne, Margaret   *Died* 26 Dec 1878   *Aged* 33 yrs   Native of Co. Kerry, Ireland
    Also his beloved wife (Husband is Patrick **Beirne**)
    May their souls rest in peace, Amen.   Erected by their Niece and Son
Beirne, Terence   *Died* 26 Jan 1867   *Aged* 2 yrs 6 mos

CATHOLIC CEMETERY LEGENDS

Beirne, Bartholomew   *Died* 24 Feb 1867   *Aged* 4 weeks
     Beloved children of Patrick and Margaret **Beirne**
Beirne, Charles   *Died* 13 Oct 1866   *Aged* 33 yrs   Native of Co. Leitrim, Ireland
Beirne, Patrick J.   *Died* 18 May 1902   *Aged* 29 yrs   May his soul rest in peace, Amen.

Gunning Owen   1891 – 1982   U. S. Navy WWI
Geraghty, John T.   1880 – 1940
Geraghty, Alvina   1887 – 1976
Smith, Gloria J.   23 Mar 1930 – 3 Feb 1994   In loving Memory
Gunning, Samuel O.   1835 – 1910   Father   Native of Co. Sligo, Ireland
Gunning, Mary   No dates given
Gunning, Ellen   No dates given
Gunning, Baby   No dates found
Gunning, Fremon   *Died* 25 Oct 1886   *Aged* 70 yrs

Kerrigan, Mary   *Aged* 66 yrs
Kerrigan, Thomas J.   *Aged* 28 yrs
Kerrigan, Patric   *Aged* 59 yrs
Kerrigan, Peter   *Died* 21 Aug 1873   *Aged* 54 yrs 9 mos   May he rest in peace Amen
Kerrigan, Ambrose F.   *Died* 19 Aug 1873   *Aged* 1 day
Kerrigan, Mary J.   *Died* 20 May 1866   *Aged* 3 yrs 11 mos ?dys
     Daughter of **Peter** & Mary **Kerrigan**
Kerrigan, Patrick K.   *Died* 9 Dec 1922   *Aged* 62yrs

Keegan, Annie L.   *Died* 7 Jun 1867   *Aged* 9 mos 5 dys
Keegan, Annie Louise   *Died* 1 Jan 1873   *Aged* 5 yrs 2 mos 19 dys
Keegan, Charles M.   *Died* 7 Jun 1873   *Aged* 6 weeks
     Children of Jas and Mary J. **Keegan**
Keegan, Ann J.   31 Feb 1884 – 31 Aug 1886
Keegan, Infant   *Died* 25 Oct 1887   *Aged* 1 yrs 8 mos
     Children of James & Mary **Keegan**

Sheehan, Julia   1833 – 1921   Sister
Sheehan, Daniel F.   1861 – 1922   Brother
Sheehan, Johanna C.   *Died* 18 Nov 1941   Mother

McCoughlan, Peter   *Died* 27 Jan 1869   *Aged* 38 yrs   Native of Co. Londonderry, Ireland
Coughlan, William   *Died* 4 Jul 1904   *Aged* 79 yrs   Native of Ireland

Doyle, James   No dates given
Doyle, Mary   *Died* 2 Jul 1908

Kerrigan   No dates or names given

Mullin, Mary Ellen   *Died* 22 Mar 1872   Daughter of J. G. and Julia **Mullin**
Mullin, Mary   *Died* 8 Jan 1879   *Aged* 59 yrs   Native of Ireland

Melody, Bridget   *Died* 18 Dec 1869   In memory of   Wife of Thos M **Melody**

Cain, Bridget   *Died* 11 Feb 1869   *Aged* 33 yrs   Sacred to the memory of
Cain, Daniel   1826 – 1909

CATHOLIC CEMETERY LEGENDS

Curran, Mark   *Died* 19 Jul 1873   *Aged* 54 yrs   May he rest in peace Amen
     Native of Deaney Parish of Nuns Reavis Co. Kidare, Ireland
     Farewell my wife and children all, From you a father Christ doth call
     Mourn not for me it is in vain, To call me to your sight again.
Curran, John J.   *Died* 1 Aug 1893
Curran, Mark   *Died* 5 Feb 1886
Curran, Mark   *Died* 4 Nov 1896
Curran, Sarah   *Died* 16 May 1889
Curran, Susan   *Died* 5 Feb 1886

Stanton, William   *Died* 2 Nov 1872   *Aged* 42 yrs
     Native of Kirney Co., Ireland   In memory of his beloved wife ??

Mellarkey, John J.   *Died* 25 Jan 1900   *Aged* 26 yrs 4 mos 17 dys
Mellarkey, James   *Died* 1 Apr 1923
Mellarkey, Edward   12 May 1814 – 14 Feb 1909
Mellarkey, Mary   27 Nov 1824 – 2 Apr 1915

Ward, Mary E.   *Died* 22 Aug 1875   *Aged* 22 yrs   Wife of W. W. **Ward**
Ward, Mary E.   No dates given   Infant daughter of Mary E. & W. W. **Ward**
Ward, Infant son   No dates given   Infant daughter of Mary E. & W. W. **Ward**
Ward, Caroline Hazel   *Died* 7 Jul 1885   *Aged* 3 yrs 1 mos 14 dys

O'Brien, Bridget   *Died* 13 Jan 1866   *Aged* 39 yrs 9 mos 7 dys
     Wife of Philip **O'Brien**   Native of Roscommon, Ireland

Smith, Anthony   1826 – 1898
Smith, Margaret   1827 – 1910
Smith, Agnes   1862 – 1864
Smith, Thomas   1865 – 1870
Smith, Margaret J.   1858 – 1873
Smith, Susan T.   1870 – 1886
Smith, John P.   1856 – 1897
Smith, William C.   1864 – 1926

Flynn, Morris   *Died* 27 Aug 1867   In memory of   Son of P. and S. **Flynn**
Flynn, Patrick   *Died* 15 Sep 1893

Hughes, Winifred   *Died* 18 Feb 1874   *Aged* 35 yrs   Wife of Michael **Hughes**
Hughes, Alois   17 Dec 1880 – 19 Apr 1955

Royer, A.   *Died* 20 Jul 1865
Royer, W. J.   13 Aug 1862 – 12 Jul 1865
Royer   Name and date unreadable

Miller, George W.   23 Feb 1862 – 20 Mar 1865
Miller, Joseph   10 May 1866 – 8 Jul 1869
     Sons of Geo and Jane **Miller**
Miller, Mary Alice   11 Sep 1865 – 19 Sep 1869   Daughter of Geo and Jane **Miller**
Miller, John R.   17 May 1864 – 26 Mar 1865   Son of Geo and Jane **Miller**
Miller, May   4 Sep 1870 – 27 May 1880
Miller, John   *Died* 24 Feb 1860   *Aged* 23 yrs

CATHOLIC CEMETERY LEGENDS

Denehey, Mary	*Died* 7 Jun 1872	*Aged* 28 yrs	Wife of Cornelius **Denehey**
Dougherty, Grace	1867 – 1931		
Dougherty, Katie	1869 – 1916		
Dougherty, Daniel	1830 – 1914		
Dougherty, Daniel	1876 – 1913		
Dougherty, Annie	1885 – 1906		
Dougherty, Catherine	Nov 1869 – 6 May 1916		
Dougherty, John	*Died* 8 Nov 1887		
Dougherty, Thomas	*Died* 4 Jul 1896		Native of Ireland
Pettit, Catherine T.	1856 – 1927		
Pettit, John Joseph	*Died* 5 Sep 1876		*Aged* 1 yrs 6 mos
Pettit, Nicholas J.	1850 – 1881		
Looney, Catherine	1827 – 1917		

Havey Family	No names or dates
Havey, Gertrude	*Died at* 7 yrs
Havey, John C.	16 May 1849 – 6 Jan 1931
Havey, Louis Patrick	21 Feb 1891 – 15 Oct 1922
Havey, Lucille	13 Feb 1896 – 20 Dec 1906
Havey, Margaret	*Died* 26 Jun 1918
Havey, Veronica Amanda	25 Feb 1884 – 4 Sep 1889
Havey, Mary	*Died* 29 Jan 1897 *Aged* 83 yrs Native of Ireland

McWilliams, Bridget   *Died* 3 Jan 1900   *Aged* 28 yrs   Native of Co. Derry, Ireland

Dillon, Peter   *Died* 12 Jun 1879   *Aged* 19 yrs 11 mos   Native of Ireland

Sweeney, Michael   4 May 1826 – 12 Apr 1877   Native of Easkey Co., Sligo, Ireland
    May his soul rest in peace   Erected by his wife **Ann**
Sweeney, Annie   *Died* 3 Mar 1908
Sweeney, John   *Died* 20 May 1893   Native of Tralee, County Kerry, Ireland

Doyle, Olive R.   *Died* 6 Oct 1875   *Aged* 2 yrs 7 mos   Daughter of E. &. A. **Doyle**
Doyle, Alexander   *Died* 24 Jul 1877   *Aged* 58 yrs   In memory of
Doyle, Elizabeth J.   1829 – 14 Nov 1886
Doyle, Alexander, Jr.   *Died* Feb 1891   *Aged* 30 yrs 3 mos 4 dys

Campbell, Daniel   *Died* 31 Dec 1878   *Aged* 44 yrs   Native of Co. Down, Ireland
Campbell, Daniel T.   2 Nov 1874 – 6 Nov 1875
Campbell, Daniel   *Died* 22 Feb 1860   *Age* 35 yrs   Native of Deneglan Co., Ireland
Campbell, Infant   *Died* 15 Aug 1878
Campbell, Mrs.   No dates given
Campbell, James   *Died* 6 Nov 1918   *Aged* 75 yrs

Coopmann, Fredrick F.   14 Nov 1822 – 10 May 1876   Native of Germany

Walsh, Richard   *Died* 6 Jan 1879   *Aged* 32 yrs   Erected by his affectionate Wife

Conry, Mary   *Died* 23 Apr 1898
Conry, Thomas   *Died* 14 Dec 1898
Lee, Agnes   5 Aug 1885 – 13 Dec 1903
Lee, Thomas R.   *Died* 20 Oct 1906

CATHOLIC CEMETERY LEGENDS

Lee, James R.   *Died* 27 Jul 1882   *Aged* 1 mos

Burns, John   18?1 – 1911
Burns, Margaret   No dates given
Burns, James   *Died* 17 Jan 1861

Buckley, Michael Sr.   5 Jan 1927
Buckley, Ann   Oct 1884
Buckley, Michael Jr.   27 Apr 1912
Buckley, Patrick H.   *Died* 14 Mar 1875   *Aged* 8 yrs 4 mos 10 dys Son of M. & A. **Buckley**

O'Sullivan, Michael   *Died* 5 Feb 1881   *Aged* 43 yrs 4 mos May he rest in peace
Early, Mary A.   *Died* 22 Jan 1879   *Aged* 43 yrs   Wife of Timothy **Early**
O'Sullivan, Margaret A.   *Died* 25 Apr 1905 *Aged* 69 yrs   Wife of Simon **O'Sullivan**
O'Sullivan, Simon   *Died* 14 Mar 1869 *Aged* 43 yrs 3 mos
     In memory of the children of P. & E. **O'Sullivan** of Co. Kerry, Ireland
Early, Margaret Ann   *Died* 9 Mar 1873 *Aged* 4 yrs 9 mos 4 dys Daughter of T & M **Early**

Mulligan, James   *Died* 30 Aug 1875   *Aged* 78 yrs   Native of Co. Roscommon, Ireland

Silva, Louis J. Sr.   1942 – No death date Loving Husband & Father
Silva, Erna Jeanne   1942 – 1991   "E. J."   Loving Wife & Mother

Creden, Daniel   *Died* 7 Dec 1877 *Aged* 28 yrs   Native of Macroom Co., Cork, Ireland

Divver, Charles   No dates given
Divver, Mary A.   No dates given
Divver, William   No dates given
Divver, George W.   No dates given
Dougherty, Sadie   No dates given
Divver, Charles   *Died* 12 Aug 1900   *Aged* 33 yrs

Leigh, Kenna   *Died* 20 Apr 1993

McPhillips, Barney   *Died* 8 Feb 1892   *Aged* 65 yrs   Native of Ireland
McPhillips, Owen   *Died* 27 Mar 1861   *Aged* 42 yrs

MacWilliams, Joe   25 Apr 1948 – 27 May 1994   U. S. Army

Williams, Joseph Matthew   13 Jun 1968
Williams, James Paul   7 Nov 1971
Williams, Juliana Rebecca   28 Jan 1975
     The Williams children   Gone to be with Jesus March 11, 1976

Vinsonhaler, Howard F.   8 Sep 1914 – 20 Sep 1993   U. S. Air Force WWII

Quilty, Patrick J.   1886 – 1964
Quilty, Thelma M.   1904 – 1986

Burke, William B.   No dates given
Burke, Genevieve   No dates given
Burke, Frank   *Died* 1 Aug 1890   *Aged* 84 yrs   Native of Ireland

CATHOLIC CEMETERY LEGENDS

Burke, Ann   *Died* 25 Dec 1888   *Aged* 76 yrs

McGanney, Edward James   1891 – 1918   U. S. A. 30 Infantry
     Killed in action, Argonne, France 5 Oct 1918   He died that nations might be free
McGanney, James F.   No dates given
McGanney, Ned   No dates given
McGanney, Edward J.   No dates given
McGanney, Annie   No dates given
McGanney, Father   No dates given
McGanney, Mother   No dates given
McGanney, Baby   No dates given
McGanney, Anna   No dates given
Cramsie, John E.   17 Feb 1937
McGanney, Georgie Gray   1879 – 1953
Cramsie, Joseph   *Died* 6 Dec 1911
Cramsie, William   15 Jun 1837 – 18 Mar 1912

Research indicates that these individuals are buried in this cemetery, without headstones.
Calaghan, Michael   *Died* 16 Nov 1918
Calaghan, Patrick   3 Mar 1845 – 22 Sep 1907
Conboy, Margaret   *Died* 2 Jun 1897
Dewan, Charles   *Died* 28 Oct 1881
Dewan, Ellen   1841 – 1 Jan 1913
Dewan, James Henry   4 Jan 1876 – 8 Jul 1918
Dewan, John   *Died* 15 Nov 1894
Dewan, John D.   *Died* 9 Jun 1895
Dewan, William E.   *Died* 29 May 1901
Ford, Patrick   *Died* 23 Feb 1915
French, Joseph   6 Feb 1854 – 2 Jan 1947
Gaffney, Ellen   23 May 1867 – 1 May 1958
Henderson, Bridgett   *Died* 16 Jul 1899
Kelly, James   *Died* 29 Aug 1909
Lavelle, Mary   No dates given
Leahy, Michael   1841 – 15 May 1906
McClure, James Alexander   27 Jul 1878 – 24 Jan 1951
McClure, Joseph Patrick   7 Apr 1886 – 4 Dec 1951
McClure, Alexander Campbell   *Died* 13 Nov 1901   *Aged* 70 yrs   Native of Ireland
McCoughlin, Margaret   *Died* 5 Aug 1903
McKenna, John   No dates found
McKenna, Elizabeth T.   No dates found
Pryor, Mary   *Died* 28 Nov 1895
Pryor, Mary Ann   21 Oct 1855 – 1934
Pryor, Michael   *Died* 28 Jul 1892
Whalen, Jeremiah, Mrs.   No dates given
Young, John   1841 – 27 Jun 1908

Timbuctoo Cemetery Legends

Radworth, Charles Jay   *Died* 20 May 1871   *Aged* 1 yr 3 mos 8 dys
     Son of Marcus J. & A. W.   **Radworth**   Of such is the Kingdom of Heaven

Robinson, Sarah A.   *Died* 28 Nov 1874   *Aged* 40 yrs
     Blessed are the dead, which die in the Lord

Michaels, Walter C.   12 Jan 1916 – 10 Jul 1987   RADM US Navy   WWII

Wooden Cross on grave   No information readable

Carpenter C.   8 Sep 1822 – 23 Jan 1874   Pioneer   Mason sign F A & G

Buck, Elijah   *Died* 5 Feb 1878   *Aged* 50 yrs

Carlson, Ric   6 Feb 1946 – 26 Aug 1976   SP4 US Army   Vietnam
Carlson, Kim Loring   28 Feb 1951 – 19 Dec 1971   Peace

Farish, John Bolton   1854 – 1929
Farish, Etta Paddock   1858 – 1919
Farish, Helen Ruth   1892 – 1903
Farish, Mary Wren Prather   1810 – 1870
Farish, Adam Thomas   1810 – 1865
Collbran, Mabel Farish   1884 – 1965
Collbran, John Stuart   1883 – 1972
Collbran, Virginia Bachelder   1911 – 1993
Collbran, James Farish   1848 – 1993
Collbran, Arthur, Jr.   1911 – 1982
Collbran, Florence Farish   1882 – 1936
Perez-Triana, Margaret Collbran   1874 – 1943
Collbran, Arthur Harry   1876 - 1943

Morton, Norman Barclay   1873 – 1950

Kershaw, Theodore Gourdin   23 May 1906 – 12 May 1975   Brig Gen US Air Force WWII

Boyer, Lizzy   *Died* 28 Jul 1869   *Aged* 9 mos & 10 dys   In Memory Of
Boyer, John   *Died* 24 Dec 1866   Infant
     Children of John & Eliza **Boyer**
Boyer, Eliza Lenihan   *Died* 25 May 1872   *Aged* 37 yrs 22 dys   Wife of John **Boyer**
     Native of Bullinnamona, Cowly Cork, Ireland   May her soul rest in peace

Calvin, John   Not readable   Farewell
Calvin, Marg (rest unreadable)   *Died* unreadable 1870

Wright, Mabel Bennett   1903 – 2001

Kershaw, Henrietta Collbran   22 Feb 1913 – 31 Mar 1986

Kershaw, Diane C.   14 Nov 1931 – 29 Oct 1996   In Memory of my Beloved
     She is in a better place

TIMBUCTOO CEMETERY LEGENDS

Kershaw, Theodore G.　　　23 Apr 1934 – No death date given
　　　This is a double stone with above
Pritchett, Susanna R.　　　*Died* 9 Jul 1866　*Aged* 47 yrs　Native of Pennsylvania
　　　Wife of Jacob **Pritchett** The bright eye is sealed, and the soft lip is closed
　　　Who art once love and feeling so, sweetly reposed
Pritchett, Jacob H.　　　1819 – 1899

Mohn, Mary H.　*Died* 11 Apr 1857　*Aged* 16 yrs　Native of Philadelphia, Pennsylvania
　　　Wife of Joseph **Mohn** Yet again we hope to meet thee, when the day of life is fled.
　　　Then in heaven with joy to greet thee where no farewell tears is shed

Holmes, Thatcher B.　　　*Died* 15 Jul 1873　　*Aged* 36 yrs
Holmes, Winslow　　　19 Dec 1865 – 1 Mar 1866

Sims, Louise Bundschu　　　1876 – 1956
Sims, Richard Maury　　　1875 – 1935
Sims, Richard Maury, Jr.　　　20 Sep 1910 – 19 Nov 1985　Berkley, Mt. Tamalpais, CA
Sims, Jane Prichard Loomis　　22 Apr 1914 – 20 Feb 1996
　　　Enjoy yourself it's later than you think
Sims, Robert Lee　　　24 May 1912 – 3 Dec 1972

Cranshaw, James　*Died* 17 Dec 1877　　*Aged* 70 yrs　　　Native of England

Congdon, Lucretia　　　*Died* 4 Mar 1878　　*Aged* 79 yrs 6 mos

Congdon, George N.　　*Died* 17 Dec 1860　　*Aged* 38 yrs 9 mos 5 dys
Congdon, Infant Daughter　　No dates given
　　　Daughter of George N. & Amelia A. **Congdon**

Congdon, Sarah E.　*Died* 1 Jan 1870　*Aged* 12 yrs 14 dys　Daughter of F & L **Congdon**
　　　Rest daughter thy struggle oer, the last fond look is given,
　　　Thy weary spirit soared above, and sweetly rests in heaven

Flint, George W.　　　*Died* 19 May 1873　　*Aged* 8 mos 12 dys

McAllis, Fannie　　*Died* 1 Aug 1867　*Aged* 2 mos　Daughter of J. & F. McAllis

Birmingham, Abbie C. L.　　*Died* 3 Jun 1885　　*Aged* 31 yrs 2 mos 12 dys
　　　Dau of G. S. **Birmingham**
Birmingham, Ida May　　　*Died* 22 Jul 1864　　Aged 8 yrs 6 mos 21 dys
　　　Dau of G. S. **Birmingham**
　　　The little one has passed away and where is love,
　　　Her Heavenly Father wanted her and took her home above

Stone, Charlie E.　　　*Died* 24 May 1866　　　*Aged* 5 yrs 6 mos 29 dys
Stone, Willie A.　　　*Died* 10 Jan 1865　　　*Aged* 6 yrs 6 mos 3 dys
　　Sons of W. H. & P. L. **Stone**
　　　The Lord gave & the Lord tooketh them away, Blessed be the name of the Lord forever.

Govne, John C.　　　*Died* 7 Mar 1867　　　*Aged* 33 yrs
Govne, Ann　　　*Died* 12 Oct 1866　　　*Aged* 28 yrs
Govne, Thomas H.　　　*Died* 23 Jan 1865　　　*Aged* 1 yr 11 mos 4 dys
Govne, Maria　　　*Died* 20 Aug 1865　　　*Aged* 22 yrs

TIMBUCTOO CEMETERY LEGENDS

Govne, William *Died* 14 Dec 1865 Aged 38 yrs

Hapgood, James Mortimer 28 Mar 1872 – 4 Apr 1959
Hapgood, Fannie Marple 28 Mar 1862 – 25 Apr 1946
Marple, Harry 1876 – 1933
Marple, George 1871 – 1950
Marple, Whelton 1822 – 1907
Marple, Elizabeth 1836 – 1860
Marple, Samuel 1878 – 1949
Marple, John W. 1856 – 1860
Marple, Brant 1860 – 1876
Marple, Fred 1872 – 1929
Marple, Infant 2 footstones with Infant on them (records show 5 infants)

Wilson, Theodore *Died* 20 Dec 1863 *Aged* 34 yrs Native of Washington, Maine

Miller, John L. *Died* 8 Dec 1855 *Aged* 24 yrs Rest is unreadable

Jones, Simon G. *Died* 6 Jan 1860 *Aged* 29 yrs A Native of Waterville, Maine

Estmon, Alden *Died* 23 Mar 1861 *Aged* 27 yrs 6 mos 5 dys

Chadwick, Jacob F. *Died* 20 Dec 1864 *Aged* 26 yrs Native of Palermo, Maine
  And shall we in another state, know the dear friends whom here we know, Then truly will the joy be great, to meet within that heavenly gate, those whom we loved here below
Linscott, James O. *Died* 4 Feb 1869 *Aged* 26 yrs Native of Palermo, Maine
  Dearest brother thou hast left us, and thy loss we deeply feel
  But it's God that hast breft us, He can all our sorrows heal

Ah, Day Toon *Died* 12 Mar 1908 *Aged* 45 yrs

Wah, Ying Yim *Died* 6 Jan 1909 *Aged* 75 yrs

Housekeeper, Unknown Name Only a wooden cross to mark her grave

Crawford, Charles Jay 12 Feb 1870 – 20 May 1871

McClaskey, John Calvin 9 Sep 1856 – 31 Jul 1870

Pioneer Cemetery Legends AKA Davies Cemetery

Moody, Ellen R. *Died* 9 Jul 1856   *Aged* 32 yrs   A native of Mississippi
  Gone, but not forgotten

Simpson, John Clayton   29 Mar 1860 – 11 Mar 1865
  Only son of Thomas B. & Mary H. **Simpson**
  Jesus said, suffer little children to come unto me & forbid them not.
  For of such is the Kingdom of Heaven.

Dittmer, John   *Died* 2 Jul 1871   Aged 40 yrs 10 mos   Farewell Husband
  Dear wife, farewell, I go to dwell with Jesus Christ on high.
  There, to sing praise to my King to all Eternity.
Dittmer, James Henry   *Died* 15 Sep 1869   *Aged* 16 mos 13 dys
  Son of **John** & Katie **Dittmer**
Dittmer, Mary Elizabeth   *Died* 25 Sep 1869   *Aged* 6 yrs 10 mos
  Daughter of **John** & Katie **Dittmer**

Cory, Wm H.   *Died* 10 Oct 1873   *Aged* 20 yrs

Cory, Daniel   *Died* 8 Jan 1899   *Aged* 70 yrs

Cory, Elizabeth   *Died* 20 Nov 1868   *Aged* 36 yrs

Bowman, William   *Died* 29 Aug 1867   *Aged* 49 yrs
  Farewell unto my wife and children
  A call from Jesus Christ   Mourn not?

Lambert Sheldon   *Died* 30 Dec   *Aged* 4 yrs

Brickell, Catherine Estelle   *Died* 21 Dec 1866   *Aged* 6 mos
  Daughter of **John** & Jennie **Brickell**
Waldron, Henry   *Died* 8 Nov 1867   *Aged* 44 yrs   Native of Vermont

Newspaper Information

11 Jul 1856; Marysville Daily Appeal; Page 2 Col 3; DIED - At Empire Ranch, on the 9th inst., Mrs. Ellen R., wife of Tom B. **Moody**, Esq., aged 84 years.

3 Jul 1860; Marysville Daily Appeal; Page 2 Col 4; Married – On the 2d inst., in this city (Marysville), by Rev. Mr. E. B. Walsworth, Mr. George **Jeffery** to Miss Maria L. **Davis**, both of Empire Ranch. (Smartsville).

17 Jan 1861 (Thursday); Marysville Daily Appeal; Page 2 Col 1; Mining Accident at Sucker Flat. – On Tuesday between 9 and 10 o'clock a.m. Mr. James **Burns** was standing near an old stump of a tree on the bank of the Dead Rabbit claims, at Sucker Flat, when the stump gave way, with some of the bank, falling with him to the bottom of the claim, between fifty and sixty feet, killing him instantly. Mr. **Burns** was a young man, in high standing and has left behind him a great many friends.

25 Jan 1861 (Friday); Marysville Daily Appeal; Page 3 Col 2; CITY SEXTON'S REPORT. Of the Mortality of the City of Marysville, for the year ending December 31, 1860. 30 Jan – Alice **Beatty**, 4 yrs 10 ms. 3 ds., born in California; 28 Dec – Infant of Patrick **Daly**, 8 mos., born Cal.; 30 Aug – James **Daly**, 64 yrs., from Ireland; 4 Feb-John T. **McCarty**, 31 yrs. 5 ms. 11 ds., born Indiana.; 24 Feb – John **Miller**, 23 yrs., born Ireland.

11 Apr 1861 (Thursday); Marysville Weekly Appeal; Page 3 Col 2; MORTALITY OF THE CITY OF MARYSVILLE, FOR THE MONTH ENDING MARCH 31st, 1861. Prepared for the Appeal by E. Hamilton, City Sexton. 27 Mar – Owen **McPhillipps**, 42, born Ireland.

29 Dec 1866 (Saturday); Marysville Weekly Appeal; Page 3 Col 4; Died – In Smartsville, December 21st, **Catherine** Estelle, daughter of **John** and Jennie A. **Brickell**, aged six months.

19 May 1867: Page 2 Col 3; Died – At Mooney flat, May 18th, the wife of T.P. **Otis**, aged 34 years. The funeral will take place at the residence of the bereaved husband tomorrow afternoon at 1 o'clock. Friends of the family are invited to attend.

21 May 1867: Page 2 Col 3; Died – At Mooney Flat, May 18th, 1867, Mrs. T. P. **Otis**, aged 34 years. At the rising of the sun, she was cheerful and happy in the midst of her family; ere the meridian, her spirit took wing. Deeply do her friends and neighbors mourn her loss, for truly did she love her friends and neighbors. [New York papers please copy.]

25 May 1867 (Saturday); Marysville Weekly Appeal; Page 3 Col 8; Died- At Mooney's Flat, May 18th, the wife of T. P. **Otis**, aged 34 years. At the rising of the sun she was cheerful and happy in the midst of her family; ere the meridian, her

NEWSPAPER INFORMATION

spirit took wing. Deeply do her friends and neighbors mourn her loss, for truly did she love her friends and neighbors. New York papers please copy.

8 Jun 1867 (Saturday); Marysville Weekly Appeal; Page 3 Col 7; Died- At Empire Ranch; Yuba County, June 2d, Samuel **Davies**, formerly of Philadelphia, aged 65 years. Funeral will take place from his late residence, on Tuesday June 4th, at 2 o'clock.

10 Aug 1867 (Saturday); Marysville Weekly Appeal; Page 3 Col 8; Died- At Timbuctoo, August 1st, **Fannie**, infant daughter of **John** and Fannie **McAllis**, aged 2 months. Sweet babe, rest in peace. Brooklyn (N. Y.) papers please copy.

7 Sep 1867 (Saturday); Marysville Weekly Appeal; Page 3 Col 7; Born- In Timbuctoo, August 29th, to Captain A. **Robinson** and wife, a son.

16 Nov 1867 (Saturday); Marysville Weekly Appeal; Page 3 Col 8; Died – At Sucker Flat, Yuba County, Nov. 8th, Henry **Waldron**, a native of Vermont, aged 44 years. Vermont papers please copy.

12 Aug 1868: Page 2 Col 4; Died – At French Corral, Aug 10th, Mrs. Elizabeth **Daniel**, aged 55 years. Her body sleeps beside her daughter and grandson near Smartsville. Deceased was born in Scotland. She was a descendant of the family of Robert **Bruce**, and was a member of the Old School Presbyterian Church. Her home in Mooney Flat is filled with natural curiosities, especially geological and botanical specimens collected and arranged with her own hands, indicating a mind of high culture. She had set her "house in order." She waited for death as a welcome visitor, and died. "Like one who draws the drapery of his couch about him, and lies down to pleasant dreams."

2 Oct 1869 (Saturday); Marysville Weekly Appeal; Page 3 Col 6; Died – In Timbuctoo, September 15th, **James** Henry, son of **John** and Katie **Dittmer**, aged 16 months and 13 days. Sleep, Harry, Sleep! Not in thy cradle bed, Nor on thy mother's breast, But with the quiet dead.

2 Oct 1869 (Saturday); Marysville Weekly Appeal; Page 3 Col 6; Died – In Timbuctoo, September 25th, **Mary** Elizabeth, daughter of **John** and Katie **Dittmer**, aged 6 years and 10 months. We weep for you, Mary, A shadow o'er our pass way crossed; About one little spot of earth, We mourn our beautiful, our lost. Oh, strange at first our darkened lives, Our struggle with that sorrow cold; And sweet at last to know our lambs, Are safe within the Shepherd's fold. Wheeling (West Virginia) papers please copy.

2 Apr 1873: Page 2 Col 3; Died – In Smartsville, March 28th, Henry **Clay**, only son of E. H. and Elizabeth **Jewell**, aged 15 yrs and 10 mos.

20 Jul 1873 (Sunday); Page 3 Col 2; MINING ACCIDENT – An inquest was held at Timbuctoo on the 18th inst., on the body of Thatcher B. **Holmes**, a well known

NEWSPAPER INFORMATION

and respected inhabitant of that place, who was unfortunately killed on the evening of the 15th by the caving of a bank in the Babb claim. The jury returned a verdict that the deceased came to his death by the bank in the Babb claim falling upon him. Deceased left a widow and four young children to regret his untimely fate.

22 Jul 1873 (Tuesday); Page 3 Col 2; Providing for the Widow. – The people of Smartsville and vicinity have given another instance of their liberality. Last week Thatcher **Holmes** was killed by a cave in the Babb mine, leaving a widow and several children. A subscription was started for their benefit, and $1,000 was donated. Such liberality is beyond all praise.

22 Jul 1873 (Tuesday); Page 3 Col 2; Fatal Accident – Mark **Curran**, an old resident of Timbuctoo, was found dead near that place yesterday morning. On Sunday, deceased started for Spencerville, distant from Timbuctoo some five or six miles, to have some repairs made to his wagon. He had not proceeded far on his way, when it is supposed his horses became frightened and ran away, throwing him from the wagon. The two horses and wagon were found in a gulch on Monday, the horses lying on their backs and the wagon upside down. The body of **Curran** was discovered about three hundred yards distant. Deceased was a worthy citizen of Timbuctoo, and leaves a wife and four children to mourn their sudden loss. An inquest was held upon the body yesterday.

23 Aug 1873 (Saturday); Page 2 Col 2; Died- In Sutter county, August 21st, Peter **Kerrigan**, aged 50 years. (The funeral will take place from his late residence this morning at 7 o'clock, and will arrive at St. Joseph's Cathedral, Marysville, at 9 a.m. Funeral services over, from thence to Smartsville, to be buried in the family plot. Friends and acquaintances are invited to attend.)

25 Sep 1873; Page 3 Col 2; An affray occurred at the Oregon House, on Monday the 15th instant, between I. H. Foulks and Thomas **Haworth**, which resulted fatally to the latter last Saturday morning. The parties met at the Oregon House on the day mentioned, when a quarrel originated between them in relation to a lawsuit about a wagon belonging to Thomas Skehan, in which suit Foulks had been a witness, and during this wrangle of words the deceased accused Foulks with testifying to that which was not true. This accusation led to a personal encounter, in which the parties fought till they reached the street, and there the fight terminated, by a stunning blow by Foulks, which broke the bridge of Haworth's nose, and drove a piece of the bone into the brain. Haworth's injuries were considered slight until the day before his death, and on a fatal termination, Foulks surrendered himself to Justice Soward, who set his examination for today, when District Attorney **Davis** will appear for the People and N. E. Whiteside for the Prisoner. Both Haworth and Foulks have had the reputation of being sober, orderly and quiet citizens. Haworth was a farmer, and brother-in-law of L. B. **Clark**, and Foulks is a laboring man in the employ of George P. **Housh**. The deceased was about thirty-five years of age.

11 Oct 1873; Page 3 Col 1; Death of Wm. Henry **Cory** – Died in this city (Marysville), October 10th, Wm. Henry **Cory**, son of Daniel and Elizabeth **Cory**,

NEWSPAPER INFORMATION

aged 20 years and 4 months. The remains will be taken to Smartsville this morning, and the funeral will take place this afternoon from his late residence. Friends are invited to attend.

11 Dec 1873 (Thursday); Page 2 Col 3; MARRIED – In this city (Marysville), December 10th, by Rev. Father Farley, Mr. Morris **Murphy**, of this city, to Miss Katie **Havey**, of Smartsville. (The happy bridegroom remembered the Appeal by a liberal display of Heldsieck, and the young bride was the recipient of many complimentary toasts.

24 Aug 1875 (Tuesday); Page 2 Col 2; DIED – At Smartsville, August 22d, Mrs. Mary E. **Ward**, wife of W. W. **Ward**, aged 22 years.

13 Oct 1875; Page 2 Col 2; Died – Near Smartsville, October 12th, Charles D. **Roberts**, aged 37 yrs, 7 mos and 18 dys.

30 May 1876; Page 2 Col 2; Married – In Smartsville, May 24th, by the Rev. Geo. R. Davis. Allen **Presley** to Miss Rachael R. E. **Flint**, both of Smartsville.

13 Jun 1876 (Tuesday); Marysville Daily Appeal; Page 2 Col 2; Married. In Smartsville, June 11th, by E. H. **Jewell**, J. P., Jacob W. **Foreman** to Ida J. **Gassaway**.

8 Jul 1876 (Saturday); Page 2 Col 3; DIED. In Linda Township, June 7th, **Louisa**, youngest daughter of **Riley** and Louisa **Lane**, aged 2 years and 7 months. Funeral will take place from the residence of the parents, on the Smartsville road eight miles east of this city, this morning at 10 o'clock. Friends and acquaintances are respectfully invited to attend.

13 Dec 1876 (Wednesday); Marysville Daily Appeal; Page 3 Col 2; Died at Smartsville – Thomas Jeffrey **Spargo**, aged 32 years, a native of Cornwall, England, died at Smartsville on the 11th instant, and the funeral of deceased took place in that town on Tuesday, 12 instant, under the auspices of Lavonia Lodge, I. O. O. F.

26 Jun 1877; Page 2 Col 2; Died – In Smartsville, June 24th, N. **Trevena**, aged 57 years.

25 Jul 1877 (Wednesday); Marysville Daily Democrat; Page 2 Col 2; Died – In Smartsville, July 24th, Alexander **Doyle**, aged 58 years, a native of Longford, Ireland. Indiana papers please copy.

7 Dec 1877 (Friday); Page 2 Col 3; DIED – In Smartsville, December 5th, David (Daniel) **Creedon**, aged 28 years. Funeral this day at 2 o'clock p.m. from the residence of his sister. Friends of the family are invited to attend.

NEWSPAPER INFORMATION

16 Dec 1877; Page 3 Col 1; Fatal Accident at Smartsville – By special telegram to the Daily Appeal, received from Smartsville last evening about 7 o'clock, we learn that James **Cranshaw**, an aged citizen, was thrown from his wagon yesterday in front of the "Excelsior Store", and received injuries from which he died a few hours after the accident. The deceased was a native of England, and 69 years of age. Our telegram gave no further particulars of the sad affair.

9 Jan 1878 (Wednesday); Page 2 Col 3; MARRIED. At the residence of the bride's brother, January 7th, Mr. William W. **Ward**, of Smartsville, to Miss Lizzie **McCune**, of Sutter county, formerly of New York. At 8 o'clock the ceremony was performed by the Rev. Father **Hines**, in his usual pious and graceful manner, in the presence of a large number of their friends and acquaintances, who enjoyed a pleasant and sociable event, as well as the splendid supper which was served immediately after the ceremony. After the party dispersed the happy couple went to Marysville to take the coach for San Francisco on their bridal trip. May their trip through life be pleasant and happy and replete with every blessing. A Spectator, Long may you live, Oh gentle Bride, To grace your husband's home, As through this world so wide, You and he may roam. May he cherish long the love, Which he has happily won, And may he bless the Father above, Who has made you and him as one

23 May 1878; Page 2 Col 3; Died – At Smartsville, May 22d, Mrs. M. J. **Perkins**, aged 62 years, 4 months and 10 days. (Peru, Indiana, papers please print.)

10 Nov 1878; Page 2 Col 2; Died – In Smartsville, November 9th, Wm. **Meredith**, aged 42 yrs. The funeral will take place from his late residence near the Empire Ranch, this afternoon at 3 o'clock. Friends invited to attend.

8 Dec 1878 (Sunday); Page 2 Col 2; DIED. At Smartsville, December 6th, Richard **Walsh**, a native of Dungarran, County of Waterford, Ireland, aged 32 years. North Hampton (Mass.) papers please copy.

1 Jan 1879 (Wednesday); Page 2 Col 3; DIED. In this city (Marysville), December 31st, Daniel **Campbell**, aged 45 years. The funeral will bake place from his late residence on Maiden lane, between Second and Third streets this morning, at 8 o'clock, when the remains will be taken to Smartsville and buried in the family grounds.

8 Jan 1879 (Wednesday); Page 2 Col 4; DIED. In this city (Marysville, January 6th, Mrs. Mary **Mullins**, a native of Ireland, aged 59 years. (Boston, Mass., papers please copy).

2 Mar 1879; Page 2 Col 4; Died – In Smartsville, February 28th, Betty, wife of Thos. **Bridge**, a native of England, aged 65 years, 1 month, 11 days. Friends and acquaintances, and those of her son John W. **Bridge**, are respectfully to attend the funeral from the Episcopal Church at Smartsville today (Sunday) at 2 o'clock. The Rev. Thomas Smith will officiate.

NEWSPAPER INFORMATION

23 Sep 1879; Page 2 Col 4; Died – At Timbuctoo, Sept. 20 of typhoid fever, James Ogdon **Williams**, a native of Missouri, aged 29 years, 6 months and 29 days.

30 Dec 1879 (Tuesday); Page 2 Col 2; MARRIED. At the residence of the brides' father, in the Foothills, Yuba co., Dec 24, by the Rev P. L. **Carden**, Mr. James **Mellarkey** of Smartsville to Miss Mary **Huffman**, also of Yuba County.

31 Mar 1880 (Wednesday); Page 2 Col 3; MARRIED. At Smartsville, Mar 29, by Rev. Father **Coleman**, John **McKenna** to Elizabeth T. **Murphy**, both of Smartsville.

13 Apr 1880; Page 2 Col 2; Married – In this city (Marysville), April 7th, Mr. John **Walsh** to Miss Margaret **McClaskey**.

30 Sep 1880 (Thursday); Page 2 Col 3; DIED. Near Smartsville, Yuba Co., September 27th, Jerry (Jeremiah) **Lyons**, a native of the parish of Ballyvourham, County Cork, Ireland, aged 53 years

13 Aug 1881 (Saturday); Page 3 Col 1; MATRIMONIAL – The Sacramento Bee of last evening says: "Wm. W. **Ward**, the well-known proprietor of the Railroad eating house at Marysville, and Miss Amanda **Ryan**, also of Marysville, were last evening united in marriage, at the residence of C. L. **Warner**, corner of Sixth and K streets, in this city. Rev. H **Rice** performed the ceremony, and Miss Minnie **Ryan** (a sister of the bride) officiated as bridesmaid, with James O. **Welch** as groomsman. A number of friends of the couple were present and numerous handsome gifts were bestowed on the newly married pair. Subsequent to the ceremony there was feasting and general merry-making."

1 Sep 1881; Page 2 Col 2; Died – Drowned in the Yuba river, at Parks Bar, Aug 24th, Annie **Elizabeth**, daughter of Eugene D. and Elizabeth **Hapgood** aged 16 years and 15 days.

10 Nov 1881; Page 2 Col 3; Died – In Marysville, November 9th, Samuel **Reid**, aged 55 years. Funeral today at Smartsville.

15 Mar 1882; Page 2 Col 2; Married – At the United States Hotel in Marysville, March 14th, by N. Sewell, J.P., James **Welch** to Miss Lillie E. **Jones**, both of Yuba County.

29 Jul 1882 (Saturday); Page 2 Col 3; DIED. In Smartsville, July 27th, **James** R., infant son of **Mary** Ann and Thomas R. **Lee**, aged one month.

22 Aug 1882 (Tuesday); Page 2 Col 3; DIED. In Linda township, Yuba county, August 20th, **Mary** Jane, daughter of **Winifred** and Peter **Casey**, aged 14 years and 6 months.

NEWSPAPER INFORMATION

27 Sep 1882 (Wednesday); Page 2 Col 3; DIED. Near Smartsville, September 26th, Ezekiel C. **Dougherty**, a native of Virginia, aged 81 years, 10 months and 17 days.

30 Nov 1882 (Thursday); Page 2 Col 2; DIED In Smartsville, November 29th, **Ella** (Ellen), daughter of **Patrick** and Johanna **Ryan**, aged 3 years and 3 months.

6 Apr 1884; Page 2 Col 2; Died – At Mooney's Ranch, near Smartsville, April 5th, Thomas **Mooney** aged 63 years. The funeral will take place from his late residence this (Sunday) afternoon at 2 o'clock. Friends invited to attend.

28 Oct 1884; Page 2 Col 3; Died – At the residence of G. W. **Sutliff**, Linda Township, October 27th, Wm. **Hartley**, aged 69 years. Services will be held at the house this morning at 8 o'clock.

28 Oct 1884; Page 3 Col 1; Died – Yesterday at 6 p.m. Patrick **Havey**, of Smartsville. The funeral party will leave here at 7 a.m. today. The funeral services will be at the Catholic Church in Smartsville, upon the arrival of the party there.

29 Oct 1884; Page 3 Col 2; Funeral – The funeral of the late Patrick **Havey** took place yesterday. The remains were taken from this city (Marysville) to Smartsville where the Rev. Father **Coleman** officiated, both at the church and at the grave. The following gentlemen acted as pall- bearers; Michael **Havey**, Peter **Earner**, Wm. **Kelley**, Thos. **Kennedy**, Mark **Curran** and Michael **Pryor**.

9 Jul 1885 (Thursday); Page 2 Col 4; DIED – In this city, July 7th, **Caroline** Hazel, eldest daughter of **William** W. and Amanda **Ward**, aged 3 years, 1 month and 14 days. (Sacramento and San Francisco papers please copy.) Friends and acquaintances are respectfully invited to attend the funeral from the parents' residence on B. street, between Sixth and Seventh, this afternoon at 2 o'clock.

6 Aug 1885 (Thursday); Marysville Daily Appeal; Page 2 Col 4; Died – At her residence two miles south of Smartsville, August 5th, Mrs. G. W. **Thursh**, aged 65 years. The funeral will take place from her late residence at 1 p.m. today.

17 Jan 1886; Page 2 Col 4; Died – At his late residence near this city (Marysville), January 15th, L. B. **Clark**, aged 62 yrs, 11 mos and 2 days. Interment to day at Smartsville.

7 Feb 1886 (Sunday); Page 2 Col 3; DIED. Near Smartsville, February 5th, Mrs. Susan **Curran**, aged 53 years, 3 months and 3 days. Friends and acquaintances are respectfully invited to attend the funeral from the Catholic church, Smartsville, Monday at 2 o'clock p. m. Interment Catholic Cemetery at Smartsville.

17 Sep 1886 (Friday); Page 2 Col 3; MARRIAGES. In this city (Marysville), September 12th, by the Rev. Father **Callan**, Samuel O. **Gunning** to Alvina **Lotsen**.

NEWSPAPER INFORMATION

27 Oct 1886 (Wednesday); Page 2 Col 3; DEATHS. In Timbuctoo, October 25th, Fremon **Gunning**, a native of France, aged 70 years. Funeral from Smartsville this afternoon at 2 o'clock. Interment, Smartsville cemetery.

16 Nov 1886 (Tuesday); Page 2 Col 4; DEATHS. In this city (Marysville), November 14th, Mrs. Elizabeth J. **Doyle**, a native of Kentucky, aged 57 years. Friends and acquaintances are respectfully invited to attend the funeral from her late residence, Seventh street, between C and D, this morning at 8 o'clock. Interment, Catholic cemetery at Smartsville this afternoon at 1 o'clock.

7 Oct 1887 (Friday); Page 2 Col 2; Young Ben (Thomas Benjamin) **Conlin** Frightfully Crushed by a wagon. On Wednesday afternoon about 4 o'clock Ben Conlin, a young man who has teamed on the road between this city and mountain towns for five years past, was driving his heavily-laden freight wagon, to which was attached four horses, toward Smartsville. At the top of the hill some distance east of the Empire ranch, he let an old man named **Mellarkey** off the seat, so that the latter could reach his home near by, through a footpath. A few seconds' later young Conlin was driving down the steep eastern incline of the hill, humming a song. He was in plain sight of all the people in Smartsville who were out of doors. Many were looking toward the hill, and when the young driver was down but a few feet they were stricken with fear at the sight which they had witnessed. The large wagon was seen to give a lurch forward and then go to one side, the horses being pulled about but making no effort to run. As the wagon went over of the outside of the grade which slopes gradually, young **Conlin** fell, all the time retaining possession of the lines. His body was caught by the brake, and he was forced down so that the left front wheel passed over his right thigh and left leg just below the knee. When he was reached by those who hurriedly went to his assistance he was sitting in the middle of the road, the horses being perfectly quiet. "Are you badly hurt, Ben?" said someone one. "No, but one of my legs is broken, I guess." He replied, without showing any signs of pain. He was then conveyed to his home, where it was found that the right thigh had been terribly mashed and the bones broken; also that his left leg had been broken, and his left wrist and hand considerably cut. The kneecap of the right leg was crushed. The doctors found that amputation offered the only hopes of saving his life, and stated that the operation would be attended by great risk. Yesterday morning at about 11 o'clock Doctors Brown and Powell put the young man under the influence of ether; and proceeded with the operation on the right leg at the thigh, the left leg having been bandaged. It took about an hour to perform the operation. After the patient had partially recovered consciousness he gave two or three gasps, asked for some water and died without showing signs of pain. The cause of the wagon getting out of **Conlin's** control was the breaking of an iron on the break. The deceased was a son of Thomas **Conlin**, aged 23 years, was a very popular young man in Smartsville and with those by whom he was known on the road. He was a member of Yuba Parlor, N. S. G. W. His cousin, Joe P. **McQuaid** of this city, was at his bedside until his death.

NEWSPAPER INFORMATION

8 Oct 1887 (Saturday); Page 2 Col 2; DEATHS. In Smartsville, October 6th, Thomas Benjamin **Conlin**, aged 22 years, 8 months and 17 days. Friends and acquaintances are respectfully invited to attend the funeral from the Catholic Church in Smartsville this afternoon at 2 o'clock.

25 Oct 1887; Page 2 Col 2; Deaths – In Yuba City, October 23d, only **child** of John and Mary **Keegan**, aged 1 year and 8 months. The remains were taken to Smartsville yesterday for interment.

9 Nov 1887 (Wednesday); Page 3 Col 3; An awful Fall. San Andreas. November 8th – John **Dougherty** last night fell 700 feet in the Sheep Ranch mine and was dashed to pieces.

10 Nov 1887 (Thursday); Page 3 Col 1; He Was From Smartsville. John **Dougherty**, the young man who met a fearful death by falling 700 feet down a mining shaft near San Andreas, which fact was related in the specials of the Appeal yesterday, was formerly a resident of Smartsville, where his parents now reside. He married a Marysville young lady named **Divver**.

23 Nov 1887; Page 2 Col 2; Marriages – In Smartsville, November 21st, by Rev. W. R. Willis, Wallace J. **Sanford** to Eva C. **Jones**.

24 Nov 1887 (Thursday); Page 2 Col 2; MARRIAGES. In Smartsville, November 23d, at the residence of the bride's parents, Geo. B. **Allen** to Miss Laura D. **Brown**, both of Smartsville.

27 Dec 1887; Page 2 Col 2; Died – In Smartsville, December 25th, Alice M. **Tifft**, aged 14 years. Funeral this morning at 11 o'clock in Smartsville.

28 Dec 1887 (Wednesday); Page 2 Col 2; MARRIAGES. In Smartsville, December 27th, by Rev. Father **Dalton** of Grass Valley, John J. **McGrath** of Linda to Miss Jane **Cain** of Smartsville.

3 Aug 1888 (Friday); Page 3 Col 2; John L. **Murphy** received telegrams from Smartsville yesterday announcing the accidental death of his nephew **Charlie**, son of Morris **Murphy**. No particulars were given further than that the boy was killed in the Pittsburgh mine.

4 Aug 1888 (Saturday); Page 3 Col 1; The death of Morris **Murphy's** son of Smartsville was due to the fall of a mass of earth upon him in an old mining tunnel where he was playing. He was found dead, when the earth was removed.

5 Aug 1888 (Sunday); Page 3 Col 2; Charles E. **Murphy**, the boy recently killed at Smartsville, was buried on Friday evening last. Thomas **Kelly**, Henry **Keegan**, John **Mellarkey**, John **Kelly**, Daniel **Dougherty** and James **McCarty** acted as pallbearers.

NEWSPAPER INFORMATION

7 Aug 1888 (Tuesday); Page 3 Col 4; Mrs. Jeremiah **Whalen** died at Smartsville last Saturday afternoon after a long illness, at the age of 82 years. She leaves a husband and many children and grandchildren, and was much revered by all those who knew her. The funeral took place at Smartsville yesterday afternoon, and was largely attended.

27 Dec 1888 (Thursday); Page 3 Col 1; Mrs. Ann **Burke**, one of the oldest residents of Smartsville, died on last Tuesday afternoon at her home. She was 76 years old.

17 May 1889 (Friday); Page 2 Col 2; Deaths. In Smartville, May 16th, Sarah A. **Curran**, aged 25 years and 8 months. Friends and acquaintances are respectfully invited to attend the funeral from St. Joseph's Cathedral, Sunday morning, at 10 o'clock, when high mass will be held.

17 May 1889 (Friday); Page 3 Col 1; SMARTSVILLE, May 16th – The sad intelligence was received in town at an early hour this morning of the sudden death of Miss Sarah Ann **Curran**, of Timbuctoo. For some time past she has not been in the best of health, being a sufferer from lung troubles, but it was not thought there was any immediate danger from her malady. This morning she arose and prepared breakfast for her brother, **Dennis**, who works in one of the mines, and started him off to his work. A short time after, she took a spell of violent coughing, which ended in a severe hemorrhage, from which death resulted in a few moments. "**Sallie**," as she was familiarly known, leaves a large circle of friends to grieve her parting, and three brothers, two of which are here. She was aged about 27 years.

2 Jul 1889 (Tuesday); Page 2 Col 2; Deaths – In Marysville, July 1st, Thomas **Walsh**, a native of Ireland, aged 55 years. Friends & acquaintances are respectfully invited to attend the funeral from St. Joseph's Cathedral tomorrow afternoon at 3 o'clock.

22 Aug 1889 (Thursday); Marysville Daily Appeal; Page 2 Col 2; Marriages – In Smartsville, August 21st, at the residence of the bride's parents, by Rev. J. H. **Jones**, Simon A. **Davis** and Julia A. **Tifft**.

5 Sep 1889 (Thursday); Page 2 Col 2; DIED. In Smartsville, September 4th, Amanda **Havey**, aged 5 years, 5 months and 8 days. Funeral from residence of John **Havey**, Smartsville, Friday at 1 p.m.

5 Nov 1889 (Tuesday); Marysville Daily Appeal; Page 1 Col e; Marriages – In Smartsville, November 4th, by the Rev. J. H. **Jones**, at the residence of the bride's parents, Robert **Beatty** and Frances E. **Tifft**.

12 Nov 1889 (Tuesday); Page 3 Col 3; Frank **McQuaid**, of Smartsville, was married Sunday morning to Miss Kate **McNamara**, of Sweetland, Nevada County. They left by early train from Marysville Monday morning, for a two weeks sojourn to San Francisco.

NEWSPAPER INFORMATION

6 Jan 1890; Page 3 Col 1; DIED. Name of the Man Who Died at the County Hospital. The man who was taken to the county Hospital suffering from a wound in the head, died Sunday morning. An Autopsy disclosed the fact that the imamate cause of death was a clot of blood at the base of the brain, due to rupture of a blood vessel, which produced paralysis of the left side. The deceased has been identified as William **McCarty** whose brother resides at Red Bluff.

25 May 1890; Page 2 Col 2; DEATHS – In Smartsville, May 23d, Mary **Presley**, a native of San Francisco, aged 22 years, 9 months, and 2 days.

26 Apr 1890 (Saturday); Page 4 Col 3; DIED.– In Marysville, April 26, 1890, Mrs. Mary A. **Divver**, a native of Ireland, aged 54 years, 4 months and 1 day. Friends and acquaintances are respectfully invited to attend the funeral from her late residence on Second street, east of A. on Monday, April 28th at 7:15 o'clock a.m. Interment at Catholic cemetery at Smartsville. Services at Catholic Church, Smartsville, at 11 o'clock a.m.

17 Jul 1890 (Thursday); Page 3 Col 1; Inquest on the Remains of Thomas H. **Curtis**. Coroner Bevan returned yesterday morning from Smartsville, where he was called to hold an inquest over the remains of a man found in the Yuba River near that place. The inquest was held late on Tuesday night. After hearing the evidence in the case, the jury rendered a verdict to the effect that the name of the deceased was Thomas H. **Curtis**, a former resident of Nevada City, and between 35 and 40 years of age: that he came to his death from drowning about the 9th of July, his body having been found in the Yuba river, near the Parks Bar Bridge, on July 15th. Whether it was a case of accidental death or suicide they were unable to say. At 1 o'clock a.m. yesterday the remains were consigned to their last resting place in the I. O. O. F. cemetery at Smartsville. The Rev. A. D. Wheaton officiated, and the following gentlemen acted as pall-bearers: Thomas **Mooney**, Wm. **Pierce**, A. **Mooney**, Mark **Curran**, P. **McGovern** and Robert **Jeffrey**.

5 Aug 1890 (Tuesday); Page 2 Col 2; Deaths. At the Yuba County Hospital, August 1st, Patrick **Burke**, a native of Ireland, aged 84 years.

7 Oct 1890 (Tuesday); Page 1 Col 1; Demise of an Old Citizen of Smartsville, Monday. Lawrence **Daly**, who has resided in Smartsville for over a quarter of a century, died at an early hour Monday morning of paralysis. He was an old time Democrat, straightforward and honest in all his dealings with the public, and was respected by all who knew him. The interment took place this afternoon.

8 Oct 1890 (Wednesday); Page 4 Col 2; Died. Daley – In Smartsville, October 6, 1890, Lawrence **Daly**, a native of Ireland, aged 58 years.

4 Nov 1890 (Tuesday); Page 2 Col 2; Deaths – In Smartsville, November 2d, Mrs. Elisa **Gratiot**, a native of Canada, aged 66 years. The deceased was a sister-in-law of ex-Minister to France, E. B. **Washburn**. (Dubuque, Iowa, papers please copy).

NEWSPAPER INFORMATION

16 Nov 1890; Page 2 Col 3; Deaths – In Smartsville, November 15th, Harry Clinton **Jeffery**, youngest son of Mr. And Mrs. George **Jeffery**, aged 13 years 10 months and 18 days.

15 February 1891 (Sunday); Page 2 Col 2; Died- In Marysville, February 14, 1891, Alexander **Doyle**, Aged 30 years, 3 months and 4 days. Funeral from the residence of Mrs. Charles **Armstrong** in Smartsville, this afternoon at 2 o'clock.

28 Feb 1891; Page 4 Col 2; Died – Magonigal – At his residence, near Smartsville, February 28, 1891, William **Magonigal**, a native of Ireland, aged 72 years. Friends and acquaintances are invited to attend the funeral from his late residence tomorrow, at 11 o'clock a.m. Interment in Fraternal cemetery.

1 Mar 1891; Page 1 Col 2; Funeral of Wm. **Magonigal** – The funeral of William Magonigal, took place at Smartsville, from Union Church, at 1:30 yesterday afternoon and was very largely attended. Rev. J. T. **Vineyard** was the officiating clergyman. The following gentlemen acted as pallbearers: John **Saunders**, Orlo **Whiteside**, Miles **Wellman**, T. A. **Newbert**, Samuel **Kuster**, and William **Marple**.

26 Mar 1891 (Thursday); Page 2 Col 2; DEATHS. In Smartsville, March 24th, Martin **Fortune**, a native of Ireland, aged 74 years.

17 Apr 1891 (Friday); Page 2 Col 2; DIED. In Smartsville, April 15tth, Mary Ellen **Driscoll**, aged 19 years, 4 months and 23 days. Friends and acquaintances are respectfully invited to attend the funeral this afternoon at 2 o'clock.

1 Jul 1891 (Wednesday); Marysville Daily Democrat; Page 1 Col 2; Death of Stephen **Murphy**. Stephen **Murphy** died in Paso Robles last evening, aged 75 years. He was the father of John L. **Murphy**, the efficient police officer of this city. Mr. **Murphy** had a large circle of friends in Yuba county, having lived in Smartsville several years. He was considered well off and will leave his relatives some property.

10 Jul 1891 (Friday); Page 2 Col 2; DEATHS In Smartsville, July 9th, John **McGovern**, aged 64 years.

10 Jul 1891 (Friday); Page 2 Col 2; DEATHS. Near Smartsville, July 9th, Alice Elizabeth **Vineyard**, beloved wife of William **Vineyard**, aged 37 years, 2 months and 22 days. Friends and acquaintances are respectfully invited to the funeral from her late residence this afternoon at 2 o'clock.

3 Nov 1891; Page 2 Col 2; Married – In this City, November first, at the Methodist Episcopal Parsonage, by the Rev. J. P. Macanlay, William **Bushby** and Miss Fannie **Owen**, both of Yuba County.

17 Dec 1891; Page 2 Col 2; Died – Near Smartsville, December 16th, Ellen **Wallace**, a native of Scotland, aged 88 years.

NEWSPAPER INFORMATION

18 Jan 1892 (Monday); Marysville Daily Democrat; Page 1 Col 3; Death of Mrs. (Rosana) Stephen **Murphy**. A telegram was received this afternoon by D. P. **Donahue** announcing the death of Mrs. Stephen **Murphy**, mother of John L. **Murphy**, at Gonzales, San Luis Obispo county. The deceased was a resident of Smartsville for a number of years and was well known in this vicinity. The remains will arrive on the Oregon Express tonight and the funeral will take place from the undertaking rooms of R. E. **Bevan** tomorrow morning. Interment at Smartsville.

26 Jan 1892; Page 2 Col 2; Died – In Smartsville, January 24th, Lucretia **Congdon**, wife of F. D. **Congdon**, a native of Connecticut, aged 61 years, 1 month and 29 days. Friends and acquaintances are respectfully invited to attend the funeral from her late residence in Smartsville this afternoon at 2 o'clock.

13 Nov 1892; Page 1 Col 5; The funeral of Nancy **Magonigal** took place at Smartsville yesterday. The funeral was held at the Union Church, Rev. John T. **Vineyard** officiating. The pallbearers were: Samuel **Kuster**, Thomas **Bonnie**, Purk **Hutchinson**, Orio **Whitesides**, Miles **Wellman** and George **Bushby**. The funeral was largely attended.

29 Jul 1892 (Friday); Marysville Daily Democrat; Page 2 Col 2; DIED. Pryor – In Smartsville, July 28, 1892, Michael **Pryor**, aged 60 years. Friends and acquaintances are respectfully invited to attend the funeral from the Catholic church at Smartsville tomorrow morning at 10 o'clock.

30 Jul 1892 (Saturday); Marysville Daily Democrat; Page 1 Col 1; DEATH OF W. E. **SMITH**. He Passes Away Early This Morning While Visiting Friends. Word was received in this city today from Smartsville, stating that W. E. **Smith**, mention of whose serious illness was made in the Democrat last evening. Died there at an early hour this morning. Mr. **Smith** was a member of Marysville Encampment, No. 6, and Oriental Lodge, No. 45, I. O. O. F.; Marysville Parlor, No. 6, N. S. G. W., and Marysville Lodge, No. 1,656, K. of H. Deceased was well known and had many friends throughout the county. He was employed at the meat market of Jenkins & Crowell during the past twelve years, and the members of the firm say that he was one of the most trustworthy employees they have ever had, and was always to be depended upon when wanted. Mr. **Smith** is the first member of the Marysville Parlor, No. 6, N. S. G. W., that has died since the Parlor was organized September 10, 1880. He was married in Smartsville November 11, 1891, to Miss Eugenia **Hapgood**, of that place, and it was at her parent's home that he died. His life was insured for $2,000 in the Knights of Honor, and he also had a policy in an Odd Fellows' Insurance Company, but the amount of it could not be learned. Coroner Bevan, who was in Smartsville at the time of his death, telegraphed that he would bring the remains to this city tonight, and that the funeral would take place Tuesday..

29 Jul 1892 (Friday); Page 2 Col 2; DIED. In Smartsville, July 28th, Michael **Pryor**, aged 60 years. Friends and acquaintances are respectfully invited to attend

NEWSPAPER INFORMATION

the funeral, which will take place from the Catholic church at Smartsville, Saturday morning at 10 o'clock.

24 Feb 1893 (Friday); Marysville Daily Democrat, Page 2 Col 2; THE TOMB. **McGovern** – At Timbuctoo, Yuba County, February 23, 1893, Mrs. John **McGovern**, a native of Ireland, aged 60 years.

28 Feb 1893 (Tuesday); Page 1 Col 5; A well known practitioner of this county died Saturday. Dr. Ray Wilkins **Tifft** died rather suddenly last Saturday evening at his residence at Smartsville of neuralgia of the heart. The doctor had not been in good health lately and was recovering from an attack of la grippe, but was able to be around and attend to his patients. He was feeling better on Saturday than he had been for some time and accompanied by his wife he visited a patient at Pleasant Valley. On his return home he ate a hearty dinner and soon after was taken with a pain in the region of the heart, suffered intensely for a few minutes and then died. He has been a resident of Yuba county for about thirty five years and of the State for nearly forty years. He had resided in Smartsville for the past eighteen years and was much respected by all classes. He leaves a wife and four daughters to mourn his loss namely, Mrs. J. W. **Hill**, Mrs. **Harris**, Mrs. S. A. **Davis** and Mrs. Robert **Beatty**. He was a native of Ohio and about 61 years of age. He was a member of Roses' Bar Lodge, No. 89, F. & A. M., of which he was Worthy Master; of Yuba Lodge, No. 5, I. O. O. F., of this city (Marysville), and of Smartsville Lodge, No. 119, A. O. U. W. The funeral will be held under the auspices of Roses' Bar Lodge this morning at 10 o'clock. Services will be held at his late residence, by the Rev. Edward L. **Allen** of this city (Marysville). Coroner **Bevan** did not think it necessary to hold an inquest as the attending physician certified that he had died of neuralgia of the heart.

1 Mar 1893; Page 1 Col 6; A large Attendance and Impressive Service at Smartsville. The funeral of the late Dr. Ray Wilkins **Tifft** took place in Smartsville yesterday and was the largest funeral seen in that section for many years. It was held under the auspices of Roses' Bar Lodge, No. 89, F. & A. M. The following other Masonic Lodges were represented; Nicolas Lodge, of Wheatland, No. 129; Yuba Lodge, No. 39; Corinthian Lodge, No. 9; Nevada Lodge, No. 13; Madison Lodge, No. 23, of Grass Valley. Smartsville Lodge, No. 119, A. O. U. W. was also well represented as wee several Odd Fellow lodges. The procession from the residence to the cemetery, was a very large one, there being sixty Masons, Thirty Odd Fellows and about twenty United Workmen in line. The beautiful Masonic services were read at the grave by the Hon. E. M. **Preston** and by the Rev. Edward L. **Allen**, who acted as chaplain. Some very handsome floral offerings were placed on the casket. The following acted as pall bearers; Ed **Hapgood** and S. **Davy**, representing the Masonic Lodges, T. A. **Newbert** and Oro **Whitesides**, the Odd Fellows and W. W. **Chamberlain** and O. **Woebler**, the United Workmen. There were seventy vehicles in line. The interment took place in the Smartsville cemetery.

NEWSPAPER INFORMATION

22 May 1893 (Monday); Marysville Daily Democrat; Page 1 Col 3; Death of an Irish Gentleman. John **Sweeney**, a native of Tralee, County Kerry, Ireland, died in this city at an early hour on Saturday evening. He was suffering from an attack of la grippe, contracted while on a recent visit to San Francisco, which resulted in his death. He leaves three sons, **John, Richard** and **Robert**, and a married daughter, Mrs. P. **Nugent**, of Live Oak. The funeral took place this morning at 10 o'clock from St Joseph's church, the following gentlemen acted as pall bearers: Harry **Dalton**, Patrick **Corcoran**, Malachy **Carew** and Samuel **Gunning**. Rev. Father **Coleman** officiated at the church and grave.

22 May 1893 (Monday); Marysville Daily Democrat; Page 3 Col 2; THE TOMB. In this city, May 20th, John **Sweeney** a native of Tralee, County Kerry, Ireland.

2 Aug 1893 (Wednesday); Marysville Daily Democrat; Page 1 Col 4; Death of a Smartsville Young man. Several telegrams conveyed the news to this city (Marysville), last evening, of the death, at Yreka, of John J. **Curran**, an native of Timbuctoo, a suburb of Smartsville. The young man died yesterday and it is surmised that his early taking way was caused by a hemorrhage from the lungs, as he was subject to these attacks. Coroner **Bevan** went to Yreka on last night's express train and will take charge of the remains, conveying them to this city, hence to Smartsville, where the interment will take place. The funeral will probably be held tomorrow. Two brothers of the deceased young man are in this city today, awaiting the arrival of their brother's remains. John J. **Curran** was 27 years and 13 days old at the time of his demise.

2 Aug 1893 (Wednesday), Marysville Daily Democrat; Page 1 Col 5; THE TOMB. **Curran** – In Yreka, August 1, 1893, John J. **Curran**, a native of Smartsville, Cal., aged 27 years and 13 days. Funeral notice hereafter.

23 Nov 1893; Daily Democrat Newspaper; Page 1 Col 4; The aged wife of R. E. **Welch**, and mother of James **Thomas** and Joseph **Welch**, of Cabbage Patch, died at the family residence at 2 o'clock yesterday afternoon, after an illness extending over six years. The deceased lady was a native of Missouri, and 67 years of age. The funeral will take place from the late residence at 2 o'clock tomorrow afternoon. The interment to take place at the Empire cemetery (Fraternal), near Smartsville.

24 Nov 1893 (Friday); Page 1 Col 4; Death of Mrs. R. E. **Welch**. Mrs. R. E. **Welch** died at her residence at the Cabbage Patch on Wednesday afternoon. The deceased was a native of Missouri and sixty seven years of age. The funeral will take place from the late residence at 2 o'clock this afternoon. The interment will take place at the Empire cemetery (Fraternal) near Smartsville.

5 Jan 1894 (Friday); Marysville Daily Democrat; Page 1 Col 1; BONANZA JIM. A Story of a Yuba County Man Who Farms, Mines, Produces Oranges and Wool. And Tells Stories. James **O'Brien**, the subject of this sketch, is a native of County Cork, Ireland, says the San Francisco Chronicle. Being of that energetic and ambitious nature which has ever characterized him through his later life, he at an

NEWSPAPER INFORMATION

early age, resolved to try his fortune in America. Accordingly, in 1848, he sailed for the United States. After a voyage marked only by the usual events, he landed in Westfield, Mass., late in the same year. Here he remained until 1858, at which time the rush to the gold fields of California had created such an excitement that thousands were leaving daily by every available route for the new El Dorado. Young **O'Brien**, with all the impulses of youth and health, resolved to cast his lot with the rest of those dauntless pioneers many of whose names have since become historical in the annals of this State. The close of 1858 found him at Smartsville, Cal., with his rocker, pan and long tom, enduring all the hardship attendant with the life of the early miner, sleeping upon the ground with only the stars for a roof and doing his own cooking and washing. How bountiful has been his reward in return for the manner in which he endured will be shown by the following: At Rose's bar, on the Yuba, near Smartsville, Mr. **O'Brien** first began his mining venture, n what was known in those days as river mining. As this character of mining could only be carried on during the summer months he spent the remainder of each year in prospecting the mountain regions. After six years of untiring labor, working his rocker in summer and prospecting in winter, he was at last rewarded by the discovery of the "Pickles Hydraulic Mine." Out of this and other adjacent mines over $13,000,000 has been taken in thirty years. In 1883 all the hydraulic mines were closed down and Mr. **O'Brien's** mine, like many others, has been closed ever since. Loath to be idle, however, Mr. **O'Brien** at once turned his attention to stock raising, and his success in this direction has been equally as great as it was in mining. He owns in divided tracts about 10,000 acres out of which he has 3,000 acres in grain. At Smartsville he owns a ten acre orange grove, which is considered one of the finest in the State. He is also extensively engaged in stock raising, having at the present time 3,000 sheep, 140 horses and 300 cattle, all of a very fine breed. Notwithstanding the many duties, which are naturally incumbent upon Mr. **O'Brien** from such a diversity of interests, he still finds time to take an active interest in the political affairs of the State. He has attended at least ten State conventions and has always been a conscientious and valuable worker in the cause of the Hydraulic miner. He is much esteemed by all who know him for his business integrity and his unbiased opinions toward both the miners and the anti-debris people. Mr. **O'Brien** is also the owner of a large and valuable hydraulic mine in Plumas county, for which he has a concession to operate under the system of dams. He now has a large force of men at work upon the construction of these dams, and by the opening of spring the mine will be running in accordance with the requirements of the law. Mr. **O'Brien** was married in 1860 to Miss Mary **Kirby** of Massachusetts. He is the father of eight children, six girls and two boys. Surrounded by his happy family in his comfortable home at Smartsville, he will, it is hoped, have many more happy years of health, wealth and pleasure to look forward to

13 Nov 1894 (Tuesday); Marysville Daily Democrat; Page 1 Col 4; Mrs. James **O'Brien** Departs this Life at Her Smartsville Home. At a few minutes past 11 o'clock today Mrs. James **O'Brien** breathed her last at her Smartsville home surrounded by those whom she loved best. The lady, who was one of the most respected residents of this section, has been a patient sufferer for many months with

NEWSPAPER INFORMATION

that dreaded complaint, Bright's disease, and her death was not unexpected. She leaves two sons, six daughters and a husband to mourn her demise. Undertaker Bevan went to Smartsville this afternoon to take charge of the remains. At the time of going to press the hour for the funeral had not been decided upon, but it will be on Thursday.

14 Nov 1894 (Wednesday); Marysville Daily Democrat; Page 4 Col 2; The Tomb. At Smartsville, November 13, 1894, Mary Kirby **O'Brien**, a native of Ireland, aged 62 years, 6 months and 28 days. The funeral will take place from the Catholic Church, Smartsville, tomorrow (Thursday) morning at 11 o'clock. Interment, Smartsville cemetery. No flowers. (San Francisco papers please copy)

14 Nov 1894 (Wednesday); Marysville Daily Democrat; Page 4 Col 2; The Tomb. Fisher – At Smartsville, November 12, 1894, Gracie **Fisher**, a native of California, aged 14 years.

24 Nov 1894 (Saturday); Marysville Daily Democrat; Page 1 Col 1; As Esteemed Christian Lady Breathes Her Last at an Advanced Age. At the advanced age of 66 years, 8 months and 29 days, Mrs. Margaret **McKenna** departed this life at the home of her son-in-law, Daniel P. **Donahue**, at 10 o'clock this morning. Fro several weeks the aged lady had been failing in health and she was prepared for the summons. Her life has ever been one of usefulness, gentle ness and forbearance and she has well earned the crown of glory that has been awarded her. A general breaking down and heart failure brought about her demise. She leaves three children, Mrs. **Donahue**, **Thomas A.**, and James **McKenna**, to mourn their loss. Mrs. **McKenna** was a native of Ireland but had resided in this city since 1855. The funeral will take place from St.; Joseph's church on next Monday morning at 10 o'clock. The interment will take place in the Catholic cemetery.

12 May 1895 (Sunday); Page 1 Col 4; The Passing of a Pioneer Smartsville Miner. County Clerk Gordon **Bowman** telephones to the Appeal office from Smartsville about 8 o'clock last evening that Daniel **McGanney** had died about a half an hour previous from a stroke of apoplexy. The deceased was an old and respected citizen of Smartsville, where he has resided since 1856. On his first arrival in Smartsville he worked in some mining claims and was a partner of Tom **Conlin** for years. He afterwards embarked in the Blue Point mine, and with his partner, J. P. Pierce, made considerable money, which he invested in land. When a few years ago had investments made him seek the protection of the courts and go into insolvency it was a great surprise, as he was considered as safe in money matters in his section of the country as the bank. The deceased was a native of Ireland and about 67 years of age. He had a large circle of acquaintances in this city that will regret to hear of his death. He was one of the old landmarks of Smartsville, and his death will also be regretted by his old friends and neighbors. He leaves a wife, four sons, **Edward**, who is an attorney in New York, **James**, **Frank** and **Dan**, who reside in Smartsville, and two daughters, **Annie** and **May**, to mourn his loss. Coroner Bevan left for Smartsville last night, but did not know if it would be necessary to hold an inquest as no particulars were received.

NEWSPAPER INFORMATION

4 May 1895 (Tuesday); Page 1 Col 8; Funeral of Daniel **McGanney**. The funeral of the late Daniel **McGanney**, Sr. took place at Smartsville yesterday and was largely attended. Father Twomey officiated at the services at the Catholic Church and at the Smartsville Cemetery, where the interment took place. Some handsome floral pieces were laid on the casket. The following gentlemen acted as pall bearers: John **McQuaid**, William **Cramsie**, J. **Marlin**, J. **Miller**, Dale **Sanders**, T. A. **Newbert**, Wm. **Riley** and M. **Breenan**.

11 Jun 1895 (Tuesday); Page 1 Col 6; To be Married Today. Dr. E. F. **Holbrook** and Miss Kate (Catherine) A. **O'Brien**, both residents of Smartsville, will be married this afternoon at the residence of Mrs. R. F. **Pierce**, Hyde Street, San Francisco. The nuptial know will be tied by Father **Otis** of St. Mary's. Dr. Geo. S. **Holbrook** will act in the capacity of best man, and Miss Agnes **O'Brien** will be bridesmaid for her sister. The Doctor is well known in this city, which he frequently visits, and where he has many friends. Miss **O'Brien** is the daughter of James **O'Brien** of Smartsville, one of the best-known residents of Yuba county, and is a most estimable young lady who will make a charming wife for the young physician.

11 Jun 1895 (Tuesday); Page 1 Col 7; John **Dewan** died at his home in Smartsville on Sunday where he has resided for the past seven years. He resided for a number of years at Sweetland and was much respected in the community. The deceased was a member of Pactolus Council, No. 134, Order Chosen Friends, which loses its first member since its organization twelve and one half years ago. His family will receive $2,000 from the lodge. He leaves a wife, five daughters, two of whom reside in this city, and two sons, to mourn his loss. The funeral will take place from the Catholic church at Smartsville at 9 o'clock this morning.

11 Jun 1895 (Tuesday); Page 3 Col 3; Died. In Smartsville, June 9th, John **Dewan**, a native of Ireland, aged 57 years and 9 months. Funeral from the Smartsville Catholic church at 9 o'clock this morning. Interment, Catholic cemetery, Smartsville.

6 Nov 1895 (Wednesday); Page 1 Col 5; Mrs. John (Elizabeth) **Doubt**, the wife of an old and respected inhabitant of Mooney Flat died yesterday at the advanced age of seventy years. The deceased was a native of Ireland. The funeral services will be held at the Catholic church at Smartsville this afternoon.

7 Nov 1895 (Thursday); Page 1 Col 4; After many years of gradual declining health Eliza (Elizabeth) **Doubt** passes away into the great beyond from whence no traveler returns. Mrs. **Doubt** died at her late residence in Mooney Flat yesterday morning about three o'clock. The deceased has been a great and long sufferer and on many occasions, undoubtedly, death to her, would not have been an unwelcome guest. But notwithstanding her many bodily pains and years of general debilitation the messenger delayed it's coming until she had lived the allotted time. Three score years and ten. She leaves a husband and one child, a son by adoption, to mourn her loss. The funeral took place today, services in the Catholic church at 2 o'clock.

NEWSPAPER INFORMATION

The interment was made in the Catholic cemetery. Mr. **Doubt**, husband of the deceased, has been confined to his bed for several days with intermittent fever.

8 Nov 1895 (Friday); Page 4 Col 2; The Remains Arrive. Allen **Presley**, the deceased San Quentin Guard, to be Buried Today. Allen **Presley** who formerly resided at Smartsville died at San Quentin Wednesday. Mr. **Presley** was a guard at the penitentiary. He has acted in that capacity for about two years. For many years deceased was employed as a blacksmith by the Excelsior company. He was forty nine years of age, was a native of Ireland and leaves a wife and three children. The remains arrived on the north bound Oregon express this morning and were taken in charge by undertaker Bevan. The funeral will take place at 2 o'clock today under the auspices of Rose Bar Lodge No. 89, F. & A. M. Interment will be made in the fraternal cemetery.

30 Nov 1895 (Saturday); Page 1 Col 7; Pneumonia Quickly Carries off Mrs. Mary **Pryor** of Smartsville. At 8 o'clock Thursday morning died at her home in Smartsville of pneumonia. She had been ill but a few hours, Wednesday night, she caught a cold, retired early and in the early morning her condition being worse a physician was called. It was too late. Deceased was the widow of Michael **Pryor**, was sixty years of age, and a native of Ireland. She leaves four daughters. They are Mrs. John **Havey** of Smartsville, Mrs. James **Hanley** of Nevada City, Mrs. John L. **Murphy** of San Francisco and Mrs. Thomas **Heggerty** of this city. The funeral procession leaves the home of Thomas **Heggerty** of this city at 8 this morning. It will at Smartsville where services will be held in the Catholic church and interment will take place in the Smartsville cemetery.

1 Dec 1895 (Sunday); Page 1 Col 4; The funeral of Mrs. Mary **Pryor** took place yesterday from the residence of Thomas **Hegertty**. A large number of friends followed the remains to the cemetery at Smartsville. Rev. Father Twomey of the Smartsville Catholic church officiated at the funeral services, which were held in that town. Flowers were numerous. The pallbearers were Daniel **Sanders**, Wm. O. **Reilly**, Mike **Horan**, and Jas. **Burns**, Anthony **Smith** and Andrew **McCarty**.

7 Jan 1896 (Tuesday); Page 1 Col 3; Supervisor Conrath Unexpectedly Loses His Mother-In-Law. Mrs. Elizabeth **Quinkert** the mother-in-law of Supervisor **Conrath** died at his home yesterday afternoon. The death was sudden and unexpected. When the Supervisor left his home to attend the meeting of the Board the lady was apparently quite well but in the early part of the afternoon he received a message notifying him of the death. She was seventy years of age and a native of Germany. She came here a short time ago from Detroit, Michigan, to visit her daughter, Mrs. **Conrath**. The Supervisor left for Smartsville early in the afternoon with Undertaker Bevan.

8 Jan 1896 (Wednesday); Page 4 Col 2; At 10 o'clock this forenoon the remains of Mrs. (Elizabeth) **Quinkert** the mother-in-law of Supervisor Conrath will be buried. The funeral will take place from the Smartsville Catholic Church, the

NEWSPAPER INFORMATION

interment to take place at the Smartsville cemetery. Mrs. **Quinkert** was also the mother of Mrs. Jas. **McGanney**.

29 Jan 1896 (Wednesday); Page 1 Col 2 & 3; Miss Margaret **Casey** died at the residence of her father Peter **Casey** in Rose Bar Township on Monday, after an illness extending over two years. The deceased young lady resided in this city for a few years before her health failed her, and made many friends, as she was a most estimable person whose amiable ways and genial disposition were admired. She was a native California, aged twenty-four years and ten months. Consumption and lung trouble was the cause of her death. The funeral will take place this morning from the Catholic Church, Smartsville, where services will be held.

30 Jan 1896 (Thursday); Page 1 Col 6; The funeral of Margaret **Casey** took place yesterday morning services being held at the Catholic church, Smartsville, where Father **Twomey** officiated. The following gentlemen acted as pallbearers; Charles **McWilliams**, John **Toland**, Richard **Cain**, Edward **Bushby**, William **Boardman** and William **Toland**. A large cortege followed the remains to the Catholic cemetery where the last prayer was said and then the interment of a most estimable young lady took place.

28 Feb 1896; Page 1 Col 5; Death at Empire Ranch – An estimable young lady succumbs to the dread La Grippe – From Smartsville comes the sad news of the death of Miss Nellie McMurry **Mooney**, the twenty one year old daughter of the late Thomas **Mooney** of Empire Ranch. A protracted illness caused by a third attach of La Grippe was the cause of death. She departed this life at a late hour last evening. Besides her mother, five sisters and two brothers survive the deceased, who was a twin to one of the latter. At the time of going to press the hour of the funeral had not been decided upon.

7 Jul 1896 (Tuesday); Page 4 Col 2; Whiskey and Sunstroke attributed as the Cause. An overdose of Whiskey and sunstroke caused the death of Tom **Dougherty**, a miner, near the Chinese store, Timbuctoo, on Saturday. After leaving Smartsville, where he had sold some bold dust, he went to Timbuctoo and commenced to celebrate on Chinese gin. His legs became weary and his head muddled, he went to sleep out in the sun, and when found later in the day was dead. Coroner **Bevan** was notified and after the local physician had examined the body and a coroner's jury had viewed the remains, they were brought to this city. Deceased was a native of Ireland and forty-five years of age. He resided at one time at the Oregon House. When the announcement of his death was made at Smartsville it was thought that there had been foul play on account of his having had about $620 in his possession earlier in the day, and having only $1 when his clothes were searched after death. Justice McNutt being ill, Coroner Bevan went to Smartsville yesterday to hold the inquest. Nothing new was developed at the inquest. Dr. **Holbrook's** testimony showed that death was caused primarily from cerebral congestion, and secondarily from undue exposure to the sun while in a debilitated condition from the effects of sickness and alcohol.

NEWSPAPER INFORMATION

5 Aug 1896; Page 2 Col 2; Died – In Smartsville, August 4th, Mrs. Alice **McNutt**, a native of England, aged 77 yrs. Funeral notice hereafter.

6 Aug 1896; Page 1 Col 2; Mrs. Alice **McNutt** died at her late home yesterday at 3:30 p.m. The deceased was well and favorably known throughout this entire neighborhood and by her best of friends elsewhere as "Auntie". Although not feeling well for some time her death was very unexpected as well as sudden. "Auntie" was one of the most active, nimble and agile women in Yuba County. At the time of her death she was seventy-five years, one month and nineteen days of age, and was a native of England. She leaves a husband to mourn her loss. The funeral will take place tomorrow at 1 p.m. Services in the Union church. J. T. **Vineyard** will officiate.

27 Oct 1896; Page 1 Col 4; The angel of death visited a happy home in Smartsville on Sunday night. Mrs. Frances E. **Beatty**, wife of Robert **Beatty**, and daughter of the late Dr. R. W. **Tifft**, died very suddenly. She was taken suddenly ill at her home about 7:30 o'clock in the evening, and died two hours later, after suffering considerable pain. Dr. E. F. Holbrook, who attended her, stated that death resulted from internal hemorrhage, some of the internal organs having been ruptured. She leaves a husband to whom she was devotedly attached, a son five years old, a mother and three sisters to mourn her loss. She was born in this county, and was thirty-five years of age. The funeral will take place from the family residence tomorrow afternoon at 2 o'clock. Interment Smartsville Cemetery. Mrs. **Beatty** was a member of Sodality Rebekah Lodge, No. 170, of Smartsville, under whose auspices her funeral will take place.

5 Nov 1896 (Thursday); Page 1 Col 6; Broke His Neck In A Scuffle. Mark **Curran** a former resident of Smartsville engaged in a scuffle with another man at the Magalia mine, Butte county, yesterday morning, which resulted in **Curran** breaking his neck, dying a few seconds later. He was a native of California and forty years of age. His brother Daniel **Curran**, lives in Smartsville. He was not employed in the mine. Undertaker Bevan went to Magalia last evening to take charge of the remains. The funeral will take place in Smartsville.

6 Nov 1896 (Friday); Page 1 Col 6; He Was Too Slow in Getting Up to the Bar. Coroner R. E. Bevan arrived from Magalia yesterday with the remains of Mark **Curran**. The funeral will take place from the residence of Mrs. Smith at Timbuctoo this morning at 10 o'clock. Interment Smartsville cemetery. At the Coroner's inquest held in Butte county the Jury found that the deceased met his death in a playful scuffle with Al Jordan at Magalia. The Oroville Mercury in speaking of the affair says: From reliable information we learn that **Curran** met his death, not in a wrestling match but from being pulled off his feet and landed on his head. The report says that a crowd of miners gathered at a saloon after the polls were closed and proceeded to indulge in the cup that is red. Mr. **Curran** was invited to take a drink, and a man by the name of Jordan thinking that **Curran** did not respond as quick as he would have him do so grabbed **Curran** by the neck and gave him a sudden jerk, throwing Mr. **Curran** to the floor heavily. Friends noting

NEWSPAPER INFORMATION

that he did not rise, rushed to his aid and found that his neck had been broken by the fall. He died in five minutes after the accident.

7 Nov 1896 (Saturday); Page 1 Col 6; The funeral of Mark **Curran** who met with a fatal accident at Magalia, took place yesterday from Mrs. **Smith's** residence at Timbuctoo. Father Twomey officiated at the house and at the Smartsville cemetery, where the interment took place. The following acted as pallbearers; William R. **McConnell**, John H. **Lehman**, John **McKeel**, M. H. **McCarty**, T. **Smith** and John **O'Connor**. There was a good attendance as the deceased had many friends in the vicinity of Smartsville.

30 Jan 1897 (Saturday); Marysville Daily Democrat; Page 1 Col 3; "Grandma" **Havey** of Smartsville Called to Eternal Rest. In the eighty third year of her age, Mrs. Mary **Havey**, relict of the late Patrick **Havey**, entered into rest at an early hour this morning at the home of her son-in-law, William **Cramsie**, in Sicard Flat, a suburb of Smartsville. "Grandma" **Havey** was a native of Ireland. She located at Smartsville with her husband in the days when the mining camp was at its zenith, and she saw many years of usefulness. Four daughters, Mrs. Wm. **Cramsie**, Mrs. James **Byrne**, Mrs. Morris **Murphy** and Miss Maggie **Havey**, and one son, John **Havey** of Timbuctoo, survive the old lady. Her grandchildren, among whom are **John**, **Will** and Joe **Cramsie**, and Miss Adie **Byrne**, well known here, number twenty-six. One son, Patrick **Havey**, died in this city several years ago. The funeral services will be conducted in the Smartsville Catholic church at 2 o'clock tomorrow afternoon, the interment to take place in the Smartsville cemetery.

31 Jan 1897 (Sunday); Page 1 Col 5; At Her Smartsville Home on Last Friday Night. Mrs. Mary **Havey**, and respected resident of Smartsville, where she has resided since 1860, died at the residence of her son-inn-law, William **Cramsie**, on Friday evening. The deceased was a native of Ireland and eighty-five years of age. She was the relict of the late Patrick **Havey**. Five daughters, Mrs. William **Cramsie**, Mrs. James **Byrne**, Mrs. Morris **Murphy**, Mrs. Martin **Jennings** and Miss Margaret **Havey** and one son, John **Havey** of Timbuctoo, besides numerous grandchildren, survive to mourn her loss. One other son, Patrick **Havey**, died in this city several years ago. The funeral services will take place in the Smartsville Catholic church at 2 o'clock this afternoon. Interment in Smartsville cemetery.

5 May 1897 (Wednesday); Page 4 Col 3; MARRRIED. At Bonanza Ranch, Smartsville, Yuba County, Mar 2, by Father Twomey, F. J. **McGanney** and Miss Georgia **Gray**.

8 May 1897 (Saturday); Page 2 Col 1; Died at Smartsville. Grass Valley Union: Maurice **O'Connell** received a telephone message yesterday afternoon appraising him of the death of John **Smith**, who formerly worked for him in this city. Death resulted from consumption. Two children – boys ten and thirteen years of age – survive him. Deceased was well know here, having been employed for a number of years by the late Dennis **Meager** and Mr. **O'Connell**. The funeral will take place on Saturday afternoon at Smartsville.

NEWSPAPER INFORMATION

8 May 1897 (Saturday); Marysville Daily Democrat; Page 1 Col 4; John **Smith** who died at his home in Smartsville Thursday, was buried this afternoon in the Smartsville cemetery. Deceased was an old resident of Rose Bar Township and had been employed part of the time in Grass Valley. He had been ill with consumption a long time and was a widower. Two children survive him, one ten years of age and the other thirteen.

2 Jun 1897 (Wednesday); Marysville Daily Democrat; Page 4 Col 1; After a lingering illness extending aver a period of several months, Mrs. Michael (Margaret) **Conboy** died at her home in Smartsville this morning. Her husband, who has been receiving treatment at the county hospital on account of injuries sustained in a fall at the mines, has gone home to attend the funeral, which will take place tomorrow afternoon at 2 o'clock.

3 Jun 1897 (Thursday); Marysville Daily Democrat; Page 1 Col 3; Bright's Disease was the Primary Cause of the Lady's Death. From Dr. E. F. **Holbrook**, who was here from Smartsville today, it is learned that chronic Bright's Disease and uremia were the cause leading to the death of Mrs. Margaret **Conboy** of that place, which occurred yesterday at 6 a.m. The funeral took place from the Catholic Church at Smartsville at 2 o'clock this afternoon, Rev. Father **Twomey** presiding at the services, which were conducted in the presence of a large concourse of sorrowing friends. In their hour of trial the bereft husband and daughter have the heartfelt sympathy of the community.

4 Jun 1897 (Friday); Marysville Daily Democrat; Page 1 Col 3; The Remains to Arrive from Port Costa For Burial Here. A private letter received by Daniel **McDonald** this morning conveyed the news of the death of Patrick **Daly**, a former resident of Marysville, which occurred at Port Costa last night. Daley lived in this city in the early sixties, following his trade, that of shoemaker. On moving to Smartsville, where he resided for twenty-five years, he was similarly engaged, amassing a comfortable sum in the palmist days of the mining camp. When the mines closed down he removed to Port Costa with his wife and four daughters. The young ladies have since married. Deceased was a native of Ireland and about 65 years of age. The remains will arrive tonight on the Knights Landing train or Oregon express, the interment to take place in the family plot at the Catholic cemetery tomorrow forenoon.

4 Jun 1897 (Friday); Marysville Daily Democrat; Page 4 Col 2; The Tomb. Daley – At Port Costa, June 4, 1897, Patrick **Daly**, a former resident of Marysville, a native of Ireland, aged 65 years.

5 Jun 1897 (Saturday); Marysville Daily Democrat; Page 1 Col 3; LAID TO REST. The Last Tributes of Respect Paid The Late Patrick **Daly**. The remains of the late Patrick **Daley**, a former resident of this city, who died at Port Costa Thursday, arrived on the Oregon Express train this morning. At 8 a.m. the body was conveyed from the Bevan undertaking parlors to St. Joseph's Church, where requiem low mass was read by Father **Kirley**. The interment was in the family

NEWSPAPER INFORMATION

plot, the following acting as pallbearers: P. **Lucey** and H. Boyle **Crockert**; S. O. **Gunning** and C. C. **Slattery**; Marysville: James **Driscoll** and John **Driscoll**, Mooney Flat.

20 Jul 1897; Page 1 Col 3; The funeral of Samuel **Magonigal**, who died at Cisco, on Monday morning, took place yesterday from the old family homestead, six miles south of Smartsville, where Rev. J. T. **Vineyard** officiated. The interment took place at the Smartsville cemetery, where the Odd Fellows services were read by John Toland, N. G., of Fredonia Lodge, No. 188, of which deceased was a member. The following acted as pall bearers: D. L. **Cantlin**, John **Kuster**, N. A. **Hortang**, W. **Huling**, W. **Luke** and J. **Navary**. The funeral was well attended considering the short notice.

23 Sep 1897 (Thursday); Page 1 Col 4; After a lingering illness of several months' duration from Bright's disease of the kidneys the spirit of Michael **Horan** left its tenement of clay last Saturday morning at 10 o'clock. The deceased was highly respected throughout this community, where he had lived and labored for many years. He was a native of Ireland, and aged 42 years. He leaves a wife and five little children to mourn his loss. The funeral services were held in the Catholic church at 2 o'clock p.m. Sunday. A large procession of vehicles and pedestrians followed the remains to the Catholic cemetery, where the interment was made. The following gentlemen were the pallbearers: Thomas **Conry**, John **McQuaid**, John **Burns**, Timothy **Linehan**, William **Dempsey** and William **O'Connor**.

9 Jan 1898 (Saturday); Page 4 Col 3; Died – In Smartsville, January 28th, John **Bach**, a native of Germany, aged 67 years, 7 months and 9 days. The funeral will take place Sunday afternoon at 2 o'clock under the auspices of Smartsville Lodge of Odd Fellows. Interment Smartsville cemetery.

15 Nov 1898 (Tuesday); Page 4 Col 2; Funeral and Interment took Place at Smartsville Sunday. The remains of John J. **Daly**, who was killed by masked robbers while acting as an election officer at Butte, Montana, arrived here on the Oregon express Sunday morning and where taken to Smartsville, where the funeral and interment were held. There were a large number of the sorrowing relatives and friends present at the church, where requiem high mass was said by Rev. Father Hines. The interment was made in the Smartsville cemetery. The following young men acted as pall bearers: Will **Cramsie**, W. **Jeffery**, Will **Conlin**, John **Cramsie**, J. H. **McQuaid** and J. **Linehan**.

17 Nov 1898 (Thursday); Page 1 Col 2 & 3; The remains of John J. **Daly**, the victim of the assassin's bullet last Wednesday morning in Butte City, Montana, the particulars of which have already appeared in the Appeal, were laid to its final rest in the Catholic cemetery last Sunday at 12:30 p.m. The large number who came from far and near to pay the last tribute of respect to the memory of the deceased well attested to the high esteem in which he was held. Probably in no place was he better known than in Smartsville, where he with his parents resided over twenty years, and was well beloved by all. His sudden, untimely and unnatural death is

NEWSPAPER INFORMATION

greatly deplored and the sincere sympathy of this entire neighborhood is extended to the heartbroken widowed mother, the two sorrowing brothers and three sisters who are left to mourn his loss.

18 Dec 1898 (Sunday); Page 1 Col 7; The funeral of Thomas **Conroy** who died in Smartsville on Wednesday took place on Friday afternoon, the services being held in the Catholic church. Father Hynes officiated in the absence of Father Twomey, the pastor of the church. The deceased was a pioneer resident of Yuba county, having been admitted to citizenship in the District Court of Yuba County on August 14, 1855. He leaves one daughter and two sons to mourn his demise. The interment took place in the Smartsville Catholic cemetery.

3 Jan 1899 (Tuesday); Page 4 Col 2; John C. **Walker** Falls from His Wagon and Is Killed. Coroner A. B. **Hopkins** was notified yesterday afternoon that a man had been found dead near Smartsville, and he at once left for that place to hold an inquest. The following particulars concerning the death were received at the Appeal office over the telephone last evening: A man named Jim **Walsh** while driving along the Wheatland road one mile and a half from Smartsville found a six-horse team and high wagon standing in the road. He made search for the driver, and soon found him a short distance away with his head badly crushed, and an examination showed that he was dead. From all appearances he had fallen off the wagon, and one of the wheels must have passed over his head. It was ascertained that the name of the deceased was John C. **Walker**, he was about 40 years of age, and was a former resident of Smartsville. He has been working for Billy **Freeman** on the McGanney tract, and left for his home on Saturday. On account of the heavy rain on Sunday he stopped all day at the Brady ranch and started out again yesterday morning. E. B. **Walker**, a brother of the deceased, arrived in this city on the Knight's Landing train last night, accompanied by his wife from Sacramento. He had been notified of the death of his brother and was anxious to get particulars regarding the fatal accident. An inquest was to be held last night.

6 Jan 1899 (Friday); Page 1 Col 4; Funeral of John C. **Walker**. The funeral of the late John C. **Walker**, who met with a fatal accident last Monday, took place Wednesday. Rev. Mr. **Vineyard** officiated at the Union Cemetery, Smartsville where the interment took place. The following gentlemen acted as pall bearers: A. G. **Wheaton**, John **Bach**, S. A. **Newbert**, John **Peardon**, Robert **Beatty** and Robert **Woodroff**.

12 Jan 1899; Page 4 Col 2; The remains of the late Daniel **Cory** arrived from Pentz, Butte County, yesterday and will be conveyed to Smartsville for burial Saturday. In the meantime the body will lie in state at Trinity Chapel in this City.

12 Jan 1899; Page 8 Col 3; The Tomb – Cory – At Pentz, Butte County, January 8, 1899, Daniel **Cory**, a native of England, aged 69 years. Funeral at Smartsville next Saturday at 11 a.m. Interment family plot, Smartsville.

NEWSPAPER INFORMATION

4 Feb 1899 (Saturday); Page 1 Col 5; Thomas J. **Kerrigan** died yesterday afternoon at fifteen minutes to 4 o'clock from heart trouble. He has been in poor health ever since the election, and his death has been momentarily expected for several days. He was born in Sutter county twenty-eight years ago, and is a son of Mrs. Mary **Kerrigan**, who resides on a ranch about six miles southwest of Yuba City. He leaves three sisters, Kate Annie and Sister Mary Eugenia, S. N. D., and two brothers, **P**. H. and P. A. **Kerrigan**. Tom **Kerrigan** was one of the best known and most popular of Sutter county's young men. He had a large number of friends also in Marysville and Yuba county who will be sincerely sorry to hear of his death. He was a candidate for the position of County Clerk at the last election on the Democratic ticket. The funeral will take place next Sunday morning at 10:30 o'clock from St. Joseph's church, Marysville, where requiem mass will be celebrated. The interment will be made in the Catholic cemetery.

4 Feb 1899 (Saturday); Page 4 Col 3; DIED. In Sutter county, February 3d, Thomas J. **Kerrigan**, a native of California, aged 18 years. The funeral will take place from St. Joseph's church tomorrow (Sunday). Requiem mass will be celebrated at 10:30 a.m. Interment Catholic cemetery.

7 Mar 1899; Page 5 Col 2; Funeral of Judge **McNutt** – A Delegation of Marysville Masons attended the Obsequies. The following Masons left for Smartsville this morning to attend the funeral of the late Judge J. F. **McNutt**, who was a member of Yuba Lodge and Washington Chapter of this city; A. H. **Suggett**, F. H. **Day**, R. E. **Bevan**, J. F. **Eastman**, C. J. **Covillaud**, F. A. **Crook**, H. **Burner**, W. T. **Henn**, G. W. **Hamerly**, F. W. **Potter**, I. G. **Cohn**, G. H. **Baird**, G. W. **Hall** and Sol. **Lewek**. The obsequies were held in Masonic Hall in Smartsville at 2 o'clock this afternoon. The burial was in Fraternal cemetery.

9 Apr 1899 (Sunday); Page 1 Col 3; John Rose Dead. One of Yuba county's Oldest Pioneers Pays Nature's Debt. A long and honorable life. He came to the state when San Francisco was Yerba Buena with but eleven houses. John **Rose**, a pioneer resident of Yuba county, died at his home two miles southwest of Smartsville a little before midnight Friday night. Be his death one of the oldest landmarks of this county has been removed. The deceased was born at Leith, Scotland, on March 17, 1817, and left London on July 4, 1837, for the United States. He came by sail boat via Cape Horn, stopping en route to trade along the coast of South America, arriving at Yerba Buena (now San Francisco) in October 1839. The business of the Mexican barque, Clarita, on which he arrived was trading in hides and tallow. At the time of his arrival the streets of what is now San Francisco were only paths or tracks, the city consisting of eleven frame houses and stores. The Mission Dolores was about three miles out of town. The Alcalde of Yerba Buena at that time was Jesus **Noe**, a native of California, who administered law and Mexican justice soon after his arrival Mr. Rose made a visit in charge of the ships' boats to Sonoma, to receive a cargo of coin, at which place he met for the first time, General Vallejo, who was then a young man, and who was the first California Ranchero whom he met could speak English. He was at that time commandaunte of Alta California. After remaining at Yerba Buena a year he

NEWSPAPER INFORMATION

embarked in a voyage along the coast to Peru and Chili, returning to Monterey in 1843. he remained at Monterey a year, and then embarked in ship carpentering at Yerba Buena with two others, Wm. J. **Reynolds** and a Mr. **Davis**. The partners began the construction of a vessel, but were compelled to discontinue it as the requisite lumber had to be cut in Oregon, and both it and other necessary materials could not be obtained cheaply enough. The partners were building a grist mill for General **Vallejo** when gold was discovered, most of the men in their employ left at once, but a few were persuaded to remain and finish the mill by the agreement. To take them to Sutter's mill in a wagon. This was done, the party arriving on the American river in June 1848. The next month the men who had been working for Mr. **Rose** went to the Yuba river, and located on what was afterwards known as Rose's Bar, the diggings being worked on shares. This well known bar has the distinction of being the first point on the Yuba river were gold was discovered. **Rose** and his partner started a store at the bar, and as **Rose** did the buying at Sacramento, in that way the locality became known as Roses' Bar. A company of fifty men among whom was the late William H. **Parks**, commenced to dam the Yuba river so as to mine the bed about that time, Roses' store furnishing the supplies. They completed the dam and commenced to work early in October. The rains set in on the eighth and in two days the water overflowed the dam and washed it away. In a few days work they had taken one thousand dollars each during the year the bar became very populous, and in 1850 there were two thousand men working there. The deceased, who was one of the first white settlers at Rose's Bar, owned at one time considerable land Rose and his partners bought the grass land along the south bank of the Yuba river from Michael C. **Nye** in the Spring of 1849. They kept a large number of cattle which they grazed on the plains and from their banks supplied the miners with meat. **Rose** and **Kenlock** lived at the old house where the town of Linda was afterwards built. In December 1849, the Linda company arrived at the ranch in the steamer "Linda", and they were so well pleased with the location and beauty of the spot, that they advised Rose to lay out a town. They promised to take or sell enough lots to pay him for any outlay. The partnership of **Rose**, **Reynolds** and **Kenlock** was dissolved in the spring of 1850, Rose keeping the ranch as his portion. In the spring **Rose** laid out a town containing about one square mile, and named it Linda in honor of the company, and the little pioneer steamer. The "Linda" brought up a load of Marysville people, and the new town was inaugurated and christened over many a bottle of wine. **Rose** established a ferry across the river, a store was opened and a few dwellings erected. This was the condition of the place for two years, when all expectation of building a town was given up. In 1856 a bridge was built across the river at this point, and was carried away by the flood in December 1861. The site of the old town is now covered with slickens and overgrown with willows. From 1848 to 1870 the deceased mined mostly on the Yuba river, and was the first promoter and largest owner of the Enterprise Hydraulic Mine at Smartsville. He was naturalized by the treaty of Queretaro. The deceased was a brother-in-law of Lee **Newbert**, Sr. at this city (Marysville), and leaves a wife, three sons, **James**, **Frank** and **Oscar**, and three daughters, Miss Belle **Rose** of San Francisco, Mrs. Joe **Emery** and Mrs. Ethel **Rose**, the latter of whom teaches at Prairie School district, to mourn his loss. His son **James** is well known in Marysville having been employed as night watchman

NEWSPAPER INFORMATION

at the Buckeye Mills for several years. Mr. **Rose** was one of the invited guests at the Admission Day celebration in 1891, and also attended the Native Son's celebration in San Francisco a few years ago, having received a special invitation he being one of California's oldest pioneers. What a transformation scene he witnessed from the day he first landed at Yerba Buena, full of life and hope for a bright future. Although not possessed of much wealth in the declining years of his life, he has left to his family that which is far more to be prized, an honorable record, which has long since become part of the State's history. The funeral takes place this afternoon at 2:30. Interment, Fraternal cemetery, Smartsville.

26 Apr 1899; Page 1 Col 3; Mrs. Jane **Bowman** dead. An old and respected resident of this county gone to her last rest. Mrs. Jane Bowman, the aged mother of County Clerk Gordon **Bowman**, died at 3:45 yesterday afternoon at her home fourteen miles southeast of Marysville. She had been on the sick list for the past two weeks and the attending physician attributed her death to old age as much as anything, she having attained the ripe age of seventy-three years. The deceased had resided in this county for many years, and was much respected by her neighbors and many friends. She was a devoted mother and was possessed of other attributes that go to constitute a good Christian Woman. She leaves two sons **James**, who resides at Spenceville, and Gordon **Bowman**, County clerk of this county, who resides in this city; also two daughters, who have resided in the ranch with their mother. The relatives of the deceased have the sympathy of their many friends, for no family can suffer a greater loss than that which occurs when the mother is taken.

16 Apr 1899 (Wednesday); Page 1 Col 4 & 5; Mrs. Ann **Beatty**, an old and much respected resident of Smartsville, died at her home yesterday morning regretted by her many friends and acquaintances. The deceased was the wife of Richard **Beatty**, and the mother of **John** and Robert **Beatty**, and had reached the ripe old age of 82 years. The funeral will take place from the Smartsville Catholic church at 2 o'clock, tomorrow (Thursday) afternoon.

27 Apr 1899 (Thursday); Page 4 Col 3; The funeral of the late Mrs. Jane **Bowman** will take place at 10 o'clock this morning, from her late residence, fourteen miles southeast of Marysville. The interment will take place at the Fraternal cemetery, Smartsville. The deceased, who was a native of Ireland, has resided in Yuba county since 1862.

27 Apr 1899 (Thursday); Page 1 Col 3; Mrs. Richard (Ann) **Beatty** after several months of feeble health, with slow and gradual decline, breathed her last Monday morning at 4 o'clock. Her death was not unexpected as the family physician had several days ago pronounced her case a hopeless one. Mrs. **Beatty** was a good Christian woman, from whom the sing of death had been long since removed, and the "deep valley," that is so dark to some, was beautifully illuminated by the presence of her Lord. The deceased was a native of Ireland and 82 years of age. She leaves to mourn her loss a husband and two sons, **Robert**, the general manager of the Excelsior store, and **John**, of Santa Ana. The funeral will take place tomorrow from the Catholic church and the interment in the Catholic cemetery.

NEWSPAPER INFORMATION

27 Apr 1899 (Thursday); Page 2 Col 4; DIED. In Smartsville, April 25th, Mrs. Ann **Beatty**, wife of Richard **Beatty**, and mother of John and Robert Beatty, a native of Ireland, aged 82 years. The funeral services will be held at the Smartsville Catholic church this afternoon at 2 o'clock. Interment Smartsville Catholic cemetery.

27 Apr 1899 (Thursday); Page 2 Col 4; DIED. At her residence in Roses Bar Township, April 25th, Mrs. Jane **Bowman**, mother of County Clerk Gordon **Bowman**, a native of Ireland, aged 73 years, 3 months and 28 days. The funeral will take place at 10 o'clock this morning from her late residence where services will be held. Interment Fraternal cemetery, Smartsville. Rev. J. R. Vineyard will officiate.

21 Jun 1899 (Wednesday); Page 2 Col 3; DIED At Smartsville, June 20th, Mrs. John (Mary S.) **McQuaid**, a native of County Galway, Ireland, aged 65 years.

21 June 1899 (Wednesday); Page 4 Col 2; Mrs. John (Mary S.) **McQuaid**, who has been a patient sufferer of Bright's disease for some time past, died at her home at Smartsville yesterday, regretted by her neighbors and many friends. The deceased was a native of Ireland, aged 65 years, had been a resident of Smartsville over 40 years, and her death has cast a gloom over the entire community. She was a dutiful wife, a kind and indulgent mother, a good neighbor and a true friend. She leaves to mourn her loss besides her husband, John **McQuaid**, to whom she was married September 29, 1861, four sons, **Frank**, **Tom** and Joe **McQuaid**, who are residents of San Francisco, and John H. **McQuaid**, who resided in Smartsville, and three daughters, Mary **McQuaid** and Mrs. William **Carey**, who resides in San Francisco, and Margaret **McQuaid**, who resides in Smartsville.. Mrs. Thomas **Conlin** of Smartsville is a sister of the deceased. She was married September 29, 1861. Three sons arrived from San Francisco on the noon train yesterday expecting to see their mother before she passed away, but soon after reaching this city they were notified of her death. The funeral services will be held in the Smartsville Catholic Church tomorrow (Thursday).

10 Jul 1899; Page 1 Col 3; Samuel **Kay**, a member of oriental lodge, I. O. O. F. of this city, who was sick a long time, died at his home in Smartsville today. The hour of the funeral will be announced later.

11 Jul 1899; Page 4 Col 2; The funeral of the late Samuel **Kay**, whose death was announced in the Democrat yesterday, will take place from Smith's Hotel in Mooney Flat tomorrow afternoon at 2 o'clock. The interment will be in I. O. O. F. cemetery Smartsville. Deceased was a native of England and 68 years of age.

18 July 1899 (Tuesday); Page 4 Col 5; DIED In Smartsville, July 16th, Mrs. Bridget **Henderson**, a native of County Temporary, Ireland, aged 57 years. Services at the Smartsville Catholic Church this morning at 10:30. Interment Smartsville Catholic cemetery.

NEWSPAPER INFORMATION

1 Aug 1899 (Tuesday); Page 1 Col 6; A pioneer of one of Sutter county's best known families has been stricken down, and in the death of Mrs. Mary **Kerrigan** the community suffers a great loss. The affliction is made doubly sad because of the passing of her son, **Thomas**, which occurred last March, and to the family thus terribly bereaved their friends and acquaintances extend sincere and heartfelt sympathy. The deceased was the widow of the late Peter **Kerrigan**, and she breathed her last Sunday evening about 6 o'clock at the family home a few miles below Yuba City. Her death was directly the result of Bright's disease, though she also suffered from other afflictions, She was born August 15, 1833, in County Sligo, Ireland, and had she lived would have been 66 years old on the 15th of this month. Mrs. **Kerrigan** was one of the pioneer women of California, coming to this State in 1853. She settled at first in San Francisco, and went from there to Smartsville, where she married. Early in the sixties she located with her husband and family in Sutter county, and resided here continuously up to the time of her death. She leaves two sons, **Peter** A. and P. H. **Kerrigan**, and three daughters, **Katherine**, **Annie** and Sister M. **Eugenia** of the San Francisco Convent. The funeral will take place from St. Joseph's Church, Marysville, Wednesday morning at 10 o'clock. At which time requiem mass will be read. The burial will be made in the Catholic cemetery. For the benefit of friends and neighbors it can be stated that the family will leave the house at 8 o'clock and proceed to the church. Mrs. **Kerrigan** carried a life insurance policy of $3,000 in the beneficiary fund of the Rainbow Council of Chosen Friends of Marysville, of which order she was a member.

1 Aug 1899 (Tuesday); Page 4 Col 5; DIED. At her home in Sutter county, July 30th, Mrs. Mary **Kerrigan**, relict of the late Peter **Kerrigan**, a native of County Sligo, Ireland, aged 65 years, 11 months and 15 days. The funeral will leave the family residence tomorrow (Wednesday) morning at 8 o'clock and proceed to St Joseph's church where requiem mass will be celebrated. Interment Catholic Church cemetery.

13 Sep 1899; Page 5 Col 2; In the prime of life William **Colling** succumbed to a dread disease. William H. **Colling**, a member of the local lodge A. O. U. W. and Charter member of the Smartsville lodge of Foresters, died last night, of consumption. Until a year and a half ago, when his health began to fail, deceased was employed by the Excelsior Company. He was a native of this state and 44 years of age. Beside the widow, three sons and a daughter survive. A brother resides at Paso Robles. The funeral will take place from the family residence at 11 a.m. tomorrow, under the auspices of the above named societies. Interment I. O. O. F. cemetery.

14 Sep 1899; Page 2 Col 2; SMARTVILLE SEP 1899. Miss Effie **Black** has returned to her place of business at Oakland. Robert **Beatty** has returned from a several weeks' visit at the Bay much benefited by the trip. William H. **Colling**, after a long illness with consumption, passed in the great beyond 8:30 last night. Mr. **Colling** came to this place when a mere lad and was always known for his good traits of character and manly uprightness. His business during long residence

NEWSPAPER INFORMATION

was the care of cattle. He was employed in the capacity for a number of years by the late D. **McGanney**. About thirteen years ago he took a position in the same line with the Ex. W. and M. Company which he held up to the time of his indisposition about twelve months ago. The deceased was a member of the A.O.U.W. and Court Gem of Yuba, F. of A. He was a native of California and 44 years of age. He leaves a wife and four children and one brother, Charles **Colling**, in Lower California. The funeral will take place Thursday morning at 11 o'clock. The interment will be made in the Family Plot in Fraternal cemetery.

14 Sep 1899 (Thursday); Page 3 Col 3; Died – At the Empire Ranch, near Smartsville, September 12th, William Henry **Colling**, a native of California, aged 44 years, 11 months and 3 days. The funeral will take place at 11 o'clock this morning from his late residence and will be under the auspices of Marysville Lodge, A. O. U. W. and the F. of A. Interment Fraternal cemetery, Smartsville.

14 Sep 1899 (Thursday); Page 1 Col 2; William H. **Colling**, after along illness with consumption, passed in the great beyond 8:30 last night. Mr. Colling came to this place when a mere lad and was always known for his good traits of character and manly uprightness. His business during his long residence was the care of cattle. He was employed in the capacity for a number of years by the late D. **McGanney**. About thirteen years ago he took a position in the same line with the Ex. W. and M. Company which he held up to the time of his indisposition about twelve months ago. The deceased was a member of the A. O. U. W. and Court Gem of Yuba; F. of A. He was a native of California and 44 years of age. He leaves a wife and four children and one brother, **Charles**, in Lower California. The funeral will take place Thursday morning at 11 o'clock. The interment will be made in the family plot in fraternal cemetery.

17 Sep 1899 (Sunday); Page 1 Col 5; Mrs. John (Ann Breslin) **Dempsey**, who died at her home two miles south of Smartsville on Friday evening, was a native of county Galway, Ireland, and about 60 years of age. She had been a resident of this county about twenty-six years, and previous to that time resided in Nevada County. She leaves a husband, a well known resident of this county, 4 sons and 1 daughter, Mrs. T. B. **Henderson** to mourn her loss Mrs. Margaret **Ruth** of Dobbins is a sister of the deceased. The funeral cortege will leave her late residence at 9:30 this morning and proceed to the Catholic Church at Smartsville where mass will be celebrated. The interment will take place in the Catholic cemetery.

17 Sep 1899 (Sunday); Page 4 Col 4; DIED. Near Smartsville, September 15th, **Annie**, the beloved wife of John **Dempsey**, a native of County Galway, Ireland, aged 60 years. The funeral cortege will leave her late residence at 9:30 this morning and proceed to the Smartsville Catholic church where services will be held. Interment Catholic cemetery.

5 Nov 1899 (Sunday); Page 4 Col 4; Death of An Aged Man. J. H. **Pritchett**, a resident of the Fourth Ward, passed away to his eternal rest last night after a life of over eighty years on this terrestrial globe. He was a member of the Smartsville

NEWSPAPER INFORMATION

Lodge of Chosen Friends, and the interment will be made in the Fraternal cemetery at that place. Due notice of the funeral will be published later as the details had not been arranged last night.

7 Nov 1899 (Tuesday); Page 4 Col 4; The funeral of the late Jacob H. **Pritchett** who died in this city (Marysville) last Saturday evening, will take place today. The cortege will leave the late residence on Yuba Street at 7 o'clock this morning, and the interment will take place at Timbuctoo cemetery.

8 Nov 1899 (Wednesday); Page 1 Col 5; Funeral of Jacob H. **Pritchett**. The funeral of the late Jacob H. **Pritchett** who died in this city (Marysville) last Saturday, took place at Timbuctoo yesterday. The ritual of the chosen Friends was read at the grave by H. **Wheaton** and Mrs. George **Allen**. The following gentlemen acted as pall bearers; J. W. **Foreman**, W. **Marple**, Andrew **McCarthy**, M. **Wheaton**, E. D. **Hapgood**, and P. **Lemon**.

2 Dec 1899 (Saturday); Page 1 Col 4; On thanksgiving Day at Butte, Montana, Phillip **Walsh**, son of John **Walsh** and wife of Smartsville, died, aged 34 years. The remains will be brought here and later interred in the cemetery at Smartsville. Due notice will be given regarding the funeral arrangements. Mrs. J. J. **McGrath** and Ed. **Bushby** were related to the deceased. Two sisters and three brothers remain to mourn his loss.

3 Dec 1899 (Sunday); Page 4 Col 5; DIED In Butte City, Montana, November 30th, Phillip **Walsh**, son of Mr. and Mrs. John **Walsh** well known residents of Smartsville, a native of California, aged 33 years 5 months and 17 days. The remains will arrive in this city Monday evening at 5:15. The interment will take place in Smartsville Tuesday.

7 Dec 1899 (Thursday); Page 4 Col 4; DIED In Smartsville, December 5th, Mrs. Mary **Rooney**, a native of New York City, aged 43 years, 4 months, and 18 days. The funeral will take place from the Catholic Church, Smartsville, this morning.

27 Dec 1899 (Wednesday); Page 1 Col 6; Horace C. **Newbert** and Miss Ella **Foreman** were married in the U. S. Hotel parlors at 2 o'clock Sunday afternoon by Rev. Dr. J. C. Chase, pastor of the M. E. Church. Frank H. Cramsie was the best man and Miss Rose Newbert officiated as the bridesmaid. The contracting parties are well known residents of Smartsville and vicinity. Their many friends join in wishing them a long and happy married life.

4 Jan 1900 (Thursday); Page 1 Col 5; Ex-Supervisor Wm. B. **Filcher**, a well known resident of Plumas District, was married at 11 o'clock yesterday morning at the parlors of the Western Hotel to Mrs. Eugenia **Hapgood Smith** of Smartsville, one of Yuba county's best known school teachers. The nuptial knot was tied by Rev. L. A. **Green** of the M. E. Church, Wheatland, in the presence of the relatives of the nuptial twain and invited guests. At the close of the ceremony those present were invited to the dining room where a choice menu for the wedding dinner had been

NEWSPAPER INFORMATION

prepared. The contracting parties are well known in this city and county where their many friends will gladly extend to them the best wishes for a long and happy married life. The Appeal acknowledges the receipt of wedding cake and extends congratulations. Mr. and Mrs. **Filcher** left for San Francisco on the afternoon train where they will spend their honeymoon.

6 Jan 1900 (Saturday); Page 1 Col 5; The funeral of the late Bridget **McWilliams** took place at Smartsville yesterday, Father Twomey officiating at the Catholic cemetery where the interment took place. The following gentlemen acted as pall bearers: Wm. **Dewan**, M. A. **McCarty**, Peter Augustus **Hanley**, J. F. **Doyle**, A. J. **McCarty**, and Thomas **Smith**. Several floral pieces were placed on the casket

28 Jan 1900 (Sunday); Page 1 Col 7; John **Mellarkey** a former resident of Smartsville, died at his home in Randsburg a few days ago. About a month ago he was in Marysville with his wife, en route to his old home to visit his aged parents. The remains arrived in this city last evening on the Knights Landing train and were removed to the undertaking parlors of R. E. **Bevan**, who will proceed with them to Smartsville this morning, where the funeral and interment will take place.

30 Jan 1900 (Tuesday); Page 1 Col 7; The remains of the late John J. **Mellarkey**, who died at Randsburg last week, were taken to Smartsville Sunday morning by undertaker R. E. **Bevan** for interment. The services, which were held in the Catholic Church by Father Twomey, were largely attended. The following friends of the deceased acted as pall bearers: T. A. **Newbert**, J. E. **Cramsie**, Robert **Byrne**, E. W. **Doyle**, J. D. **Byrne** and N. L. **Newbert**. A large cortege followed the remains to the Catholic cemetery, where the last prayer was said; after which the interment took place in the family plot. Several handsome floral pieces were sent by friends of the deceased. He was a member of the A. O. U. W. Lodge at Randsburg.

20 Apr 1900 (Friday); Page 1 Col 7; Mrs. Richard (Elizabeth) **Beasley**, who died in San Francisco on Tuesday, was the oldest daughter of the late **Patrick** and Bridget **Dugan** of Smartsville, and was 39 years of age. Besides her husband she leaves six children to mourn her demise. The funeral services were held at St. Rose's Church yesterday and the interment took place in the Holy Cross cemetery.

20 Apr 1900 (Friday); Page 4 Col 5; DIED. In San Francisco, April 17th, **Elizabeth** E., the beloved wife of Richard **Beasley**, eldest daughter of the late **Patrick** and Bridget **Dugan** of Smartsville, aged 39 years.

22 May 1900 (Tuesday); Page 1 Col 6; **Mary**, the beloved wife of Thomas **McNally**, died at her home at Sucker Flat, near Smartsville, at 7 o'clock Sunday evening, after having been on the sick list for several months. The deceased was a native of Ireland, and about 55 years of age, and had resided at Sucker Flat with her husband, who is a miner, for the past twenty-five years. She leaves to mourn her loss a husband and the following children: Mrs. **Fevier**, **Kate**, **Tom** and **James** McNally, who reside in San Francisco, and Annie, who resided with her parents.

NEWSPAPER INFORMATION

The funeral services will be held at 2 o'clock this afternoon at the Catholic Church, Smartsville, where Rev. A. Twomey will officiate. The interment will take place in the Catholic cemetery Smartsville.

25 Jul 1900 (Wednesday); Page 4 Col 3; DIED. Near Smartsville, July 24th, Ella **Newbert**, wife of Horace **Newbert**, daughter of Mr. and Mrs. **Foreman**, sister of **Pearl**, **William** and Frank **Foreman**, aged 18 years, 11 months and 13 days. Funeral from her late residence Thursday afternoon at 2 o'clock. Interment I. O. O. F. cemetery, Smartsville.

26 Jul 1900 (Thursday); Page 1 Col 7; After many months of lingering illness with disease, and ever hoping against hope, the young life of Mrs. Ella **Newbert** yesterday, at the hour of 12:20 o'clock, at the home of her parents, Mr. and Mrs. J. **Foreman**, fled its earthly tenement of clay. Death no matter when or where it breaks into the family circle, especially when that circle is bound about by the holy bonds of fond affection, darkness hovers o'er that home and sadness takes possession of every surviving heart, yet in the case before us that darkness becomes more intense and that sadness pronounced. Not because the dread messenger came unannounced or there was not sufficient time for a preparation for the great change, but more especially because of the fewness of her years. Eighteen years and eleven months ago Ella was born. Seven months yesterday in Marysville the solemn words were said that made H. C. **Newbert** and the deceased, man and wife. But notwithstanding the solemnity of that event, the occasion was characterized by joy and gladness. Tomorrow solemn words will again be said but not a spark of joy, not a ray of gladness, saves in the hope of one day and that to be without reunion. The deceased leaves besides a husband, father, mother, two brothers, one sister and many relatives to mourn her loss.

27 Jul 1900 (Friday); Marysville Daily Democrat; Page 8 Col 3; The Tomb – Peckham – In Linda township, July 26, 1900, at the residence of Thomas **Hammon**, Mrs. Marian **Peckham**, a native of England, aged 77 years, 7 months and 12 days. Funeral will take place from the residence of Thomas **Hammon** tomorrow (Saturday), at 9:30 a.m. Interment in Smartsville Cemetery.

9 Aug 1900 (Thursday); Page 1 Col 7; We are in receipt of the sad intelligence of the death of George **Schmidt** at Mooney Flat, where, with his large family, he had resided for many years. Mr. **Schmidt** lived to the advanced age of 87 years. The time for the funeral had not been set at this writing, but it will probably take place on Friday.

9 Aug 1900 (Thursday); Page 4 Col 4; DIED. At Mooney Flat, August 8th, George **Schmidt**, a native of Bavaria, Germany, aged 87 years and 1 day. The funeral services will be held at the Catholic Church, Smartsville, Friday afternoon at 2 o'clock. Interment, Catholic cemetery.

14 Aug 1900 (Tuesday); Page 1 Col 6; Charles **Divver** died about 12:30 Sunday afternoon at the home of his sister, Mrs. Margaret **Hopkins**, in the St. Nicholas

NEWSPAPER INFORMATION

Building, from the effects of kidney troubles, after an illness of about two months. The deceased was born at Bullards Bar, Yuba county, thirty three years ago, and during his father's lifetime, assisted him as janitor of the public schools of the city, and after his death, filled the position for a short time, until his brother George was appointed. He worked for a number of years on the Chandon ranch, in Marysville township, and was a faithful employee. He persisted in staying with his work even after he became sick, and when some members of his family visited him at Otto Rubel's place last Thursday they had some difficulty in securing his consent to come to town and be properly cared for. He leaves one sister, Mrs. Margaret **Hopkins**, and three brothers, **George**, **William** and **Bernard**, to mourn his demise. The funeral cortege will leave his late residence in this city at 6 o'clock this morning and proceed to Smartsville, where the services will be held at St. Rose's Church at 10:30 a.m. The interment will take place in the family plot at the Catholic Cemetery.

14 Aug 1900 (Tuesday); Page 4 Col 3; DIED. In this city, August 12th, Charles **Divver**, a native of Yuba County, aged 33 years. The funeral cortege will leave his late residence at 6 o'clock this morning and proceed to Smartsville where the services will be held in the Catholic Church at 10:30. Interment family plot Smartsville Catholic cemetery.

27 Oct 1900 (Saturday); Page 1 Col 6; John **Quinkert** died at his home at Smartsville at 5:30 last evening, regretted by his neighbors and many friends. The deceased was a native of Prussia and was 76 years of age, and came to this State from Detroit, Michigan, about the year 1895 with his wife, who died rather suddenly at Smartsville on January 6, 1896. He was naturalized on May 9, 1855, in the Circuit Court at Wayne county, Michigan. Deceased leaves three children residing in this State to mourn his demise. John **Quinkert**, Jr. and Mrs. L. **Conrath** of Smartsville, and Mrs. Mary J. E. **Plamond** of San Francisco. The deceased had been in poor health for the past few years, and his death was not unexpected. He was a man of genial disposition, who was fondly attached to his children. The funeral will take place tomorrow (Sunday) afternoon from his late residence, and the services will be held at the Catholic Church at 2 o'clock. Interment in the family plot in the Smartsville Catholic cemetery.

4 Nov 1900 (Sunday); Page 4 Col 4; Louis Perry **Foust**, a pioneer resident of Yuba county, died at his home at Smartsville yesterday. The deceased was a native of Ohio and 73 years of age, and had resided in Yuba county nearly fifty years. He was a blacksmith by trade and worked for over twenty years for the Blue Point Mining Company. He lived at Timbuctoo in the early fifties. He was an upright and honest man, whose word was as good as his bond. He leaves a wife, Elizabeth **Foust**, to mourn his demise. The deceased belonged to the Order of Chosen Friends. The funeral will take place from his late residence tomorrow (Monday) afternoon at 2 o'clock under the auspices of Pastoles Council, O. C. F. The interment will take place in the Smartsville cemetery.

NEWSPAPER INFORMATION

1 Dec 1900 (Saturday); Page 1 Col 6; Accidentally Shot. Thomas P. **Ryan** Loses His Life From the Discharge of a Gun. Thomas P. **Ryan** was killed at the Empire ranch near Smartsville on Thanksgiving morning by the accidental discharge of a shotgun, the load from which was received by him in the right breast, causing instant death. Coroner A. B. **Hopkins**, when notified of the fatal accident, proceeded immediately to Smartsville to hold an inquest. The following jury were sworn to inquire into the causes leading to his death; John H. **McQuaid**, W. J. **Dempsey**, John **Doyle**, James **Kelly**, William **Dewan** and James **Fraser**. Leo **Meade**, a 17 year old boy testified that the deceased was standing on a wagon tongue at the Empire ranch about 10 o'clock Thursday morning with his shotgun in his hand when the gun fell to the ground and was discharged, the deceased receiving the charge as above stated. Arthur **Mooney** testified that when the gun fell out of the deceased's hands that the contents of one of the barrels struck the deceased in the right breast and killed him instantly. Patrick **Ryan**, father of the deceased, testified that the deceased was 22 years and 9 months of age. The jury found that his death resulted from the accidental discharge of a shotgun. The funeral will take place in Smartsville on Sunday.

27 Dec 1900 (Thursday); Page 1 Col 7; The funeral of the late John **Barrett**, the third victim of the Pennsylvania mine disaster at Browns Valley, took place last Thursday according to a previous announcement in this column of The Appeal. Rev. A. Twomey officiated at the church and grave. The interment was made in the family plot in the Catholic cemetery and the following gentlemen acted as pallbearers: A. G. **Wheaton**, W. A. **O'Brien**, Daniel **Dempsey**, Robert **Beatty**, P. **Callahan**, and James **Kelly**.

31 Jan 1901 (Thursday); Page 1 Col 4; ELECTRIC CURRENT INFLICTS DEATH. Peculiar Accident That Befell James **Doyle**, Jr. at Smartsville. Was Killed Instantly. A Power Current Wire Intercepts the Telephone Line Transferring to it a Heavy Voltage Flow Momentarily. News was received in this city yesterday that James **Doyle**, Jr., a lineman in the employ of the Bay Counties Power Company, had been instantly killed at Smartsville from the effects of an electric shock sustained by him while endeavoring to telephone from that place to some point along the line of the company's works. The deceased, a young man twenty-seven years of age, had been connected with the power company for a year and for the latter three or four months of that time had been located at Smartsville where he had charge of the company's line between the Yuba and Bear rivers. For convenience a telephone was installed at his residence that he might be able to talk along the line at any time either day or night. The young man at about a quarter to eleven o'clock yesterday forenoon had gone to the instrument for such purpose and had but taken the receiver down and placed it to his ear when he was stricken and died within a few moments. One of the electricians connected with the power company's works at Colgate informed the Appeal correspondent at Smartsville that the accident was in all likelihood due to a wire carrying current under heavy voltage having become loosened or broken off and falling against the telephone wire thus throwing its force with the results noted. The dead man was born and raised at Smartsville where his funeral will undoubtedly take place although

NEWSPAPER INFORMATION

particulars regarding the same have not as yet come to hand. Coroner Hopkins, who was summoned shortly after the accident, has gone to the scene of the casualty, where he will empanel a jury and inquire into the causes leading up to the sad occurrence. The deceased was a brother of Mrs. J. C. **Baldwin** and Mrs. F. L. **Johnson** of Maysville, of Mrs. H. G. **Merriam** of Dobbins, and of **Edward** and John **Doyle** of Smartsville, where his aged parents reside. He had a number of friends in this city that heard the announcement of the fatal accident with regret.

1 Feb 1901 (Friday); Page 4 Col 2; Coroner's Jury Finds That He Succumbed to an Electric Shock. Coroner A. B. Hopkins held an inquest at Smartsville on Wednesday evening to ascertain the causes leading to the death of James J. **Doyle**, Jr. The following jury were sworn: A. G. **Wheaton**, George H. **Meredith**, J. H. **Warne**, Peter A. **Hanley**, James **Kelley**, Sr., and Jacob W. **Foreman**. Mrs. **Doyle**, the mother of deceased, testified that there was a telephone in front of their house, and she thought that her son was answering a call from the power house when she saw him fall; heard a snap as if something had broken. Had heard the bell ring just before the stricken man had gone to the telephone. James **Doyle**, Sr., the father of the deceased, testified that heard his son speak when he fell. He said, "Mamma do not be scared or worried – I am all right," and then expired. Thought that he had had a regular call from No. 3 bell. The deceased was breathing heavily when he got to him. Dr. E. F. **Holbrook** testified that death was due to paralysis of the centers of respiration and circulation caused by a current of electricity. It was possible for a person to articulate after such a shock. J. D. **Boyd**, an employee of the company, testified that the deceased, who was one of the linesman had a regular call of three bells. Did not know if such an accident could occur by a live wire crossing a telephone wire, and could not tell if there had been any break on the line between Smartsville and the powerhouse. The jury found that deceased came to his death from a shock of electricity, the immediate cause being paralysis of the centers of respiration and circulation, said shock being received while in the discharge of his duty as a linesman in the employ of the Bay Counties Power Company.

1 Feb 1901 (Friday); Page 4 Col 3; Died- In Smartsville, January 30th, James Joseph **Doyle** Jr., a native of California, aged 28 years. Requiem mass for the repose of his soul will be celebrated at the Catholic Church, Smartsville, on Sunday morning at 10:30 o'clock. Interment Catholic cemetery, Smartsville

16 February 1901; Page 1 Col 5; After an Extended Illness Mrs. John **Walsh** Passes to the Beyond. Mrs. John **Walsh**, an old and respected resident of Rose Bar Township, died yesterday morning at her late residence near Smartsville, after an extended illness. The deceased was a native of County Dounegal, Ireland, and was 72 years of age. She was the mother of John E., **Edward**, **Mary** and William **Walsh**, and of Mrs. George **Hymens**, and was a sister of Mrs. George **Bushby**. She was a kind-hearted and very hospitable woman, and by her death one of the old-time residents has been removed from that section of the country. The funeral cortege will leave her late residence at 1 o'clock tomorrow (Sunday) afternoon, and the services will be held in the Catholic Church, Smartsville; at 2 o'clock, after which the interment will take place in the Smartsville cemetery.

NEWSPAPER INFORMATION

16 Feb 1901 (Saturday); Page 4 Col 3; DIED At her home near Smartsville February 15th, **Mary**, the beloved wife of John **Walsh**, a native of County Donegal, Ireland, aged 72 years. The funeral cortege will leave her late residence at 1 o'clock Sunday afternoon and proceed to the Smartsville Catholic Church, where services will be held at 2 o'clock. Interment Catholic cemetery.

19 Feb 1901 (Tuesday); Page 1 Col 4; The funeral of the late Mrs. Mary **Walsh** took place on Sunday and was largely attended, the deceased having had many friends in the community where she had resided so many years. The cortege left the **Walsh** residence at 1 o'clock and proceeded to the Catholic Church at Smartsville where the services were held. Father **Twomey** officiated at the church and at the Smartsville cemetery where the interment took place in the family plot. The following gentlemen acted as pallbearers: John **McQuaid**, J. B. **Barrie**, Patrick **Ryan**, Daniel **Sanders**, William **Nutley** and C. F. **Boardman**. Among the floral pieces were a large pillow with the emblem "Rest Mother" and a large cross.

21 Feb 1901; Page 1 Col 6; A message came to this place from Reno last Monday night bearing the sad news of the death of Mrs. V. **Emery** (nee Rose). The deceased was born and educated in this place. She was married to Joseph **Emery**, February 6, 1895, and at the time of her death was 34 years of age. Mr. And Mrs. Emery resided in this place up to about 18 months ago when they went to Reno, where last Monday, after a short illness the wife passed away. The particulars of her death had not reached us at this writing. James **Rose**, a brother, has gone to meet the remains, and the funeral is appointed for this afternoon at 2 o'clock. The interment will be made in the fraternal cemetery. Mrs. Emery leaves to mourn her loss besides her husband and 3 children a mother, Mrs. J. **Rose**, three brothers, **Frank** of Stockton, **James** and **Oscar** of this place, and two sisters, Miss **Bell** of San Francisco, and Mrs. C. **Burris** of Browns Valley. Later - the funeral is postponed until tomorrow at 11 o'clock a.m.

26 Feb 1901; Page 1 Col 5; TOLAND – BACH NUPTIALS. A well Known Yuba County Educator Unites with One of Smartsville's Fair Daughters. A pretty wedding took place at 8 o'clock Sunday morning at the residence of Mrs. M. A. **Bach** at Smartsville, at which time her daughter **Elizabeth** and John **Toland** were united in the holy bonds of matrimony. The young couple stood under a floral horseshoe while the Rev. John T. **Vineyard** spoke the words that united them. John E. **Bach**, a brother of the bride, acted as groomsman and Miss Mary J. **Toland**, a sister of the bridegroom, as bridesmaid. The parlors in which the ceremony took place were decorated with violets and smilax. At the conclusion of the ceremony the guests were entertained at a wedding breakfast by Mrs. **Bach**. The young couple received many handsome and useful wedding presents. The bride is the daughter of the late John **Bach**, and has resided with her mother in Smartsville. She is a young lady of a bright and cheery disposition, well educated, and possessed of those attributes , which help to make a happy home. She will undoubtedly make a dutiful and loving wife. The groom is the son of Robert **Toland**, an old and respected resident of Rose Bar Township. He has taught school at Mooney Flat, was principal of his home school at Smartsville for several years

and also occupied the same position in the grammar grade school of this city (Maysville). He was a member of the County Board of Education for several years, and is at present employed in the U. S. Custom service in San Francisco, gaining his position after a successful competitive civil service examination. He is in all respects a model man, whose paradise will be his home. Mr. and Mrs. **Toland** arrived in this city on Sunday afternoon and left for San Francisco, via Sacramento, on the 8 o'clock train yesterday morning. They carry with them the best wishes of a large circle of friends for a long and happy married life.

12 May 1901; Page 1 Col 3; SANFORD-MURDOCK. Justice **Morrissey** Joins Two of Cupids Victims in Hymeneal Bonds. Al B. **Sanford** and Miss Emily T. **Murdock** were married in this city at 8 o'clock last evening by J. M. **Morrissey**, J. P. The nuptial know was tied in the office of the Justice and was the first marriage ceremony performed by him since taking the oath of office. The groom is the son of Benjamin **Sanford**, a pioneer resident Smartsville, and is a successful rancher of the locality. His bride has been a resident of Berkeley, and is a well-educated and handsome young lady, who will no doubt make a dutiful wife and a loving helpmate for her husband. The many friends of the contracting parties will join in wishing them a long, happy and prosperous married life.

30 May 1901 (Thursday); Page 1 Col 6; William Edward **Dewan**, a son of the proprietress of the Dewan Hotel, Smartsville, was killed yesterday by the premature discharge of a blast in the Gold Hill mine, near Grass Valley, at which place he was employed. His working partner was seriously injured. No further details were obtainable last night. The information above given was sent to the dead man's relatives at Smartsville, and the later telephoned to The Appeal by this paper's correspondent at that place. The body will arrive in Smartsville this morning. The date of the funeral had not been determined upon last night. Mr. **Dewan** was born twenty-nine years ago at Sweetland and had never married. Besides his mother there remain one brother, James **Dewan**, now employed at or near Grass Valley, and four sisters, two of whom, the Miss **Jessie** and Nellie **Dewan**, reside in San Francisco, another at Sweetland and another, Mrs. Kate **Burnes**, at Grass Valley. Mr. **Dewan** was a member of Court Gem of Yuba, No. 121, F. of A., of Smartsville. Presumably this lodge will have charge of or at least participate in the funeral ceremonies. Mr. **Dewan's** terrible death was a severe shock to the people of Smartsville, among whom he stood well and had many friends.

31 May 1901 (Friday); Page 4 Col 2; Young Dewan's Death. A Missed Shot from the Night Shift the Probable Cause. The Grass Valley Union publishes some further particulars with regard to the death of William E. **Dewan**, which occurred as the result of an accident in the Gold Hill mine at Grass Valley on Wednesday afternoon. What caused the blast to explode at the time is not positively known. It is supposed that it was a "missed hole" that had failed to connect in the morning when the night shift quit work. Coroner **Daniels** summoned a jury and held an inquest Wednesday evening. The verdict was that death was caused by an unaccountable explosion of a loaded hole, and concluded by exonerating the mining company from all blame. The deceased was a native of Birchville, Nevada

county. His widowed mother is at present conducting a hotel at Smartsville, five sisters and James **Dewan**, an only brother, are left to mourn his loss. John **Dewan**, another brother, was killed two years ago by falling over a cliff on the Yuba river. He had been a cripple for about fourteen years. Deputy Postmaster J. H. **McQuaid** and Al **Wheaton** of Smartsville, as representatives at Court Gem of Yuba, A. O. F. took charge of the remains and removed them to Smartsville, where the funeral and interment will take place today. When James **Dewan**, a brother, arrived from the W. Y. O. D. mine, where he works, he insisted on seeing the body, and as the blanket was lightly lifted he went into hysterics and his moans were heart rending to hear. Mrs. Kate **Burns**, a sister, was taken to the office of the mine, but owing to her sad cries and screams was not permitted to go near the body. John **Breen**, a cousin, came also, but was persuaded to remain away from the terrible sight. James **Ledwich**, who was in the mine with **Dewan**, was also badly injured, his left hand being shattered an d he receiving many bruises.

12 Jun 1901 (Wednesday); Page 4 Col 2; The Funeral of Mrs. Catherine **Mitchell**, the beloved wife of John **Mitchell**, who died at the family home near Smartsville on Monday morning, took place at noon yesterday. The services were held at the Catholic church, Smartsville. The deceased was about 75 years of age. The interment took place in the Smartsville Catholic cemetery, where the services were concluded by Rev. A. **Twomey**.

13 Jun 1901 (Thursday); Page 1 Col 6; All that is mortal of the late Mrs. Catherine **Mitchell** was quietly laid to its long and peaceful rest yesterday at the noon hour in the Catholic Cemetery. The deceased was a native of Longsford county, Ireland, and at the time of death was 75 years of age. In her death the Church has lost a true and faithful member and this community a noble friend and neighbor. She had at all times and places a cheerful smile, a kind work, a willing hand and a generous heart. She leaves a mother to mourn her loss, whom to leave was her greatest pain. Rev. A. **Twomey** officiated and the following were the pallbearers: John **Dempsey**, F. **McNally**, Mr. **Wilson**, Daniel **Cain**, D. **Carson** and John **Burne**. Mr. John **Mitchell** desires through the columns of the Appeal to return thanks to the many friends who were so untiring in their labors in caring for his wife during her long sickness, as well as for the many kindnesses shown toward himself during his residence among our people.

13 Jun 1901 (Thursday); Page 1 Col 4; John C. **Hall** Fails to Recover From Results of Recent Accident Near Round Tent. John Clark **Hall** (not Jos. B. **Welch**) who was so seriously injured in a runaway at the ranch of his father-in-law, D. N. **Jones** near Round Tent, died yesterday morning. The deceased was a native of Colorado, and 31 years of age. He leaves a wife and one daughter, L. **Otis** aged 7 years to mourn his loss. The funeral will take place from the residence of D. N **Jones** at 10 o'clock tomorrow (Friday) morning. The interment at the Union cemetery at Smartsville at 1 o'clock. When the first news of the accident, which resulted in the death of Mr. **Hall**, reached The Appeal office, our informant stated that it was Jos. B. **Welch** who had sustained the injuries spoken of and the notice of the occurrence so appeared in yesterday's paper. While the episode is nonetheless sad, still the

NEWSPAPER INFORMATION

friends of Mr. **Welch** well are glad to know that it was not he on whom the mishap fell.

14 Nov 1901 (Thursday); Page 1 Col 3; Alexander Campbell **McClure**, who lived at Smartsville for many years, died at the County Hospital yesterday into which place he was recently admitted. The deceased was a native of Ireland and 70 years of age. He leaves one son who resides at Smartsville. His friends have been notified of his death and the time for the funeral will be set later pending which Undertaker R. E. **Bevan** has taken charge of the remains.

16 Nov 1901 (Saturday); Page 1 Col 3; The funeral of Alexander **McClure** of Smartsville, who died in this city, took place yesterday. The services were held at St. Joseph's Church at 11 o'clock where Rev. A. **Twomey** officiated, as well as at the Catholic cemetery, where the interment took place.

3 Aug 1901 (Saturday); Page 4 Col 5; Died – In Rose Bar precinct, August 2d, Frank **Bushby**, a native of California, aged 25 years, 4 months and 5 days. The funeral will take place from his late residence Sunday morning at 8 o'clock. Interment Fraternal cemetery, Smartsville

10 Jan 1902 (Friday); Page 1 Col 5; Passed Away Last Evening Without Recovering Consciousness. A telephone message was received in this city (Marysville) about 8 o'clock last evening stating that Edward **Bushby** was dead. He was removed about noon yesterday to the French Hospital, and it was thought that he had a fighting chance, although he remained unconscious. The deceased was a native of Yuba county and about 34 years of age. He resided on the home place with his parents, Mr. and Mrs. George **Bushby**, for many years, and was a very industrious man. He came to this city highly recommended and was appointed driver of the hose carriage. His name was also at one time mentioned for appointment on the police force. After serving the city faithfully he resigned his position and went to San Francisco, where he obtained a position as grip man on the California street car line. Last Monday Ed, complained of having a severe cold, and remained in bed all day, as he was threatened with pneumonia. When his landlady went to his room to call him on Tuesday morning she found him unconscious and the gas escaping in his room. He was at once removed to the emergency hospital, but never regained consciousness. The physicians stated that the gas had more effect on him because of the condition of his lungs. The deceased was a member of the Odd Fellows Lodge of Smartsville and of the Fraternal Brotherhood of this city, in which he was insured for $3600, the policy being made out in favor of his father. His death is a very sad one – the more so because his younger brother, **Frank**, committed suicide only about six months ago, while suffering from a high fever. Besides his aged parents he leaves two brothers, **Will** and **Joe**, and three sisters to mourn his demise. The bereaved family has the sympathy of their many friends and the public generally. The remains will be brought to this city probably tonight. The father of the deceased was at Ed's bedside when he passed away.

NEWSPAPER INFORMATION

10 Jan 1902 (Tuesday); Page 1 Col 5; Died-Near Smartsville, June 9th, John **Walsh**, a native of Dublin, Ireland, aged 72 years. The funeral cortège will leave his late residence at 1 o'clock Wednesday afternoon, Services will be held at the Smartsville Catholic Church. Interment Catholic cemetery

16 Jan 1902; Page 1 Col 5; Death at Smartsville – Mrs. George **Jeffery**, An Aged and Honored Resident, Passed Away Yesterday. Maria L., the beloved wife of George **Jeffery**, died at her home at Smartsville at 3:15 yesterday morning. The deceased was a native of Pennsylvania, and about 61 years of age. She was an old and much respected resident of the community in which she had resided so many years and her death will be sincerely mourned. She leaves to mourn her demise a husband, who is over 70 years of age, and the following children; **Lizzie, Della, George, William** and Robert **Jeffery** and Bertha **Colling**. The funeral will take place from the family residence tomorrow afternoon at 2 o'clock. The interment will be made in the Fraternal cemetery.

28 Jan 1902 (Tuesday); Page 1 Col 6; Mrs. Mary **McGanney**, an old and much respected resident of Smartsville, died at 10 o'clock on Sunday morning at her home. The deceased was a native of county Tyrone, Ireland, was about 68 years of age, and was the relict of the late Daniel McGanney who died in 1895. Previous to her marriage the deceased lived at Timbuctoo, but since that time has lived in Smartsville for over forty years. She leaves the following children to mourn her demise: Four sons, **Edward, James** F., **Daniel** C. and **Frank**, and two daughters, **Annie** and **Mary**. The deceased had been an invalid for some time and was fully prepared to meet death when the final message came. She was a good mother, a model wife, and her death will cast a gloom over the community in which she had lived so long. The funeral will take place at 2 o'clock this afternoon from the Catholic Church, Smartsville, where the services will be held, interment in the family plot, Catholic cemetery.

28 Jan 1902 (Tuesday); Page 4 Col 4; Died. In Smartsville, January 26th, Mrs. Mary **McGanney**, a native of the county Tyrone, Ireland, aged 68 years. (San Francisco papers please copy). Funeral services will be held at 2 o'clock this afternoon at the Catholic church, Smartsville. Interment Smartsville Catholic cemetery.

2 Mar 1902 (Sunday); Page 1 Col 3; James **McCarty** Killed. Was Working in a Mine in Siskiyou County But Particulars Not Obtainable. Undertaker J. K. **Kelly** was notified yesterday afternoon that James **McCarty**, a former resident of Timbuctoo, had been killed near Gazelle, Siskiyou county. He was requested to go after the remains. The deceased was the youngest son of Andrew **McCarthy**, a well-known resident of Timbuctoo, and it is stated that he was either working on a dredger or in a mine at Callahan's, near Gazelle. It was impossible to get any particulars about the fatal accident, as the wires were all down last evening.

4 Mar 1902 (Tuesday); Page 1 Col 3; James **McCarty's** Death. The Poor Fellow was Caught in Mining Machinery and Perished Miserable. Undertaker J. K. **Kelly**

returned from Red Bluff yesterday morning with the remains of James **McCarty**, who was killed on Saturday morning last at Callahan's, twenty miles from Gazelle, Siskiyou county, while working on a dredger. It was ascertained that **McCarty** was working near the machinery when the end of his coat got caught in the cogwheels. He threw his left arm around to extricate the garment but instead his arm was caught and his body drawn into the machinery and firmly held down on its left side. Death resulted in a few minutes. When he was first caught he shouted "Oh Pete," to his partner, Pete **Galligan**, but the latter was unable to render him any assistance. Undertaker Kelly accompanied the remains to the old home at Timbuctoo yesterday afternoon. The funeral services will be held at the Catholic Church at Smartsville at 10:30 this morning, where Father **Twomey** will officiate. The interment will take place in the Smartsville Catholic cemetery.

4 Mar 1902 (Tuesday); Page 4 Col 5; DIED. At **Callahan's**, near Gazelle, Siskiyou county, March 1st, James **McCarthy**, a native of Yuba County, aged 29 years. Funeral services will be held at the Catholic church, Smartsville, this morning at 10:30 o'clock. Interment Catholic cemetery.

9 Mar 1902 (Sunday); Page 1 Col 4; The sudden and tragic death of Father Andrew **Twomey** yesterday came as a shock to all those who were informed of the sad event and has caused sincere sorrow. The life of the dead priest was spent in going about doing good, in living strictly up to the requirements of his high calling as a minister in God's service, and, therefore, it is but natural that his untimely taking off, should cause deep and lasting grief. The following account was telephoned to The Appeal by its Smartsville correspondent: Father Twomey left Smartsville yesterday morning between 7 and 8 o'clock for Rackerby, where he was to have preached today. He started by way of the Peoria School House, where he was to have held services at 10 o'clock yesterday morning. When he reached Dry Creek he made several ineffectual efforts to cross over, near where the road strikes the stream. Finally he desisted and went up alongside the creek in a point opposite M. V. **Hendricks'** place. Here he again entered the swollen waters and despite the swift current seemed likely to succeed in his risky attempt. Indeed his horses were within two or three feet of the opposite bank when they lost control of themselves and were whirled about and carried off their feet, headed down the stream. When this occurred, the buggy was up set,, fastening its occupant within it, as it turned turtle, thus drowning him, helpless even to strike a stroke to save himself from such a terrible ending. After a few minutes the rushing water moved the buggy in such a way as to permit the imprisoned body to escape and it was then washed down stream quite a distance, catching finally on the projecting limbs of a willow. Here it was held for several moments, eventually, however, working loose and again being washed down stream, until its progress was checked by striking ground in a comparatively shallow place. The body was recovered from this point, after having been in the water three and a half hours. The accident occurred about 9:30 o'clock yesterday morning, the body being rescued a little after one o'clock in the afternoon. It was then prepared hastily for removal to Smartsville, reaching that place about 4:30 yesterday. M. V. **Hendricks**, it is said, was an eyewitness of the shocking catastrophe and took the lead in recovering the body. The horses were

NEWSPAPER INFORMATION

gotten out of the water, not much the worse for their thrilling experience, but the buggy was wrecked. Father **Twomey** was born at Clonakilty, County Cork, Ireland, Thirty-six years ago. He studied for the priesthood at All Hallows College, Dublin, and was ordained on June 24, 1888. After two months' vacation he started for California. Soon after his arrival here he was appointed by Bishop Manogue as pastor of the Catholic Church at Smartsville, where he has resided ever since, having been the successor of Rev. M. **Coleman**. The deceased was loved by all the residents of the foothills, irrespective of creed. His field was large and his success most encouraging. He succeeded in building a church at Dobbins, among other important achievements, and had recently been deeded some land at Rackerby on which it was his intention to erect another church in the near future. Father **Twomey** was the true type of the Irish Catholic priest. He was loyal and devoted to his people, and no hardships were too great for him to endure when their welfare was at stake. He traveled through the mountain districts at all hours of the night during the winter months to give comfort and consolation to the dying, and he was ever ready to nurse a sick person and cheer him on the road to recovery. In Smartsville, which was his home for over thirteen years, his loss will be deeply deplored by all classes. He was a genial, cultured and high-minded gentleman, who was charitable to a fault; for his big heart could not see any person in need if he was in a position to render aid. He was a pulpit orator of considerable ability, and his sermons were always interesting. Father Twomey was naturalized in the Superior Court of this county on November 20, 1893, and while truly devoted to his motherland, was one of the most loyal and devoted of our citizens. He frequently visited Marysville, where he had a large circle of devoted friends. True to his church, loyal and devoted to his parishioners and his friends, faithful to the best interests of society, he has passed away to the deep regret and sorrow of the residents of Yuba county. When the news of the fatal accident reached this city Father **Coleman**, one of his best friends, left at once for Sicard Flat to assist in hunting for the remains. The arrangements for the funeral will be perfected this morning.

9 Mar 1902 (Sunday); Page 4 Col 4; Died Drowned in Dry Creek, Sicard Flat, March 8th, Rev. Andrew **Twomey**, a native of county Cork, Ireland, aged 36 years. Funeral notice hereafter.

20 Apr 1902 (Sunday); Page 1 Col 5; Miss Sadie Josephine **Dougherty** died about midnight on Friday night at the residence of her aunt, Mrs. Margaret **Hopkins**, in the St. Nicholas building, after an illness of about six weeks. The deceased was a native of Smartsville and about 24 years of age. Her father was killed in a mining accident at Angels Camp several years ago. Her mother has also been dead for several years. She had lived with her aunt for about six years and was greatly attached to her. She had a mind at peace with all. Mrs. **Hopkins** loved her as if she had been her own child. Every one that knew her could not help loving her as she was a model girl in all respects. Besides her aunt, Mrs. Margaret **Hopkins**, she leaves three uncles – **George** and Byron **Divver**, of this city, and Dan **Dougherty**, of Smartsville; also a grandfather – Michael **Dougherty**, of Stockton. She was a member of the local lodge of the Fraternal Brotherhood, and Mrs. **Hopkins**, who is

her beneficiary, will receive $500 from that order. The funeral services will be held at the Catholic church at Smartsville tomorrow afternoon at 1 o'clock and the remains will be buried in the Catholic cemetery by the side of her mother.

22 May 1902 (Thursday); Page 3 Col 5; Patrick J. **Beirne**, a painter living at 231 Shipley street, died Sunday morning at St. Luke's hospital from shock caused by the amputation of his right leg. About a week ago Mr. **Beirne** while painting the skylight of a house on Ivy avenue fell from a ladder, sustaining a compound fracture of the leg. The physicians found it necessary to amputate the limb. The deceased was the beloved son of the late **Patrick** and Margaret **Beirne**, and a nephew of Mrs. O. **Sanders**, of Smartsville, with whom he had resided a greater part of his life. His many friends here were shocked to learn of his untimely death, for Patrick was a favorite with all, and the family has the sympathy of the entire community.

10 Jun 1902 (Tuesday); Page 3 Col 5; DIED Near Smartsville, June 9th, John **Walsh**, a native of Dublin, Ireland, aged 72 years. The funeral cortege will leave his late residence at 1 o'clock Wednesday afternoon. Services will be held at the Smartsville Catholic church. Interment Catholic Cemetery.

12 Aug 1902 (Tuesday); Page 4 Col 4; John **Mitchell**, who died at Wheatland, was buried at Smartsville on Sunday afternoon. The services were held at the Catholic Church, and the interment took place in the Smartsville Catholic cemetery. The deceased was a former resident of Linda and also once resided at Fernly, Nevada county, as well as at Smartsville

14 Aug 1902 (Thursday); Page 3 Col 4; John **Mitchell** and Benjamin **Frazier** Called to Final Rest. It is our sad task this week to chronicle the death of two old pioneers. John **Mitchell** died on the 8th inst. At 3 o'clock p.m. in Wheatland, where he was taken ill while en route from San Francisco to his home in this place. Rev. J. J. **Hynes** was called to his bedside last week and administered the consolations of the church. The deceased lived in this place about three years and was always considered a kind and obliging neighbor. The remains were brought here in charge of Joseph **Gardner** of Wheatland and interred in the family plot in the Catholic cemetery. The following gentlemen acted as pallbearers: Timothy **Linehan**, Charles **Compton**, John **Burnes**, Michael **Bregnan**, James **Byrne** and Thomas **McNally**. The second death is that of Benjamin **Frazier**, who departed this life on the morning of the 8th, Mr. **Frazier** had been for several years in very feeble health and his death was not unexpected. The deceased resided in this place for a number of years and was engaged in the boot and shoe making business. His wife died in 1886, since which time he has been carrying on a saloon and small grocery business. He was considered a good tradesman, and was strictly honest in all his business transactions. The deceased was born in Bethlehem, New York, on the 24th day of March 1822. He leaves a brother in Lyons, Kansas, and a nephew in Nashly, New Hampshire. The remains were in charge of J. H. **Conlin** and were interred in the fraternal cemetery, Rev. J. T. Vineyard officiating. The following

NEWSPAPER INFORMATION

named gentlemen acted as pallbearers; Samuel **Fisher**, T. A. **Newbert**, Patrick **Callahan**, Charles **Murch**, W. C. **Huling** and A. G. **Wheaton**.

11 Sep 1902 (Thursday); Page 1 Col 3; Died – At Smartsville, early Tuesday morning, James E. **Rose**, a native of California, aged 40 years, 6 months and 2 days. Funeral this morning; interment Smartsville cemetery.

1 Oct 1902 (Wednesday); Page 4 Col 2; Rose Anna **Byrne**, the beloved wife of James **Byrne**, of Smartsville, died at 8:45 yesterday morning at the residence of her daughter, Mrs. Clarence **Owen**, in this city. The deceased was a native of New Jersey, and about 55 years of age. She was the mother of Mrs. Clarence **Owen**, Mrs. B. F. **Compton** and of **James** D., **John** Bernard, **Robert** Emmet, **Henry** J., **Sarah** and Rose **Byrne**. She was a sister of John **Harvey** and Mrs. William **Cramsie**, old and respected residents of Smartsville, and Sister Margaret **Harvey**. Mrs. **Byrne** was married about 30 years, and had lived in Smartsville for three years before her marriage. She was much respected in the community she resided in for so many years, and the announcement of her death was heard with deep regret by her neighbors and old friends. The remains were taken to Smartsville yesterday afternoon. The funeral services will take place tomorrow in the church of the Immaculate Conception, where Rev. J. J. **Hynes** will officiate. The interment will be made in the Smartsville Catholic cemetery.

10 Oct 1902 (Friday); Page 1 Col 3; The death is announced of Mrs. Thomas A. **Newbert** at her home in Timbuctoo, where it took place at an early hour yesterday morning. The maiden name of the deceased was Minnie **Jackson**, and before her marriage she resided in Lone Bar township. She had a stepbrother named Harry **Hogarth** of Angel Camp, Calaveras county. She was an old and much respected resident of Yuba county. She had suffered from cancer for some time. Besides her aged husband, she leaves six children to mourn her demise – namely, Mrs. Henry **Creps**, **Horace** C., **William** L., **Thomas** A., **Rose** and Ada Augusta **Newbert**. She was a sister in law of the late Lee **Newbert** of this city, and an aunt of. H. **Newbert** of the Marysville Gas & Electric Co., and of **Chester** and Miss Edna **Newbert**. The deceased name was Millicent E. **Newbert**. She was a native of New York, aged 52 years, 2 months and 5 days. The funeral will take place tomorrow at 11 a.m. from her late residence. The interment will be made in the Odd Fellows' Cemetery, Smartsville.

10 Oct 1902 (Friday); Page 3 Col 5; In Timbuctoo, October 9, **Millicent** E., beloved wife of T. A. **Newbert**, a native of New York, aged 52 years, 2 months, and 5 days. The funeral will take place tomorrow at 11 o'clock from her late residence. Interment Odd Fellows cemetery, Smartsville.

12 Oct 1902 (Sunday); Page 1 Col 4; Funeral of Mrs. Newbert – Mary went from here o attend the services at Timbuctoo and Smartsville. The funeral of the late Mrs. T. A. **Newbert** took place at 11 o'clock yesterday morning from her late residence at Timbuctoo, and was very largely attended. Rev. John T. **Vineyard** officiated, at the house, and at the Odd Fellows cemetery, where the interment took

NEWSPAPER INFORMATION

place. There were numerous floral pieces, including a pillow and a broken wheel. The following gentlemen acted as pallbearers; Thomas R. **Lee**, John **Peardon**, W. **Huling**, A. J. **McCarthy**, W. **Jeffrey** and W. R. **McConnell**.

16 Oct 1902 (Thursday); Page 1 Col 3; John H. **McQuaid** and Miss Katherine L. **Rearden** of Smartsville were married at 8:30 last evening at St. Josephs church, Tenth street, near Howard, San Francisco. The nuptial knot was tied by Rev. Father **Sullivan** in the presence of relatives of the contracting parties and a number of invited friends. William Roger **Conlin**, a cousin of the groom, acted in the capacity of groomsman, and Neil **Denehey** and Frederick **Fay** were the ushers. Miss Louisa I. **Rearden** made a charming bridesmaid for her sister. The bride was attired in white chiffon over white taffeta. She carried a shower bouquet of lilies of the valley, and wore a large white picture hat. The bridesmaid was attired in white chiffon over blue taffeta and wore a large black picture hat. She carried a bouquet of pink bridesmaids roses. At the conclusion of the ceremony, which was an exceptionally pretty church wedding, the wedding party went to the home of Mrs. N. **Rearden**, the mother of the bride, No. 226 Twelfth street, where a reception was held. The parlors were decorated in smilax and violets, and the dining room in pink La France roses and streamers of white satin ribbon. The groom needs no introduction to the people of Yuba county as he is the son of John **McQuaid**, a prominent resident of Smartsville, He is to be congratulated on winning such a charming bride. Mrs. **McQuaid** is equally, if not better known that her husband,, as she has won much honor in her calling, being considered one of the brightest schoolteachers in Northern California. She is a pretty and accomplished young lady whose many charms and sweet disposition have won her a host of friends in this, her native county, particularly in Marysville and her old home in Smartsville. The young couple was the recipients of many useful and beautiful, as well as costly, presents. Mr. and Mrs. **McQuaid** will leave by the steamer, State of California , for Los Angeles, and will visit Catalina Island and Southern California before returning to Smartsville, where they will be at home after November 15th. Their many Marysville friends extend congratulations and best wishes, and trust that their journey through life will be one perpetual honeymoon.

28 Jun 1903 (Sunday); Page 1 Col 6; John **Woodroffe** died at his home at Mooney Flat on Friday morning. The deceased was a pioneer resident of Nevada County and followed mining for many years. He was the father of Mrs. Charles E. **Shellinger** and Miss Annie **Woodroffe**, residents of this city (Marysville), and was about 66 years of age. The funeral will take place at Mooney Flat at 10 o'clock this morning and the interment will take place in the Smartsville cemetery.

29 Sep 1903 (Tuesday); Page 1 Col 6; John **Rickey** of Mooney Flat died at the hotel in Wheatland at 5 o'clock on Sunday evening. The deceased was a native of Nova Scotia and was about 70 years of age. He had resided at Mooney Flat for the past forty years. He was the owner of some mining property which had been bonded by G. W. **Manwell** and some other gentlemen. The deceased had gone to Wheatland to sell fruit when he was taken ill. The remains of the late John **Rickey** arrived from Wheatland on the 5 o'clock train last evening, and were taken to the

NEWSPAPER INFORMATION

undertaking parlors of J. K. Kelly. A sister of the deceased arrived from San Francisco on the Oregon Express this morning to make arrangements for the funeral

30 Sep 1903 (Wednesday); Page 1 Col 5; The Funeral Services Arranged to Take Place Tomorrow. Thomas Hamilton **Welch** died in this city yesterday. He was a native of Missouri aged 56 years. The home of the deceased was at Cabbage Patch which is about eighteen miles from Marysville, and now known as Waldo. He was a brother of Mrs. Aden **Wright, Joseph** B., **Gale** W. and James **Welch**. The announcement of his death will be heard with regret by his many friends and acquaintances. The funeral will take place tomorrow, Thursday, morning at 10 o'clock from the home of his brother, Joseph B. **Welch**, near Cabbage Patch. The interment will take place in the Smartsville cemetery.

9 Dec 1903 (Wednesday); Page 2 Col 4; Died – Bowman – At her home, in Rose Bar township, December 8, 1903, Miss Lizzie **Bowman**, a native of California and sister of County Clerk Gordon **Bowman**. The funeral will take place from her late residence to-morrow (Thursday) morning at 10 o'clock. Interment, fraternal cemetery, Smartsville.

9 Dec 1903 (Wednesday); Page 2 Col 4; Bach – At her home, in Smartsville, Dec 8, 1903, Mrs. Alice Wright **Bach**, the beloved wife of John Edward **Bach**, a native of California, aged 23 years. The funeral will take place at 2:30 tomorrow (Thursday) afternoon from her late residence. Interment Smartsville Cemetery.

15 Dec 1903 (Tuesday); Page 4 Col 3; DIED. At the Mater Miserordiae Hospital, Sacramento, December 13, 1903, **Agnes**, the beloved daughter of Mr. and Mrs. Thomas R. **Lee**, a native of California, aged 18 years, 4 months and 28 days. The funeral services will be held at 2 o'clock this afternoon at the Church of the Immaculate Conception, Smartsville, Interment, Catholic Cemetery.

10 Jan 1904 (Sunday); Page 4 Col 2; DIED. At his home near Smartsville, January 9, 1904, Edward **Walsh**, a native of California, aged 40 years and 1 day. The funeral services will be held at the Catholic Church, Smartsville, tomorrow (Monday) afternoon, at 2 o'clock. Interment, Smartsville Catholic Cemetery.

19 Jan 1904 (Tuesday); Page 5 Col 6; DIED. At his home in Smartsville, January 18, 1904, Edward Thomas **Linehan**, a native of California, aged 33 years, 7 months and 12 days. The funeral will take place at 2 o'clock tomorrow (Wednesday) afternoon from the Church of the Immaculate Conception. Interment Smartsville Catholic Cemetery.

17 Mar 1904 (Thursday); Page 5 Col 3; DIED. At his home in Smartsville, March 16, 1904, Richard **Beatty**, a native of Ireland, aged about 82 years. Funeral notice hereafter.

NEWSPAPER INFORMATION

17 Mar 1904 (Thursday); Page 7 Col 3; Richard **Beatty**, a pioneer resident of Smartsville, died at his home at 1:30 yesterday afternoon after an illness extending over one year. The deceased was a native of Ireland, about 82 years of age, and was an old and respected resident of this county. He was naturalized on October 27, 1860, in the District Court of Yuba County. He leaves two sons, Robert **Beatty**, who is Postmaster at Smartsville, and secretary of the Excelsior Water and Mining Company, and John **Beatty**, who resides at Santa Ana, Orange county, and who was at one time Supervisor in Yuba County. The deceased was a member of Rose Bar Lodge, No. 89, F. and A. M. By his death one of the old landmarks of Yuba county has been removed.

4 Jul 1904; (Monday); Page 3 Col 2; Coughlan – In the city of Stockton, July 4, 1904, William **Coughlan**, a native of Ireland, aged 79 years. The remains will arrive on this afternoon's train and will be interred in the family plot, Smartsville Catholic cemetery.

6 Jul 1904; (Wednesday); Page 1 Col 4; Married on the Fourth of July. A marriage license was issued on Sunday evening to Joseph **Bushby** aged 21, and Miss Nora **Murch**, aged 27, both residents of Rose Bar township. They were married on Monday by Rev. John T. **Vineyard** at his home. The groom is the youngest son of Mr. And Mrs. George **Bushby**, well-known residents of Rose Bar Township, while the bride for about three years has lived with the family of Dan **Caine** Jr. at his home near Smartsville.

6 Jul 1904 (Wednesday); Page 4 Col 2; DIED. In the city of Stockton, July 4, 1904, William **Coughlan (McCoghlan)**, a native of Ireland, aged 79 years. The remains will arrive on this afternoon's train and will be interred in the family plot, Catholic Cemetery.

7 Jul 1904 (Thursday); Page 1 Col 6; The remains of the late William **Coughlan (McCoghlan)**, who was a nurseryman in this city for many years and who sold out his place to the S. P. Company some years back, arrived from Stockton on Yesterday afternoon's train. They were accompanied by Rev. Father **O'Connor** and by Ex-Postmaster J. J. **White** of Oakland and Mrs. **White**, the latter who was a sister of the deceased. Undertaker R. E. **Bevan** met the remains of the depot and the cortege moved to the Catholic Cemetery, where Rev. M. **Coleman** officiated, after which the interment took place in the family lot by the side of his wife.

3 Sep 1904 (Saturday); Page 1 Col 4; Ex-Supervisor Louis **Conrath** died yesterday morning at the home of his brother in law, John **Quinkert**, at Smartsville. He had been in poor health for the past two years and the news of his passing away was not unexpected. He had gone to the mountains this summer with the hopes of being able to recruit his health. On the return trip he stopped over at Smartsville and there passed away, surrounded by those relatives that he loved best on earth. The deceased was a native of Milwaukee, Wis., and about 50 years of age. He came to this section of the country about eighteen years ago, taking up his residence on the McGanney ranch, known as the "Reservation". In Nevada county, where in

NEWSPAPER INFORMATION

company with Antone **Gerig** he was engaged in the stock business. He later moved to the Bowman ranch, near Smartsville, and afterward to the Brady ranch, and after returning to Smartsville was in the stock business with J. P. **Ehert**. He was employed for some years as superintendent of the Excelsior Water and Mining Company's interests at Smartsville. Mr. **Conrath** was elected Supervisor in the Fourth district in the fall of 1892, defeating A. R. **Forbes** by 104 votes, and was re-elected in the fall of 1896 for four more years, defeating A. F. **Folsom** by 129 votes. After completing his term of office he came to Marysville, where he made his home and engaged in the insurance business, having an office on Second street, opposite the Western Hotel. As a public servant he was very popular in the district he represented and his death will be regretted by a legion of friends. The flag at the Court House was at half mast yesterday out of respect to his memory. Louis **Conrath** was a man who had a number of very fine traits of character, prominent among with was his constant desire to see his friends swell in harmony among themselves and on many an occasion he has gone a long way out of his path to bring together two neighbors between whom some difference had arisen. Resourceful in argument his plea seldom failed to restore good feeling in cases of this nature. When things were prosperous with him no person ever knew of Louis **Conrath** being appealed to in the cause of charity that he did not respond to the call, many times himself taking the initiative in an endeavor to make the way a trifle easier for some poor acquaintance who had fallen upon evil times. The recollection of such actins will do much to preserve a pleasing remembrance of the dead man among those who were his friends and who were familiar with his really kindly nature. He leaves a devoted wife, four children, **Ludwig**, **Frank**, **Clara** and Lizzie **Conrath** and one brother, Gustav **Conrath**, to mourn his demise. The deceased was a member of Marysville Lodge No. 38, A. O. U. W., in which he held an insurance policy. The funeral services will be held tomorrow (Sunday) afternoon at 2 o'clock at the church of the Immaculate Conception, Smartsville. The interment will follow in the Catholic Cemetery.

3 Sep 1904 (Saturday); Page 4 Col 2; DIED. In Smartsville, September 2, 1904, Louis **Conrath**, a native of Milwaukee, Wis., aged 49 years, 11 months and 3 days. The funeral will take place at 2 o'clock tomorrow (Sunday) afternoon, from the Church of the Immaculate Conception, where services will be held. Interment Smartsville Catholic Cemetery.

30 Sep 1904 (Friday); Page 4 Col 2; DIED. At his home, seven miles south of Smartsville, September 29, 1904, Peter **Casey**, a native of County Galway, Ireland, aged 80 years. The funeral cortege will leave his late residence Saturday morning at 10 o'clock and arrive at the Smartsville Catholic Church at 12:30 where services will be held. Interment Catholic Cemetery.

2 Nov 1904 (Wednesday); Page 1 Col 4 & 5; Formerly Resided in Marysville and Was Well Known Here. Mrs. Susan Ann Duke **Harriman** died at her home in Sicard Flat on Monday night. The deceased was a native of Canada and was aged 69 years. She was the wife of Stephen **Harriman**, a well known resident of Yuba County, to whom she was married thirty years ago by the Rev. Mr. **Ward** of the

NEWSPAPER INFORMATION

Episcopal Church of Marysville. Before her marriage she resided in Maysville for several years. She was a good Christian woman and had the respect of all who knew her. The funeral will take place tomorrow (Thursday) morning at 10 o'clock from her late home, and the interment will take place in the Fraternal Cemetery, Smartsville.

1 Dec 1904 (Thursday); Page 4 Col 3; Died –In Smartsville, November 30, 1904, Mrs. Amelia **Saunders** wife of John **Saunders** Jr., a native of Illinois, aged 82 years. The funeral will take place tomorrow morning at 10 o'clock from her late residence. Interment in the Smartsville Cemetery.

14 Dec 1904 (Wednesday); Page 4 Col 2; Married – **Gassaway** – **Colling** – In Smartsville, December 11, 1904, by Rev. C. J. **Richard**, Robert Samuel **Gassaway** and Mrs. Bertha **Colling**, both residents of Smartsville.

17 Mar 1904; Page 5 Col 4; Died – **Beatty** – at his home in Smartsville, March 16, 1904, Richard **Beatty**, a native of Ireland, aged about 82 years. Funeral notice hereafter.

17 Mar 1904; Page 7 Col 2; Richard **Beatty** Fall of Years is Called at Smartsville – Richard **Beatty**, a pioneer resident of Smartsville, died at his home at 1:30 yesterday afternoon after an illness extending over one year. The deceased was a native of Ireland, about 82 years of age, and was an old and respected resident of this county. He was naturalized on October 27, 1860, in the District Court of Yuba County. He leaves two sons, Robert **Beatty**, who is Postmaster at Smartsville, and secretary of the Excelsior Water and Mining Company, and John **Beatty**, who reside at Santa Ana, Orange County, and who was at one time Supervisor in Yuba County. The deceased was a member of Rose Bar Lodge, No. 89, F. and A. M. By his death one of the old landmarks of Yuba County has been removed.

27 Apr 1905 (Thursday); Page 1 Col 5; Robert **Toland**, an old and much respected resident of Rose Bar township, died at 2 o'clock yesterday morning at his home, three miles south of Smartsville, from heart trouble. The deceased was a native of County Antrim, Ireland, where he was born in 1833, being 72 years of age at the time of his death. In 1856, soon after arriving at the age of manhood, he left his native land and sailed for the United States, landing in Pennsylvania. He remained there until 1859, when he decided to make his home in the Golden West, and started for this Coast, coming via the Isthmus. He made the Atlantic voyage in the Northern Light, and the Pacific in the Sonora. He spent one month in San Francisco before going to the mines in Oroville, and soon afterwards made his home in Rose Bar township, where he got married in 1868, an followed ranching up to the time of his death. He leaves to mourn his demise a wife, three sons, **William J.**, **John** and **Hugh**, one daughter, **Mary**. He was the step-father of Mrs. **Green** of Berkeley, and Mrs. **McDonald** of San Francisco, and of **Robert** and James **McKaig** of Honolulu, Hawaiian Islands. The deceased was an honest and upright man, who was much respected in the community where he resided for so many years. He was a devoted husband, a kind father, a good neighbor and a true

NEWSPAPER INFORMATION

friend. He was a member of Marysville Lodge, No. 38, A. O. U. W. The funeral services will be held at his late home tomorrow (Friday) morning at 11 o'clock, and the interment will take place in the fraternal cemetery at Smartsville.

29 Apr 1905 (Saturday); Page 2 Col 2; The funeral of the late Robert **Toland** took place from his late home near Smartsville Friday. Rev. John T. **Vineyard** officiated in the ceremonies at the family home and the pall-bearers were a. **Newbert**, W. W. **Chamberlain**, James **Perkins**, Miles **Wellman** and Thomas **Lee**, all members of the A. O. U. W., of which deceased was a member. The funeral was very largely attended.

29 Apr 1905 (Saturday); Page 1 Col 3; The funeral of the late Mrs. Margaret Ann **Sullivan** at Smartsville Friday was attended by many old friends of the lady. Rev. J. J. **Hynes** officiated in the ceremony and the body was interred in the Catholic cemetery. The pallbearers were John **McQuaid**, Andrew **McCarthy**, John **Havey**, John **Byrne**, Daniel **Saunders**, and James **Kelly**.

15 Jun 1905; Page 1 Col 3; William W. **Chamberlin** answers call of the Grim Reaper. William Wallace **Chamberlin**, whose serious illness was announced in "The Appeal", died last evening at his home in Smartsville from heart trouble. The deceased was a native of Canada, 73 years of age, and had resided in Yuba Co. for the past 40 years. He had followed mining for about 20 years, and worked in the Blue Gravel and Blue Point mines at Smartsville. He also worked for J. P. Pierce and Daniel McGradey and had an interest in mines at Sucker flat. He was also interested in quarts mines in Nevada County, and was interested in the Forlorn Hope mine, being the largest stockholder. For the last 20 years he conducted a saloon in Smartsville, and relieved a liberal patronage. He was an honorable and very industrious man and his word was as good as his bond to all who knew him. He was square and honest in all his dealings, and no man was more respected in Rose Bar Township. He married Mrs. **Otis** of Mooney Flat about twenty years ago and she survives him, as well as the stepson of the deceased, James **Otis**. No more kind or generous man lived in Smartsville, and his death will be mourned by that entire community. He was a kind and affectionate husband, one of the best of citizens. By his death an old landmark has been removed that cannot be replaced. The time for the funeral has not as yet been announced.

15 Jun 1905; Page 4 Col 3; DIED In Smartsville, June 14, 1905 W. W. **Chamberlin**, a native of Canada, aged 73 years. Funeral notice hereafter.

2 Aug 1905 (Wednesday); Page 1 Col 4; William **Cramp**, an old Grand Army man, died yesterday at his home at Sicard Flat. He was a native of Maine, and was 76 years of age. He leaves one son, Wilfred R. **Cramp**, who also resided at Sicard Flat and who is 37 years of age. The deceased was a member of the Masonic order

17 Oct 1905 (Tuesday); Page 2 Col 2; Elizabeth **Hapgood** Buried in Smartsville. Last evening the funeral of the late Mrs. E. D. **Hapgood** took place in this city, Rev. W. H. **Stoy** officiating in the religious services at Kelly's undertaking parlors.

NEWSPAPER INFORMATION

This morning at 7 o'clock the funeral procession left for Smartsville where the interment took place at 11 o'clock.

1 Nov 1905; Page 8 Col 2; Granville Clifford **Keith** died at 10:25 o'clock last night at the county hospital. Keith was a miner and was brought here from Smartsville. He was admitted to the hospital on the recommendation of Supervisor Wheaton. Keith was 74 years of age, and as far as is known he leaves no relatives.

4 Nov 1905 (Saturday); Page 1 Col 5; **Bessie**, the beloved wife of Philip H. **Smith**, died at her family home at Timbuctoo at 6 o'clock yesterday morning. The deceased was a native of Canada, aged 31 years and 23 days, and leaves six sons, the youngest of which is only two weeks old, to mourn her demise. She was the daughter of Mr. and Mrs. John **Peardon** of Smartsville and a sister of **John, Fred, William, James, Frank** and Miss Bertha **Peardon** and of Mrs. T. J. **Williams** of Marysville. She was a woman of cheerful disposition, who has many friends by all of whom she was dearly loved. The funeral services will be held at 2 o'clock tomorrow (Sunday) afternoon and the interment will take place in the fraternal cemetery.

7 Dec 1905 (Thursday); Page 3 Col 3; SMATRSVILLE. Dec 1905. Nicholas **Trevenia** died at his late home in French Corral on the 3rd instant. The deceased was for several years a respected resident of this place. In the years 1877 and 1878, Mr. and Mrs. **Trevenia**, parents of the deceased, were proprietors of the Smartsville hotel, where the father died, after which the deceased and mother moved to French Corral, where they have continued to reside up to the time of his demise The remains will be brought to this place today. The funeral to take place at one o'clock p.m. The interment will be made in the family plot in the fraternal cemetery.

29 Jan 1906 (Monday); Page 5 Col 4; Died – **Harriman**, At Sicard Flat, January 29, 1906, Elsie, beloved wife of Edward L. **Harriman**, a native of New Jersey, aged 34 years, 6 months, and 29 days. The funeral will take place from the family home Wednesday at 10 a.m.

26 May 1906 (Saturday); Page 1 Col 4; Mrs. Kate **Daly**, formerly a resident of Smartsville, but late of San Francisco, died at the bay city Wednesday morning at 8:30 o'clock. Word reached that city the same day bringing the news of the death of her daughter, Mrs. Kate **O'Donnell**, which occurred in Napa county. The body of Mrs. **Daly** arrived last evening and was taken in charge by Kelly Bros. The remains will be taken to Smartsville today, where the funeral will take place at 10 o'clock. The remains will be interred in the Catholic Cemetery at that place. John Lawrence and Will **Daly**, former residents of this city, were sons of the deceased.

6 Jun 1906 (Wednesday); Page 1 Col 6; William L. **Newbert** of Smartsville and Miss Nellie **Wright** of Waldo were married yesterday morning at 11 o'clock at the home of M. Ramsey in this city. Rev. John **Vineyard** of Smartsville tied the nuptial know in the presence of relatives and a few intimate friends. Miss Minnie

NEWSPAPER INFORMATION

Wright, a sister of the bride, and Thomas **Newbert**, a brother of the groom, acted as bridesmaid and best man. The happy young couple departed for Sacramento, where they will spend their honeymoon. They will make their home at Smartsville.

20 Jun 1906 (Wednesday); Page 1 Col 5; John Edward **Cramsie** and Miss Anna **McGanney**, two popular young people of Smartsville, were married in this city Monday evening at the Catholic Church, Rev. J. J. **Hynes** of Smartsville officiating. The ceremony was performed in the presence of only relatives and a few intimate friends. They departed yesterday morning for their home at Smartsville, where they will at once take up their residence. Miss **McGanney** is a daughter of the late Daniel **McGanney** and is a very pretty and lovable young woman, with a large circle of friends. Her father at one time was considered one of the wealthiest men in California. Mr. **Cramsie** is a son of Mr. and Mrs. William **Cramsie** of Smartsville and is a very bright and industrious young man with sterling qualities. He has a host of friends at that place and in this city. He is a trusted employee of the Excelsior Water and Mining Company at that place, having been with that concern for over eight years. He is a brother of Deputy Recorder W. P. **Cramsie** of this city.

6 Oct 1906 (Saturday); Page 1 Col 4; Mrs. John **Peardon** suddenly Passes Away at Smartsville. Mrs. John **Peardon**, a well known resident of Smartsville, died suddenly last evening at that place after an illness of only a short duration. The deceased was a native of England and aged 57 years and is survived by a husband, five sons, **John**, Jr., and employee of White, Colley & Cutts, and **Fred**, a saloon keeper of this city (Marysville); **James** of Nimshew and there employed by the Bay Counties Power Company, **Frank** and **William** of Smartsville and two daughters, Mrs. T. J. **Williams** of this city (Marysville) and Miss Bertha **Peardon** of Smartsville. She has been a resident of Smartsville for nearly thirty years and was a woman of many friends, being of a kind heart and always one ready to give a helping hand to one in need. Her passing away will be deeply regretted by the entire community in which she resided for so many years. She was of a very ambitious disposition. With her husband they have been conducting a hotel at Smartsville for many years. No arrangement has as yet been made for the funeral.

21 Oct 1906 (Sunday); Page 1 Col 3 & 4; One of the most deplorable accidents that has occurred in this vicinity for many years was on yesterday morning at 9:30 o'clock. It happened about five miles below the town of Nicolaus in Sutter county. Three men – Thomas R. **Lee**, Frank **Cunningham** and T. F. **Munday** – were at work on what was presumed to be a "dead" line. They were splicing it and were standing on the ground. The wire was grounded on each side of the men as a protection and is always done in cases of that kind when that character of work is being handled, but for some unexplained cause the power was turned on and off instantly. It is not known whether it was switched at the station or whether a swinging wire at some other point came in contact with the one on which they were working. When the unexpected current came and it carried a voltage of 60,000, the three men instantly fell enveloped in flames. They did not, however, receive the full force of this voltage as the wire was grounded. The result was, however, that

NEWSPAPER INFORMATION

their companions who were about three hundred yards away and who went immediately to their assistance thought that they were dead. **Cunningham** and **Munday** showed signs of life, but they were terribly burned, particularly **Munday** on the crown of his head. Messengers were immediately sent out and physicians were summoned from Davisville, Wheatland and Nicolaus. Dr. **Harding** was the first to arrive at the scene and did everything possible to revive the three men. He discovered that **Lee** was beyond human aid, but that there was hope for the others. Cunningham, after being in a comatose condition for more than two hours regained consciousness. Three other doctors were working on the men at noon. **Lee** in the meantime died, and the remains were taken in charge by Coroner Rowe. In the afternoon an inquest was held, but nothing developed further that that the cause of death was purely accidental. His body was brought to this city by Undertaker Kelly on the Knights Landing train last evening and will be shipped to Smartsville from this city. **Lee** was aged about 50 years and lived at Smartsville for about thirty years. He was well and favorably known. He leaves a wife and two daughters of Smartsville. **Cunningham** and **Munday** were removed to Nicolaus and today will be taken to the Sister's Hospital in Sacramento, where they will be treated. From last accounts they stand a reasonable chance for recovery, but at this writing nothing definite can be said in that regard. **Cunningham** came up from Byron in Contra Costa county and **Munday** came out from Philadelphia. Both are reputed to be expert linemen and are young men. **Lee** was a teamster and was engaged with one of the crew in reconstructing the power line. Several crews are at work removing every second pole and raising the height of the remaining poles and putting on new arms and insulators.

23 Oct 1906 (Tuesday); Page 8 Col 2 & 3; The remains of the late Thomas R. **Lee**, who was accidentally killed on Saturday on the Bay Counties Power line, were taken to Smartsville on Sunday. The funeral services took place yesterday morning at 10 o'clock from the Church of Immaculate Conception, Smartsville. Rev. J. J. Hynes officiated. The following gentlemen acted as pall bearers: William **Cramsie**, G. W. **Nelson**, W. A. **Smith**, J. E. **Bach**, J. **Kelly** and J. H. **McQuaid**. The interment took place in the Catholic Cemetery.

3 Nov 1906 (Saturday); Page 9 Col 2; Mrs. Winifred **Casey**, relict of the late Peter **Casey**, who were former residents of Rose Bar township, died at 2 o'clock yesterday afternoon at the home of her daughter, Mrs. James **McWilliams**, at 622 B street, in this city. The deceased was a native of Ireland and aged 70 years. She was much respected by her many friends. She leaves to mourn her demise two daughters, Mrs. **Holt** of Chico and Mrs. James **McWilliams** of Marysville; also four sons, **John** H. and P. H. **Casey** of Browns Valley, William **Casey** of Idaho and F. H. **Casey** of Erle, Yuba county. The funeral and interment will take place tomorrow (Sunday). The interment will be in the Catholic Cemetery at Smartsville.

24 Nov 1906 (Saturday); Page 4 Col 4; DIED. At her home near Smartsville, November 22, 1906, Mrs. Anne **Dougherty**, wife of Daniel **Dougherty**, a native of County Donegal, Ireland, aged 71 years. The funeral will take place from her late residence, Sunday morning, at 9:30; then to the Church of the Immaculate

NEWSPAPER INFORMATION

Conception at Smartsville, where mass will be celebrated for the repose of her soul, at 10 o'clock. Interment Catholic Cemetery.

16 Jan 1907 (Wednesday); Page 1 Col 5; Mrs. Ann **Trevena** died at the Grass Valley county hospital on Saturday evening, where she had been for some ten days. Old age and a general breaking down of the system was the cause of death. Mrs. Trevena was a native of England and had reached the ripe old age of over eighty-four years. For many years she resided with her deceased husband at Smartsville, after moving to French Corral, where her son, Nicholas Trevena, died a few years age. Deceased was a kindly old lady who made and held many friends. The remains have been taken to Smartsville for burial.

31 Jan 1907 (Thursday); Page 8 Col 2; W. W. Marple of Timbuctoo died yesterday as a result of New Year accident. Whelton Wharton **Marple**, a well-known pioneer miner, known as the father of Timbuctoo, died at his home at that place yesterday afternoon, at the age of 83 years, death being the direct result of injuries received in a fall on New Year Day. He was a native of Pennsylvania and located at the historic old mining camp in the '40's. The camp was not known by that name at first, but John Henry, a storekeeper at Rose Bar, after reading a history of North Africa dubbed Marple "the Sultan of Timbuctoo". There after Marples' camp was Timbuctoo and Marple was the Sultan. Marple struck pay dirt at Timbuctoo and the camp was one famed in California history. There was a Theater there in early days and the best show companies that came to the coast made it a point to play there. The miners showered gold dust upon the actors and actresses upon each visit making their trips to Timbuctoo decidedly profitable. On New Year Day the old man fell from a porch at his home and sustained injuries from which he never received. Deceased was the father of eighteen children, six of them being twins. But six of these survive, they being Mrs. J. **Hapgood** and another daughter who resides in Idaho, and four sons, **Fred**, **George**, **Barney** and Sam **Marple**. Mr. Marple was a member of the Masonic order, having been identified with the Smartsville Lodge for many years. Undertaker R. E. **Bevan** of this city has taken charge of the body.

22 Jun 1907 (Saturday); Page 5 Col 3; Word reached Wheatland Thursday afternoon that W. B. **Adkins**, one of the best known residents of the foothill section had been accidentally killed at the Dairy Farm mine by falling from the skip to the bottom of the shaft, a distance of over seventy feet. From the meager details that have reached Wheatland from the mine. It appears that Mr. Adkins was working on a drift about 70 feet from the bottom. The only signal he had to those above was a lighted candle and in attempting to extinguish it and notify those above to start the skip he was taken unawares by the sudden start and thrown out backwards to the bottom where he struck his head & shoulders the shock crushing his head to a pulp. The death of Mr. Adkins is very sad indeed. He was a jolly man and liked by everybody who knew him. He leaves a wife and a baby but 11 days old to mourn the loss of an affectionate husband and father. The funeral will be held from the home at Dairy Farm mine tomorrow morning at 10 o'clock. The interment will be in the Smartsville cemetery.

NEWSPAPER INFORMATION

5 Jul 1907 (Friday); Marysville Evening Democrat; Page 8 Col 3; James Francis **McGanney** died yesterday at Smartsville. He was a native of that town and was 44 years, 3 months and 24 days. He had been residing in San Francisco for a number of years and arrived at Smartsville last Sunday, intending to stay there awhile, for the benefit of his health. He was in very bad condition then. Deceased leaves a wife and five children as well as three brothers and two sisters. They are **Ed.**, **Frank** and Dan **McGanney** and Mrs. J. E. **Cramsie**. The funeral will be held Saturday at 10 a.m., when services will be held in the church of the Immaculate Conception at Smartsville. The interment will take place in the Catholic cemetery at Smartsville.

31 Jul 1907; Page 1 Col 5; George **Jeffery** died today after spending fifty years there. George Jeffery, a pioneer blacksmith, died at his home in Smartsville this morning at the age of 77 years. He was a native of Nova Scotia, but came here many years ago and was naturalized in this county August 25, 1858. Before coming here he was mate on a ship plying between San Francisco and Honolulu, his father being the captain. Deceased was a respected old man and was known far and wide. His death is the occasion of sorrow to a host of friends. Surviving him are three daughters, Mrs. T. J. **Tyrell** of this city (Marysville), Mrs. Robert **Gassaway** and Mrs. Ella **Jeffery**, and three sons, **William B.**, **George** and Robert M. **Jeffery**. The announcement of the funeral will be made later.

1 Aug 1907 (Thursday); Page 8 Col 3; The funeral of the late George **Jeffery** of Smartsville will be held tomorrow afternoon at 1:30 o'clock and the interment will take place in the Fraternal cemetery at Smartsville.

8 Aug 1907 (Thursday); Page 2 Col 3; Smartsville, Aug 7 – Mrs. P. (Hannah) **Ryan**, a highly respected resident of this community, died here this morning, 60 years old. She leaves a husband and a daughter. Who was a native of Ireland. She resided in Smartsville for more than 30 years and was a favorite with everyone. She was kindly in disposition and was very charitable. Undertaker J. K. **Kelly** came from Marysville and prepared the body for burial. The funeral will be held from the Catholic church in Smartsville tomorrow morning. Interment will be in the Catholic cemetery at this place.

22 Aug 1907 (Thursday); Page 1 Col 3 & 4; John **Peardon**, who has resided at Smartsville for the last 40 years and whose illness was announced in the Appeal a few days ago, died at his home at 2 o'clock yesterday morning. The deceased was a native of England, being born on August 24, 1842, and has been a resident of Smartsville about 40 years, having been the proprietor of the hotel and other business interests in that prosperous little town. He was 64 years old at the time of his death. He underwent a surgical operation in San Francisco about a year and a half ago and seemed to be getting along very nicely. He visited in Marysville about a week ago and his relatives did not notice that the final call was so near at hand. A few days ago he was taken suddenly ill and the attending physician, Dr. Hall **Vestal**, sent for Dr. **Powell** and a consultation was held over the sick man, and the physicians saw that his case was very serious. He leaves to mourn his demise two

NEWSPAPER INFORMATION

daughters, Mrs. T. J. **Williams** of Marysville and Miss Bertha **Peardon** of Smartsville, and the following sons; **J. H., Fred, James, William** and Francis **Peardon**. The deceased was a prominent member of Fedonia Lodge, No. 188, of Smartsville, and the funeral will take place under its auspices next Saturday afternoon at 2 o'clock. Mr. Peardon was a man much respected in Smartsville and all parts of Yuba county where he was known, and the announcement of his death will be received with deep regret by his many friends.

10 Oct 1907 (Thursday); Page 4 Col 2; Smartsville, Oct 9 – Mrs. V. **Schmidt** of Mooney Flat, after a long illness, passed to the great beyond this morning at about 4 o'clock. The deceased was near her ninetieth year and was much esteemed by all. She leaves two sons – **Hiney** of San Francisco and **William** of Southern California – and two daughters – Mrs. Mary **Gassaway** of San Francisco and Mrs. C. **Manasco** of this place. The arrangements for interment have not yet been made public.

10 Oct 1907 (Thursday); Page 4 Col 3; DIED. At Mooney Flat, October 9, 1907, Veronica **Schmidt**, relict of George **Schmidt**, a native of Germany, aged 81 years. Funeral notice later.

11 Dec 1907 (Wednesday); Page 1 Col 2; John **Flanagan**, native of Smartsville, aged about 38 years, died in that town yesterday afternoon. He leaves a mother, brother, and sister. The funeral has not been announced.

12 Dec 1907 (Thursday); Page 4 Col 3; DIED. At Smartsville, December 10, 1907, James Henry **Flanagan**, a native of New Jersey, and aged 38 years. The funeral will take place at Smartsville this afternoon at 1 o'clock.

12 Dec 1907 (Thursday); Page 4 Col 3; DIED. Near Mooney Flat, December 11, 1907, John **Driscoll**, a native of Ireland, aged 82 years. Funeral notice hereafter.

12 Dec 1907 (Thursday); Page 7 Col 2; Smartsville, December 11 – John **Driscoll**, Sr., a pioneer resident of this neighborhood living in Mooney Flat, died this morning at 8:30 o'clock after a protracted illness. Mr. **Driscoll** was one of the most industrious, upright and honest of men. His motto was "do unto others as you would wish others do unto you." While the community deeply regret the departure of this good citizen, yet death to Mr. **Driscoll** was a sweet relief, he having been a great sufferer for many years from rheumatism.

12 Dec 1907 (Thursday); Page 7 Col 2; Smartsville, Dec. 11 – John **Flanagan**, whose illness was reported some weeks ago in this column of The Appeal, passed to the great beyond yesterday at noon. About three months ago the deceased on account of ill health, was obliged to give up work and since that time has been rapidly growing weaker, and the end was not unexpected. He was 39 years of age and leaves a mother, one brother, Thomas J. **Flanagan**, of this place and one sister, Mrs. **Wilson**, of Idaho, who came to this place several weeks ago on account of her brother's illness. The funeral will take place tomorrow, but owing to the absence

NEWSPAPER INFORMATION

of the Rev. Father **Hynes**, who is holding services in the neighborhood of Challenge, the hour has not been decided upon.

28 Jan 1908 (Tuesday); Page 3 Col 3; **William** Everett, youngest son of Mrs. **Alpha** and the late Russell M. **Widner**, died at his home in Smartsville Sunday. Deceased was a native of California, aged 25 years. He was a brother of Mrs. Ethel **Miller** of Sacramento and of Joseph **Widner** of Smartsville. He had been in poor health for some time and his death was not unexpected. The funeral will take place from the home of his mother this afternoon at 2 o'clock and the interment will be in the Smartsville cemetery.

30 Jan 1908 (Thursday); Page 6 Col 2; The funeral of William Everett **Widener** took place Tuesday from the family home in Smartsville, Rev. John **Vineyard** officiated. The following gentlemen acted as pallbearers: William **Peardon**, Frank **Cramsie**, John **Murphy**, Jack **McKeef**, William **Mead** and Art **Money**. The floral offerings were numerous and beautiful and spoke silently of the esteem in which the young man was held.

4 Mar 1908 (Wednesday); Page 1 Col 4; Mrs. Ann **Sweeney**, who had been an inmate of the county hospital for eighteen years, died at that institution yesterday afternoon. She was a native of Ireland, aged 75 years. Requiem mass will be celebrated for the repose of her soul this morning in the chapel of the convent and services will be held in the undertaking parlors of **Kelly** Brothers at 4 o'clock this afternoon. Interment will be in Smartsville cemetery Thursday afternoon at 1 o'clock. Mrs. **Sweeney** was the widow of Michael **Sweeney**, who was killed in a mine accident in Smartsville in the year 1877.

19 Jun 1908 (Friday); Page 3 Col 3; A dispatch announcing the death of Timothy Edward **Early** at Butte City, Montana, was received by his family in this city Thursday evening and also announcing that his body would be shipped to Marysville today for interment in the grave with his late wife at Smartsville, where the burial will take place on Sunday. Timothy Edward **Early** was one of the best known old pioneers of Yuba County, having arrived with other young Argonauts at Rose's Bar, this county, in the early fifties from Newark, New Jersey and followed the business of mining in pursuit of the golden fleece and was successful during his younger days and up to a few years ago, when old age admonished a halt in his labors. On the 26th day of October, 1856, being then 26 years of age, Timothy Edward **Early** was married to Mary **Sullivan** by Rev. P. L. **Deyeart** of the Catholic church in this city. The issue of the marriage was four boys and four girls, all of whom were born in Smartsville, and all of whom, except the oldest son, are still living – two sons and two daughters at Butte City, Montanan, and two daughters and one son – Mrs. William **Reilly**, Mrs. William **Hagerty**, and Frank **Early** – residing in this city. The deceased was a most successful miner and held a high position as superintendent in the mines of James **O'Brien** at Smartsville for many years and was also superintendent of the Pittsburg mines at the same place under the management of the late John **Brickell**; and also superintendent of a mine in Plumas county for several years. After the death of his noble wife, which occurred

NEWSPAPER INFORMATION

at Smartsville on January 23, 1879, he went to Butte City, Montana, and entered the service of Marcus **Daly**, where he was employed and appreciated up to the time that death called and closed his earthly labors. He was a kind and affectionate husband, a good family provider and his bereaved family will keenly feel the great loss of this generous, noble and unselfish father.

3 Jul 1908 (Friday); Page 1 Col 4; Pioneer Resident of Smartsville Dies at Daughter's Home in this city. Mrs. James **Doyle**, a native of Queenswon, Ireland, died last evening at 7 o'clock at the residence of her daughter, Mrs. J. C. **Baldwin**, 415 D. Street, this city. At the time of her death Mrs. Doyle had reached the age of three score and ten. She met and was married to James **Doyle** in the east some fifty-two years ago, and her death is a hard blow to her aged husband who survives her. They came to Yuba county, to reside about forty years ago and have lived at Smartsville most of the time since. Mrs. **Doyle** was the mother of three sons and four daughters, five of whom survive her. They are Mrs. H. **Merriam** of Dobbins, Mrs. F. L. **Johnson** of Sacramento, Mrs. J. C. **Baldwin** of Marysville, E. W. **Doyle** of Sacramento and J. F. **Doyle** of Marysville. Five years ago her oldest son, James, met with a very sad and sudden death, being electrocuted at his own home in the presence of his aged mother. Mrs. **Doyle** never entirely recovered from the shock of his death. For the past year she had been a patient sufferer from a complication of diseases, which finally caused her death. She was brought to this city from her home in Smartsville several weeks ago and has been given the very best of care and attention at the home of her daughter. Mrs. **Doyle** was a much beloved woman and leaves many sincere friends to mourn her death. She will be laid at rest in the Catholic cemetery at Smartsville on Saturday morning, July 4th.

8 Jul 1908; Page 3 Col 4; Mrs. Mary Vineyard, wife of Rev. J. T. **Vineyard**, died yesterday morning at 7 o'clock at her home at Lone Tree, near Smartsville. The funeral will take place from the family home on Thursday morning at 10 o'clock. Interment will be in the Smartsville Cemetery. Mrs. Vineyard was well and favorably known in this and Nevada Counties, and her husband is one of the best-known pastors in the valley.

19 Jul 1908; Page 1 Col 3; Henry **Fiene** died Friday afternoon at 5 o'clock at his home at Mooney Flat. He was 83 years, 3 months and 22 days of age, and leaves to mourn his death a wife, one son, William H. **Fiene** of Mooney Flat, and three daughters, Mrs. James **Dooley** and Mrs. Charles **Marks** of Sacramento, and Mrs. George **Leslie** of San Francisco. The funeral will take place at Mooney Flat this morning at 10 o'clock, and interment will be in the Smartsville cemetery.

15 Aug 1908 (Saturday); Page 2 Col 2; Miss Sara Frances **Byrne**, daughter of **James** and the late Rose Anna **Byrne**, died at San Francisco on Thursday and the body will arrive in this city today for burial. It will be taken in charge by **Kelly** Brothers and notice of the funeral will be given hereafter. Miss **Byrne** was a native of Smartsville, aged 24 years, 1 month and 7 days. She was a sister of Mrs. H. C. **Owen**, Mrs. B. F. **Compton**, Mrs. W. B. **Wilbur**, **James** D., **John** B., **Robert** E. and Henry J. **Byrne**.

NEWSPAPER INFORMATION

14 Nov 1908 (Saturday); Page 8 Col 5; Death yesterday claimed Mrs. Margaret **Conlin** of Smartsville after an illness of extended duration. The fatal ending has been expected for some time and she passed away quietly at 7:30 o'clock in the morning at her home. The deceased was married to the late Thomas Conlin, the pioneer stage man, over forty years ago in Sierra county, and was the mother of three children – **John** H., **William** R. and M. Josephine **Conlin**. All survive her. She was a sister in law of John **McQuaid**, and leaves a sister, Mrs. Vincent **Dennis**, residing in San Francisco. She was a native of County Roscommon, Ireland, and had resided in this county for many years. She was a woman of kind heart and had many friends; was ever ready to lend a helping hand to a friend and her kindly disposition won for her many friends who will deplore her death deeply. Arrangements for the funeral have not yet been made.

15 Dec 1908 (Tuesday); Page 1 Col 5; James **Young**, a pioneer resident of this county, died Sunday morning at the home of his nephew, David Jones, Jr., near Smartsville. Death was due to paralysis. The deceased was a native of Illinois, aged 72 years, and has resided in this county for about fifty years. He was a brother of the late Mrs. D. N. **Jones**. The funeral will take place this morning at 11 o'clock from the home near Smartsville. Interment will be in the Smartsville cemetery. Undertaker Kelly of this city will conduct the funeral.

15 Feb 1909 (Monday); Marysville Daily Democrat; Page 5 Col 6; In the passing of Edward **Mellarkey** at his home in Smartsville, death has removed one of the grand old pioneers of this state. Mr. **Mellarkey** was a native of County Derry, Ireland, and had reached the ripe old age of 94 years, 9 months and 2 days. He was a kindly, generous man who had ever a good word for his friends and neighbors. Edward **Mellarkey** came to this state in 1853 and located at Smartsville, where he followed mining until hydraulick mining was stopped. Since that time he has been living at home and taking life easy. He was the father of Edward **Mellarkey** of Nevada, James **Mellarkey** of Smartsville, Mrs. T. W. **McFate** of Nevada, Mrs. W. G. **House** of Sacramento and Mrs. R. **Bennett** of Los Angeles. His wife also survives him and there are also several grandchildren, two of whom reside in Smartsville – Mrs. W. P. **Walsh** and Katie **Kildahl**; the others reside in the state of Nevada. One niece, Mrs. R. **Poirier**, resides in Colusa. Deceased was held in high esteem by all who knew him in life and his name was a byword for honesty and integrity. The funeral arrangements are in the hands of Undertaker R. E. Bevan and it is likely the hour will be as near noon Tuesday as weather permits. The interment will take place in Smartsville.

23 Jul 1909 (Friday); Page 7 Col 4; The funeral of the late Daniel **Caine** will take place at the Catholic church, Smartsville, this morning at 10 o'clock. Interment will be made in the Catholic cemetery at that place. The cortege will leave this city at 5 o'clock in the morning. The funeral will be in charge of **Kelly** Bros., the undertakers.

NEWSPAPER INFORMATION

25 Jul 1909 (Sunday); Page 7 Col 3; Eugene F. **Rooney** died at his home at Smartsville yesterday morning. The funeral will take place at Smartsville at 9 o'clock today, **Kelly** brothers of this city having charge. Mr. **Rooney** was a native of California, aged 29 years and was a nephew of **John** and Charles **Saunders**. He was a grandson of Mr. and Mrs. Dan **Saunders** of Smartsville and was well and favorably known in the Smartsville section.

3 Aug 1909 (Tuesday); Page 5 Col 3; While on his way to Sacramento yesterday to have his eyes treated by a specialist, John **Dempsey**, a pioneer resident of Smartsville and well known here, died at Wheatland, where he had been taken from the train on account of his condition. Death resulted from heart disease and came suddenly and unexpected. His son, John **Dempsey**, Jr., was with the old gentleman when the end came, the son having started to Sacramento with his father. The body was brought back to this city last evening and is in charge of the **Kelly** Brothers. It will be taken to the home at Smartsville today and the funeral will take place from the Catholic church at Smartsville Wednesday morning, at 10 o'clock. Mr. **Dempsey** was 78 years of age, a native of Ireland. He leaves one daughter, Mrs. T. D. **Henderson** of Butte, Montana, and four sons; William J. **Dempsey** of this city, Michael R. **Dempsey** of Butte, Montana, Dan **Dempsey** of Yuba City, and John **Dempsey** of Smartsville.

21 Aug 1909 (Saturday); Page 1 Col 3; After living in one house at Smartsville, Yuba County, for fifty six years, Andrew **McCarty**, one of the oldest and best known pioneers of California, passed away shortly after 4 o'clock yesterday morning at the home that had been his for over half a century. Four years before settling at Smartville, **McCarty** came to California. He was a 49'er and in every sense of the word was of that sturdy stock that makes the best of citizens. When he came to this state **McCarthy** was a traveler by way of the Isthmus of Panama, but in those days there was no other means of transportation except by foot and he walked the entire distance. He settled at Smartville, or more properly speaking, at Timbuctoo, which is almost a portion of Smartville. In the year 1853, and since that time has battled with life's problems and eked out an existence in various ways. He was a highly respected citizen and had a host of friends who will mourn his loss. He leaves a wife, two daughters, Mrs. Peter **Galligan** of this city, and Miss Rose **McCarty** of Timbuctoo, three sons, A. J. and Matthew **McCarty** of Hammonton and Robert **McCarty** of Smartville. The funeral will take place Sunday morning from the Catholic church at Smartville and interment will be in the Smartville cemetery. **Kelly** Brothers of this city will have charge of the funeral arrangements.

31 Aug 1909 (Tuesday); Page 1 Col 3; James **Kelly**, a pioneer of the Smartsville region, but who for some time had been a resident of San Francisco, died at that city yesterday morning. The body will arrive on the Overland train this morning and will be taken to Smartville by Undertaker **Kelly**. Funeral services will be held at the Smartville Catholic church this morning at 10 o'clock. Mr. **Kelly** was a native of Ireland, aged 61 years. He leaves a wife, two sons and two daughters. He

NEWSPAPER INFORMATION

was well and favorably known in this city and had lived for many years at Smartville.

10 Sep 1909 (Friday); Page 1 Col 4; The funeral of Mrs. Mary **Warne**, wife of J. H. **Warne**, took place yesterday from the Methodist church at Smartville. Rev. Joseph **Sims** of Nevada City, a lifelong friend of the deceased, conducted the services at the church and grave. The funeral was largely attended, and there were many beautiful floral offerings. The honorary pallbearers were Mrs. William **Jeffrey**, Mrs. Dr. **Vestal**, Mrs. John **Quinkert**, Mrs. William **Huling**, Miss Anna **Sanders**, and Miss Alice **Bach**. The acting pallbearers were William **Huling**, William **Jeffrey**, Martin **Robins**, Arthur **Sanford**, John **McQuaid**, and John **Quinkert**. Kelly Brothers of this city (Marysville) conducted the funeral.

13 Jan 1910 (Thursday); Page 1 Col 6; William F. **Divver**, a well known resident of this city, died early Tuesday morning and the funeral will be held from the Catholic church at Smartville this afternoon at 1 o'clock. Undertaker **Bevan**, of this city, ha charge of the funeral arrangements. The deceased was aged 40 years 11 months and 9 days, and is survived by a son, Alvin **Divver**, one sister, Mrs. Maggie **Hopkins**, of Santa Rosa, and two brothers, **Byron** and George **Divver**, of this city.

30 Jan 1910; Page 3 Col 3; Much Activity in Smartsville Section. Word was received in this place a few days ago of the serious illness of Thomas A. **Newbert**, Sr. Mr. Newbert was overtaken by a stroke of paralysis last week. He is being cared for by his daughter, Mrs. H. **Creps** in Wheatland. Little encouragement is held out by the attending physician.

18 Mar 1910 (Friday); Page 1 Col 1; Col. Isaac Newton **Hays** Passes Away at Spenceville In his 89th year. Colonel Isaac Newton **Hays** died yesterday at his home in Spenceville, where he had resided for thirty years. He was born in Virginia December 12, 1821, and consequently was in his eighty-ninth year. Only one individual in many thousands attains that great age. He had passed 59 years of his life in Nevada and California. He served with distinction in the Mexican War and lost a leg in the service. This bluff old soldier will be missed from his little hostelry just this side of Spenceville, which was a sort of headquarters or meeting place for the old pioneer miners that resided thereabouts. For many years the old veteran has presided over the little gatherings that met at his road house, and the stirring times of '49 in California and the wars and battles through which the old veterans and the old miners had passed were discussed and fought over many times. Col **Hayes** left four daughters surviving him – Mrs. Lizzie **Kelly** of Spenceville, Mrs. **Kraus** of Los Angeles, Mrs. S. **Davey** of Oakland and Mrs. **Garley**, also of Oakland. Funeral services will be held at Spenceville on Saturday, March 19th, at 9:30. Interment will be made in the Smartsville cemetery, R. E. Bevan has charge of the funeral arrangements.

28 May 1910 (Saturday); Page 1 Col 6; Proprietor of Smartsville Hotel Died of Heart Disease. John H. **Peardon**, proprietor of the Smartsville hotel, died suddenly at his home yesterday afternoon after a few hours of unconsciousness. He had been

NEWSPAPER INFORMATION

suffering for some time from valvular disease of the heart, but the suddenness of his demise was most unexpected and heart-rending to his family. At seven o'clock yesterday morning he went to the stable to harness a team preparatory to driving in to Marysville. A quarter of an hour later he was found lying in the yard with his head bleeding from a fall upon a rock, and it was at first thought that his skull was fractured as he fell in a fainting spell. Dr. Hall Vestal was at once summoned and Dr. B. Kaufman of this city was also call in consultation. He was at first conscious but later sank into a state of coma and then passed quietly into the sleep of death. Deceased was born in Pennsylvania June 14, 1872. When he was five years of age his parents came to California and for the greater part of his life he had been a resident of Smartsville. For five years he made his home in Marysville and was a trusted employee of the firm of White, Cooley & Cutts. Upon the death of his father he returned to Smartsville and conducted the hotel business there. He leaves to mourn his loss a widow, two children, **Walter** and **Carmelita**, two sisters, Mrs. T. J. Williams and Miss Bertha **Peardon** of San Francisco, and four brothers, Frank G. **Peardon** and Fred **Peardon** of Marysville, James **Peardon** of Sacramento and William B. **Peardon** of Smartsville. He was a man of generous, unselfish nature, who had won countless friends throughout the county by his winning personality. The funeral will be held on Sunday afternoon at two o'clock from the family residence in Smartsville under the auspices of the woodmen of the World, of which organization he was a member. Undertaker R. E. Bevan has charge of the arrangements.

30 Jun 1910; Page 1 Col 5; Thomas A. **Newbert** died yesterday at the home of his daughter, Mrs. Henry R. **Creps** at Wheatland. Deceased was a native of Massachusetts, 79 years, 8 months and 12 days of age. He came to California in 1860 and made his home at Timbuctoo, Yuba County. He leaves the following children: Mrs. Henry R. **Creps**, Wheatland; H. C. and T. A. **Newbert** of Marigold: Mrs. J. B. **Byrne**, Marigold; William L. **Newbert**, Smartsville; Mrs. W. B. **Peardon**, Smartsville, and one brother, James **Newbert**, in Massachusetts. The date for the funeral has not been set, but it will take place at Smartsville and the interment made in the I. O. O. F. cemetery at that place. R. E. **Bevan** is the undertaker in charge.

12 Jul 1910 (Tuesday); Page 1 Col 6; Pioneer mother is called by death. Woman who was very widely known. Husband Came to California Over Seventy Years Ago and Owned Large Grant. Another pioneer mother has passed away. Mrs. Virginia **Rose**, whose death occurred in San Francisco June 8th. She was the relict of John Rose, one of California's earliest and best known pioneers. In fact John Rose was a resident of California before his wife was born, he having come to the state the same year, 1837. At one time he was one of the largest land holders in California, being proprietor of Roses rancho, a Mexican grant embracing a tract nine miles square and including all the lands from the Yuba river on the north to Johnson's rancho on the south and from the present Nevada county line on the east to Linda township on the west. Native of Virginia. Mrs. **Rose** was a native of Richmond, Virginia, born September 28, 1837, and a member of one of the most exclusive of the F. F. V.'s, a daughter of Judge Ezekiel **Dougherty**, one of the early

NEWSPAPER INFORMATION

California justices whose reputation for quaint through just rulings are household legend. Judge **Daugherty** arrived in Marysville in the spring of 1850 and removed that fall to Rose Bar, then a thriving settlement on the Yuba River near the present site of Smartsville. His family followed him west in 1853 by way of the Isthmus and in the spring of 1857, John **Rose** and Virginia **Dougherty** were wedded in the historic mining camp named for her illustrious husband, as was Rose Bar township and Rose Bar school district. She has been a resident of Rose Bar township ever since, and was on a visit to her daughter Mrs. L. B. **Bates** in San Francisco, and apparently enjoying the best of health when she was stricken with apoplexy which caused her death the same day. Refined and Cultured. She was a refined and cultured lady of antebellum days, and possessed rare gifts of mind and grace of body. An invidious reader, with a retentive mind, and an excellent and witty conversationalist her company was sought by young and old alike. A woman of deep religious convictions she was a devout member of the Episcopalian church, but above all she was a faithful wife and devoted mother, and leaves to mourn her loss two sons and two daughters, Mrs. L. B. **Bates** of San Francisco and Mrs. Chas. **Burris** of Browns' Valley, Frank **Rose** of Stockton and Oscar John **Rose** of Smartsville. The interment took place in the family plot at Smartsville the following friends acting as pall bearers; E. D. **Hapgood**, W. L. **Newbert**, W. B. **Jeffrey** and John **Mogonigal**. A choir consisting of Miss Kate **Bach**, Mrs. W. B. **Jeffrey** and J. H. **Warne** rendered appropriate music. Kelly brothers conducted the funeral.

12 Nov 1910; Page 1 Col 6; Mrs. Louisa Dooley, wife of James **Dooley** of the City, died at Mooney Flat yesterday at the Age of 33 years and 4 months. She was a native of this state. Mr. Dooley is an employee of the Southern Pacific Railroad company. Besides her husband, a mother, Mrs. Margaret **Fiene**, and a brother, W. H. **Fiene**, survive her. The funeral services will take place at the residence of relatives at Mooney Flat at 1 pm Sunday. Interment will be at Mooney Flat.

15 Dec 1910 (Thursday); Page 1 Col 1; The toll of the death Angel has been quite heavy in the County of Yuba in the past few days. Before the last sad rites in the cemetery are said and the grave closes over the remains of Dr. **Stone**, county hospital superintendent, Samuel O. **Gunning**, county auditor and recorder, is called from time to Eternity. Mr. **Gunning** had been ill for several weeks, and for some days had been at the Rideout hospital where he could receive the best of medical attention and nursing. Last week he showed some signs of improvement, and his friends and relatives believed he would pull through, but his extreme old age, being 76, told against him, and the end came peacefully yesterday at 1 p.m. At his bedside when he passed away were the members of his family. There was no better citizen in Yuba county than S. O. **Gunning** no one who had more or truer friends. He was the soul of honor, and a man who made close friendships and kept them. He had the confidence of the entire people as is witnessed by the fact that he had held the office of Auditor and Recorder for thirty years, and ha only at the recent election been elected to serve another four years term. Deceased was born in Sligo, Ireland, in January, 1834. When only twelve years old he ran away from home, found employment on a coastline vessel and for several years followed the sea. He

NEWSPAPER INFORMATION

landed in San Francisco in October 1857, and came immediately to the gold fields of Yuba county, locating at Smartsville. For a long time he was superintendent of the Excelsior, Water and Mining Company, and for ten years had charge of the Timbuctoo property. At the time of his death he was possessed of a lot of valuable mining property in the Smartsville section. S. O. **Gunning** was married in Marysville to Miss Alvina **Lotsen**, a native of Prussia, and six children were the result of the marriage, **Alvina, Ella, Samuel, Thomas, Jennie** O., and **Robert** Emmet. The funeral will take place from his late residence, 819 G street, Saturday morning at 7 a.m., thence to the Catholic church at Smartsville, where requiem mass will be said at 11 a.m. Interment in the Catholic cemetery at Smartsville.

18 Dec 1910 (Sunday); Page 1 Col 1 & 2; One of the most popular men in Northern California was hidden from view for all time when the grave closed over County Auditor and Recorder S. O. **Gunning** at Smartsville yesterday afternoon. It was one of the largest funerals that had moved through the streets of that little foothill town for half a century. Coming to Yuba county over 50 years ago while a young man, he had married and reared a family and endeared himself to every man, woman and child who knew him. He was a diamond in the rough. In his character and make up there was neither guile nor deception. He was a rough, rugged, fearless, honest character, his word was his bond. None knew him but to love him, none named him but to praise. The remains were taken from Marysville yesterday morning and a large number of hacks and autos followed the hearse until it reached his old home town and the church of the Immaculate Conception where he in his life time often attended religious services. Requiem mass was said by Rev. Father **Hynes**. The pall bearers were the following county officers: Judge E. P. **McDaniel**, Sheriff George **Voss**, Clerk J. F. **Eastman**, Treasurer George W. **Pine**, District Attorney Fred H. **Greely** and Assessor T. E. **Bevan**.

24 Dec 1910 (Saturday); Page 3 Col 1 & 2; Patrick **Ryan**, a native of County Cork, Ireland, aged 79 years, expired at his home in this city at 7 o'clock yesterday morning after an illness of a couple of weeks. Deceased was one of the best known residents of northern California, having lived in Smartsville for forty-seven years, where he owned numerous mines. He moved to this city three years ago to reside in enjoyment of his fortune in his declining years. His estate is valued at many thousands of dollars, he having owned much land in Sutter and Yuba counties. Left to mourn him are two brothers, one in Ireland and the other in Massachusetts, and a sister, Mary Jane **Ryan**, who resides in this state.

7 Mar 1911 (Tuesday); Page 2 Col 3; John **Saunders** Jr., on Saturday last, died at the home of Henry **Schoeder** at Gold Flat, Nevada county, aged 79 years, six months and three days. He came to the state in the early fifties and for many years lived near Smartsville in Yuba county.

9 Mar 1911 (Thursday); Page 2 Col 3; SMARTSVILLE. The funeral of the late John **Saunders**, Jr. took place last Monday. Rev. J. T. **Vineyard** officiated at the graveside. The interment was made in the family plot in the fraternal cemetery.

NEWSPAPER INFORMATION

20 Apr 1911 (Thursday); Page 1 Col 1; At 2 p.m. yesterday John **Walsh** passed away at his home, No. 116 Eighth Street. He was a native of Ireland and aged 87. Coming to Marysville in 1853, he had resided in Yuba county ever since. He leaves a wife, Margaret **Walsh**, and two daughters, Mrs. Charles **Bowen** and Miss Maud **Walsh**. He followed the trucking business for many years, and firm name being Walsh & Brown. John **Walsh** was an honest man, kind father, indulgent husband and respected citizen. The funeral will be held at St. Joseph's church at 10 a.m. Friday. Interment in the Catholic cemetery. Kelly Bros. undertakers.

27 May 1911 (Saturday); Page 1 Col 3; Timothy Edward **Early**, a native of Yuba county, but for many years a resident of Butte, Mont., died at the home of his sister, Mrs. William **Haggerty**, here yesterday morning at the age of 37. Mr. **Early** has many friends who will mourn his death. He has been in Marysville three months this last time, coming here because he could not stand the climate of Montana. Four sisters, Mrs. William **Haggerty** and Mrs. William **Riley** of Marysville, and Mrs. William **Moran** and Mrs. John **Fogarty** of Butte, Mont., and two brothers, Frank **Early** of this city and John **Early** of Butte, mourn his death. The remains are at Kelly Bros.' undertaking parlors.

18 Aug 1911 (Friday); Page 1 Col 3; News comes from San Francisco of the death of Frank **McQuaid**, son of John W. **McQuaid** of Smartsville, where he was born and where he lived until he went to San Francisco and secured a place on the police force twenty two years ago. The deceased was well known and liked by a large circle of friends, both at his old home in Yuba county, and also in San Francisco. He was in his forty ninth year and leaves a wife and four daughters to mourn his loss, besides his father and two brothers, **John H.** and **Thomas**, and three sisters, Mrs. W. G. **Cady**, Mrs. **Allread** and Miss Mary **McQuaid** The funeral will take place in San Francisco, where the body will be interred. (He is not buried in Smartsville, but I included him because it shows the relationships).

20 Aug 1911 (Sunday); Page 1 Col 4; George **Divver**, formally janitor of the Marysville grammar school, had one of his legs cut off by a Southern Pacific passenger train at Dunsmuir Saturday afternoon. A. P. **Lipp**, who was at Dunsmuir and saw the accident, telephoned from Klamath Falls yesterday to his son Frank **Lipp**, and told him of the accident. It seems that **Divver** had been to Klamath Falls, where he had attended the funeral of his sister, Mrs. Margaret **Hopkins**. Particulars regarding the accident are lacking. Mr. **Lipp** also tried to get Byron **Divver**, a brother of George, who is on his way to Portland from Klamath Falls, but could not reach him. George **Divver** is receiving medical attendance at Dunsmuir.

5 Sep 1911 (Tuesday); Page 1 Col 2; George **Divver**, who had his leg run over at Dunsmuir a few weeks ago and who was brought to this city after the accident to have his leg amputated, died Sunday morning. George **Divver** was a native of Yuba County, aged 40 years. For many years he was a janitor in the public schools, as well as the Rideout bank, and through his thrift he amassed quite a little property. He is survived by one brother, Byron **Divver**, and a nephew, William

NEWSPAPER INFORMATION

Divver. The funeral is to be held from the mortuary parlors of J. K. **Kelly** this morning at 6 o'clock, and the body will be taken to the cemetery at Smartsville.

2 Dec 1911 (Saturday); Page 8 Col 5; John **Burns**, a native of Ireland and 70 years of age, died last evening at his home in Smartsville. Deceased has been a resident of that town for many years. His funeral will be held in the Catholic Church in Smartsville Sunday morning at 10 o'clock and interment will be made in the Catholic cemetery.

7 Dec 1911 (Thursday); Page 1 Col 4; The many friends of County Superintendent of Schools **Cramsie** and of Justice William **Cramsie** of Smartsville will hear with deep regret of the death of Joseph D. **Cramsie** at St. Helena yesterday. The deceased, who was a father of the professor and a son of the justice, had gone to the sanitarium for his health, as stated recently in these columns, and therefore his death was not entirely unexpected. He leaves a loving father and mother, and three brothers, the superintendent, Francis J. **Cramsie** and John C. **Cramsie** of Smartsville and a sister, Miss Sarah **Cramsie**. He was in the thirty-sixth year of his age. The funeral, which will be managed by Kelly Brothers, will be announced later, but the remains will arrive in this city at 1 p.m. today.

8 Dec 1911 (Friday); Page 8 Col 4; The funeral of Joseph **Cramsie** will be held from the Catholic church in Smartsville at 2 o'clock this afternoon and interment will take place in the cemetery at that place. Father J. J. **Hynes** will officiate at the services.

7 Jan 1912 (Sunday); Page 1 Col 6; Word was received yesterday of the death in San Francisco Jan 4 of Wm. W. **Ward**, a pioneer resident of this county, who formerly resided at Smartsville in the early '80s, where he was in business, later removing to Marysville, where he formed a partnership with J. H. **Murphy** in a business venture where S. **Ewell's** store on D. street is now located. He later moved to San Francisco, where he resided at the time of his death. Four daughters and one son survive him, Mrs. E. H. **Grandjean**, Mrs. E. **Rentoff**, Anna **Ward** and Leo J. **Ward**. Phil and Edward **McCune**, nephews of the deceased, left yesterday morning for Oakland, where the funeral was held. (This man wasn't buried here, I put his Obituary in because he is the father and husband of individuals here).

9 Jan 1912 (Tuesday); Page 8 Col 3; Mrs. T. (Ellen) **Linehan**, a pioneer resident of the valley, died yesterday morning at her home at Smartsville. She had been ill for some time and several days ago it was known that it was only a question of a few days when death would end her sufferings. Her children had been summoned and were all at her bedside when the end came. Mrs. Linehan had been a resident of Smartsville for more than forty years, settling in the town when it was a lively mining camp. She was held in the highest esteem by all who knew her. A native of Ireland and was 70 years of age. She is survived by her husband, Timothy **Linehan**, three sons, **John** of Shasta county, **Peter** of San Francisco, **Timothy** of Ventura, and two daughters, Kate **Linehan** of Smartsville, and Mary **Linehan**, who are at present training to be a nurse at San Francisco. The funeral services will be

NEWSPAPER INFORMATION

held Wednesday at the Catholic Church. Mass will be celebrated by Rev. Father J. H. **Hynes**. The interment will be in the Smartsville cemetery.

20 Mar 1912 (Wednesday); Page 1 Col 4; William **Cramsie**, one of the oldest pioneers of Smartsville and father of W. P. **Cramsie**, superintendent of schools of this city, died at his home at midnight last Monday. He had been sick for some months, and just before the end came all of his relatives were summoned to his bedside to await the final departure. He was one of those sturdy pioneers of the Golden State that made glorious history. Born in Bale Mona, Ireland, on June 15, 1837, he came to this country when but 19 years of age. He spent a year teaching school in Pennsylvania and then came to this state in 1857, landing at Rose's Bar. In 1863 he moved to Mokelumne Hill, where he engaged in mining and was a partner of Marcus **Daly** and Patrick **Reddy**. When the Fraser river excitement was at its height, he traveled through Oregon, Washington, Idaho, Montana and Utah by pack train, visiting Boise City and Helena when they were only Indian Villages. Later he moved to Smartsville and ever since then made that his home. For many years he was connected with the Blue Gravel Mining company and the Excelsior Water and Mining company. In politics he was always a staunch Democrat, and at one time was spoken of as a candidate for governor. For twenty years he was justice of the peace for Rose's Bar township. He married Miss Elizabeth **Havey** in 1870, which survives him, as do his four children, John E. **Cramsie**, W. P. **Cramsie**, F. J. **Cramsie**, three sons, a daughter, Miss Sarah F. **Cramsie**. The funeral will take place at 10 o'clock this morning from the Catholic church at Smartsville and the interment will be in the Catholic cemetery. Kelly brothers have charge of the arrangements.

8 May 1912 (Wednesday); Page 8 Col 2; Funeral Today. Funeral services over the remains of the late Ella **Jeffery** were held from her home in Smartsville this morning at 10 o'clock. Interment was made in the family plot in the Smartsville cemetery under the direction of R. E. Bevan.

16 May 1912 (Thursday); Page 3 Col 3; Jacob **Foreman** buried from Smartsville Home. The funeral of Jacob Washington **Foreman**, the Smartsville pioneer, who died Monday shortly before midnight, was held from the family residence in Smartsville at half past one o'clock yesterday. The funeral was under the auspices of the Smartsville lodge of the I. O. O. F. Interment will be in the City Cemetery at Smartsville. The funeral arrangements were under the direction of Kelly Brothers of this city.

24 Aug 1912; Page 4 Col 4; Mrs. Henry **Fiene** call to her Reward - The people of Smartsville are mourning the death of Mrs. Henry **Fiene**, one of the old and respected inhabitants of the place. For many years, with her late husband, she kept the old tollhouse at Deer Creek, near Mooney Flat. For more than a generation they catered to the traveling public and numbered their friends by the hundreds. The funeral will be held today under the supervision of Kelly Brothers of Marysville. Interment will be made in the Smartsville cemetery, where the remains of her husband rest. Mrs. **Fiene** came to the United States when a young woman

from her home in Germany in 1860. She has made her residence in the vicinity of Smartsville most of the time since. To mourn her death she leaves a son, Henry **Fiene** of Mooney Flat, and one daughter, Mrs. Charles **Marks**, of Sacramento. Many expressions of regret have been heard since her decease, for she was a woman highly esteemed in the section in which she lived.

11 Sep 1912 (Wednesday); Page 1 Col 1; Joseph's Catholic church was the scene of an interesting wedding on Saturday morning when George **Herboth** and Miss Elizabeth **Conrath**, both of Marysville, were united in Marriage. The impressive ceremony at the church was performed by the Rev. Father **Coen**. Only relatives and intimate friends of the couple were at the church on account of the time that the ceremony was performed at 5 a.m. At the conclusion of the ceremony the wedding party returned to the **Conrath** home on the corner of F and Eighth streets. Here an elaborate wedding breakfast was served. The dining room was artistically decorated with La France roses and violets. Around the wedding table the health of the bride and groom was drunk and there were many toasters for a prosperous matrimonial future. The bride, both at the church and at the home afterward, looked pretty in her traveling dress of dark blue material. She was attended by Alvina **Gunning** as bridesmaid, while the grooms man was the bride's brother, L. **Conrath**. The wedding presents were numerous and costly. They came from relatives and friends in all parts of the state. Several messages of good cheer were also received by Mr. and Mrs. **Herboth** prior to their leaving for their honeymoon. Some time after breakfast Mr. and Mrs. **Herboth** took a train for the south. They will spend most of their honeymoon in Los Angeles and other points of interest in southern California. On their return they will reside in their home in Marysville.

27 Sep 1912 (Friday); Page 1 Col 3; John H. **Conlin**, a pioneer resident of Smartsville, died at San Francisco yesterday morning, the immediate cause of death being heart trouble. Deceased had been in ill health for some time and his death was not unexpected. He was a member of the firm of **Conlin** Brothers of Smartsville and had resided in San Francisco for the past three years. He was 46 years old. He is survived by one brother, William **Conlin**, of Smartsville; one sister, Mrs. Charles **Burns**, of San Francisco. Deceased was a cousin of John **McQuaid** of Smartsville. Mr. **Conlin** was a native of Smartsville and had spent his entire life in and around that place. He was always identified with anything that was for the building up of the community. The funeral will be held at Smartsville Sunday afternoon at 1:30 from the Catholic church. Rev. Father Hynes will conduct the services.

14 Nov 1912 (Thursday); Page 8 Col 2; Died – In Marysville, November 14, 1912, at the residence of Mrs. R. D. **Moncur**, 622 E. street, Mrs. Anna **Walsh**, relict of the late Michael **Walsh**, aunt of **Harry** J., **John** K., Peter F. **Kelley** and Mrs. R. D. **Moncur**, a native of Ireland, aged 67 years.

15 Nov 1912 (Friday); Page 8 Col 3; Mrs. Anna **Walsh** Succumbs to Long Illness-Funeral Saturday. Mrs. Anna **Walsh**, relict of the late John **Walsh** of Smartsville, died at the home of her niece, Mrs. Robert **Moncur**, on E. street at an

early hour Thursday morning. The death of the aged lady was not wholly unexpected, as she had suffered for some time from a complication of ailments. Deceased was a native of Ireland and at the time of death was 67 years old. About a year ago Mrs. **Walsh** came from Arizona, where her husband had died, to Marysville to make her home. Her health was failing at the time and shortly after her arrival she was stricken with a complication of illness that grew steadily worse until the time of death. She had been almost an invalid for several months. Those who survive her are: Mrs. Robert D. **Moncur**, a niece; **Harry** J., **John** K. and Peter K. **Kelley**, nephews, all of whom are residents of this city. The funeral will be held from the Moncur residence on E street at 8:30 o'clock. Saturday morning and a requiem mass will be held at the Catholic church at 9 o'clock and the interment will be in the Catholic cemetery. Mrs. **Walsh** and her husband resided in the Smartsville district for a number of years, and they are well known to the residents of that section of the county. They were among the counties best and most substantial citizens and the death of Mrs. **Walsh** will be learned of with pain by the friends of the family.

12 Jan 1913 (Sunday); Page 5 Col 4; Mrs. Ellen **Dewan**, who has been seriously ill at the home of her daughter, Mrs. William **Cocking**, of Hammonton, for some time, passed away early Saturday morning. Deceased was 72 years of age and was a native of County Cork, Ireland. Mrs. **Dewan** resided in Smartsville about twenty years, and during that time she conducted the hotel at that place. After she disposed of her property interests in Smartsville she remained there until she became stricken with the illness that caused her death, at which time she went to Hammonton to make her home with Mrs. **Cocking**. Deceased was one of the best-known pioneers of Yuba county, and her many friends will learn with much sorrow of her death. She was a devout Catholic and lived up to the teachings of her church. Although having been afflicted in the past few months she still retained that lovable disposition that made her known to so many people of the county. Before coming to Yuba county to make her home she resided with her husband at Sweetland, in Nevada county where Mr. **Dewan** was engaged in mining. It was while residing in Nevada county that she lost a son by a mine accident. She is survived by three daughters and one son, via: Mrs. Alice **Jeffries** of Forest City, Mrs. William **Beasley** of Smartsville, Mrs. Fred **Broken** of San Francisco and James **Dewan** of Grass Valley. The funeral will be held tomorrow at Smartsville, at which place the body will be interred. Kelly brothers have charge of the funeral arrangements.

18 Jan 1913 (Saturday); Page 1 Col 3; James **Bowman**, whose serious illness was announced several days ago, laid down life's burdens at the age of 61 years, Friday afternoon, when the grim reaper, leveler of all mankind, called the pioneer resident of this section of the state to his final reward. The end came at the Bowman home in Spenceville, Nevada County, where the dead man has resided for many years. Past several days ago it was announced that the patient could not long survive, and anticipating that death was near, the relatives of the dead man were at his bedside when he succumbed to the disease that afflicted him. Mr. Bowman was a native of Lawrence County, New York, but had resided in California for several decades,

NEWSPAPER INFORMATION

having come to this section when a young man many years ago the deceased secured a farm at Spenceville and entered extensively into stock raising and agriculture, which he followed successfully until the time he was taken ill. While a resident of Spenceville, deceased, who was a brother of the late Gordon **Bowman**, for many years clerk of Yuba County, reared a large family and besides being survived by a wife leaves the following other relatives: three sons, W.H. and F. W. **Bowman** of Spenceville and J. M. **Bowman** of San Jose; five daughters, Mrs. Mary **Sheppard** of Salinas, Mrs. C. E. **Smith** of Marysville, Mrs. Lewis **Russi** of Folsom, Mrs. Frank **Taylor** of Marysville and Mrs. William **Barrie** of Hammonton: one sister, Miss Emma **Bowman** of Yuba County: four cousins, Henry **Walker**, sheriff of Nevada County; Miss Elspeth **Walker**, Grass Valley; and George and Tom **Akins** of Wheatland. Funeral services will be held in the Bowman home in Spenceville Monday morning at 10:30 o'clock. Rev. John **Vineyard**, of Smartsville, for many years a personal friend of the deceased, will officiate at the last sad rites, which will be followed by burial in the Spenceville cemetery, under the direction of Kelly Brothers.

2 Feb 1913 (Sunday); Page 8 Col 4; Miss Annie D. **Sanders** died at her home at Smartsville yesterday afternoon after a lingering illness. Deceased was a native of New Jersey and was 60 years, 2 months and 11 days old at the time of death. She is survived by her mother and father, Mr. and Mrs. Daniel **Sanders**, and two brothers, **John** and **Charles**, all of Smartsville. Miss **Sanders** was a member of one of the oldest families of the country. She was a devout Christian and was a most estimable woman. Her family has resided at Smartsville for a number of years and the deceased was known all over the county for her many good deeds. The funeral will be held at the Catholic church at Smartsville Monday morning at 10 o'clock, and interment will be in the Catholic cemetery at that place. Kelly brothers have charge of the funeral arrangements.

29 Apr 1913 (Tuesday); Page 8 Col 2; Last Thursday at San Francisco Mrs. Alvina **Gunning**, widow of the late S. O. **Gunning**, for many years one of the most prominent residents of Yuba county, was wedded to Peter Fisher of Bakersfield. The marriage comes as a surprise to the many friends of the bride. The attendants were Miss Alvina **Gunning**, daughter of the bride, and Timothy **Driscoll**, a former resident. The ceremony was performed at the Mission Dolores. Mr. and Mrs. **Gunning** will make their home at Bakersfield, which has been the residence of the groom for many years.

2 May 1913 (Friday); Page 1 Col 5; SMARTSVILLE. May 1 – Miss Kate **Dougherty**, after attending the funeral of her late brother, Dan **Dougherty,** has returned to her home in San Francisco. Daniel James **Dougherty** is called to his eternal home and reward. About seven years ago the deceased met with an accident while working for the Bay Counties power company. In stretching the copper wire it slipped and cut his face. About two years thereafter a cancer made its appearance and was operated upon. A number of operations followed without favorable results and the end came early Saturday morning. Dan was a good boy, esteemed and greatly admired by all who knew him. At the time of death he was

NEWSPAPER INFORMATION

thirty-seven years old. He leaves to mourn their loss the aged father and five sisters, Mrs. J. **Keegan**, Miss **Grace** and Kate **Dougherty** of San Francisco, Mrs. R. **Ragby** and Miss Lizzie **Dougherty** of this place. The funeral took place last Monday and was largely attended. Rev. J. J. **Hynes** officiated at the church and grave. The interment was made in the Catholic cemetery and the following acted as pall bearers: W. R. **McConnell**, W. **Meade**, J. H. **McQuaid**, B. L. **Compton**, W. B. **Jeffery** and R. **McCarty**. The family desires to return sincere thanks for the many tokens of sympathy during the years of great trouble; also for the beautiful flowers that were placed by loving hands upon the casket that contained the remains of the late beloved son and brother **Dan**.

8 Aug 1913 (Friday); Page 1 Col 4; Mrs. Catherine Loretta **Dempsey**, wife of William J. **Dempsey**, a well known business man of Marysville, died yesterday morning in San Francisco. Death was unexpected and the news came as a shock to her husband, who was sitting as a juror in the coroner's inquest over the remains of the Wheatland victims. The deceased had gone to San Francisco but a few days ago and was accompanied by her husband, who returned home believing that his wife's health was improving. She was a native of California, being born in this state thirty-one years ago and before coming to this city she resided in Smartsville. Mrs. **Dempsey** is survived by her husband, a son, **Kenneth**, two sisters, Misses **Annie** and Mary **Meade**, and three brothers, **Will**, **John**, and Leo **Meade**. The funeral arrangements, under the direction of Kelly brothers, have not been announced.

9 Aug 1913 (Saturday); Page 5 Col 6; The funeral of Mrs. Catherine **Dempsey** will be held Sunday morning at 5 o'clock from the family residence at 411 Seventh street and thence to Smartsville, where requiem mass will be said in the Smartsville Catholic church at 10 o'clock a.m. Burial will be made in the Catholic cemetery at that place. The funeral arrangements are under the direction of Kelly brothers.

5 Sep 1913 (Friday); Page 5 Col 3; Another one of the pioneer women who bore privations and braved the frontier days was called yesterday when Mrs. Catherine **Driscoll** was removed from the fast thinning ranks. She was a woman who was loved by all of her friends and spoken of with high praise by her acquaintances. For the past fifty years she had resided in the section of Smartsville, where she died. She was the widow of the late John **Driscoll**, who was a well-known Yuba county miner. Death was not unexpected, as she had been failing for a long time and when it occurred yesterday many of her relatives were there. The deceased was a native of County Cork, Ireland, and was born in that country seventy-five years ago. When but a young girl she came to the United States. Mrs. **Driscoll** is survived by six children, Mrs. Kate **Barrett** and Mrs. A. **Tufts** of Marysville, and Mrs. John **Dempsey** of Smartsville, and **John**, **Dan** and James H. **Driscoll**, all of Oregon. The funeral will take place Saturday morning at 10:30 o'clock from the Smartsville Catholic church and burial will be made in the Smartsville Catholic cemetery. The services will be preached by Father Hines. Kelly brothers have charge of the funeral arrangements.

NEWSPAPER INFORMATION

6 May 1913 (Tuesday); Page 5 Col 4; AGED COUPLE MARRY. That love is a story which never grows old was demonstrated yesterday when John L. **Vineyard**, seventy six years old, of Smartsville, and Sara Eliza **Woodroffe**, sixty-three years old, of Mooney Flat, applied to County Clerk Eastman for a marriage license. They were later married by Rev. Thomas H. **Nicholas**

14 Oct 1913 (Tuesday); Page 1 Col 2; Mrs. Margaret **Walsh** Has Answered the Last Summons. Mrs. Margaret Ellen **Walsh** a well known in this city (Marysville), where she had resided for the past thirty-two years, died Sunday morning at the family home on Eighth street, between A and B. The deceased was a native of Louisiana and aged sixty-two years. She came to California fifty-five years ago. Mrs. **Walsh** is survived by two daughters, Mrs. Charles **Bowen** of Sacramento and Miss Maud **Walsh** of this city (Marysville), and one sister, Mrs. **Fitzpatrick** of San Francisco. The funeral will be held this morning at 10 o'clock from the Catholic church, and interment will be made in the Catholic cemetery under the direction of Kelly brothers.

6 Nov 1913 (Thursday); Page 1 Col 3; John **McQuaid**, for sixty years a resident of Yuba county, died yesterday morning at his home at Smartsville following a stroke of paralysis which he suffered last Monday. He was seventy-six years old and a native of Ireland, but came to Yuba county in 1876 and located at Parks Bar Bridge, where he engaged in mining for a number of years. Later he removed to Smartsville, where he reared a large family. The deceased was one of Yuba county's best-known mountaineers and he was held in high esteem by all who knew him. He is survived by the following children: **John** H. of Smartsville and Thomas E. **McQuaid** of San Francisco; Mrs. W. J. **Carey** and Miss Mary **McQuaid**, both of San Francisco, and Mrs. William **Allread** of Smartsville. The funeral, which is under the direction of Kelly brothers, will be held from the family home next Friday morning at 10 o'clock. High requiem mass will be held at the Catholic church of that town. Interment will be made in the family plot.

27 Nov 1913 (Thursday); Page 5 Col 3; Eugene D. **Hapgood**, one of the early settlers of Yuba County, died at the home of his daughter, Mrs. Walter B. **McGinnis**, in Oakland at noon Thursday, according to information that reached Marysville this afternoon. He resided in the Smartsville district for many years, following mining, and was one of the best-known residents of the eastern section of the county. He was 76 years old, and a native of the state of Vermont. He had been in ill health for some time and had been making his home with his daughter. The surviving relatives include three daughters; Mrs. W. B. **Filcher** of Pacific Grove, Mrs. Walter **McGinnis** of Oakland and Mrs. A. **Simpson**, and one son, James **Hapgood** of Smartsville. Deceased was an esteemed member of the Rose Bar Lodge, No. 89 F. & A. M. Smartsville and under the auspices of this lodge the funeral will be held Sunday afternoon at 2 o'clock. The remains will arrive in this city (Marysville) on the noon train Friday and taken in charge by R. E. Bean & Son, who have charge of the funeral arrangements.

NEWSPAPER INFORMATION

30 Dec 1913 (Tuesday); Page 1 Col 5; Miss Alvina **Gunning** and John J. **Geraghty** were married in San Francisco Sunday at the home of the bride's mother, only relatives and a few close friends of the contracting parties witnessing the ceremony. Both of the young people are well and favorably known in this city. Miss **Gunning** has spent most of her life here and up until a few weeks ago she held the position of deputy to Recorder Greeley. The groom holds a position at the local shops of the Yuba construction company. After a honeymoon spent around the bay the couple will return to this city where they will make their home.

7 Jul 1914 (Tuesday); Page 8 Col 4; Daniel **Dougherty**, 89 years old and a native of Ireland, died at his home at Sucker Flat in the Smartsville district, early yesterday morning. Death was due to the complications of old age. He had resided at his late home since 1852. He came to the United States in 1847 and five years later he crossed the continent from New York. In the early mining days of California the deceased was a prosperous merchant and freighter, but his fortunes were lost in the mines. Daniel **Dougherty** is survived by five daughters, Mrs. John **Keegan** of San Francisco, Mrs. Albert **Rigby** of Smartsville, Mrs. John **Worl** of Grass Valley and Misses **Katie** and Grace **Dougherty** of Smartsville. The funeral will be held from the Catholic Church of Smartsville Wednesday morning at 10 o'clock. The funeral arrangements are under the direction of Kelly Brothers.

20 Sep 1914 (Sunday); Page 4 Col 1; Elmer Dixon **Bristow** will lead Miss Agnes Virginia **Havey** to hymen's altar this morning at 10 o'clock at the Catholic church at Smartsville, Rev. Father Dermody, rector of the church, officiating. While it has been generally known that the time this couple would wed, the date of the affair was not given out until yesterday, when application for a marriage license was made at the office of the Yuba County clerk. The bride to be is well know in Yuba county and has resided at Smartsville a number of years. She is the daughter of John **Havey** and wife, well known and prominent residents of the Smartsville section. She is an unusually attractive young woman of the blonde type and numbers her friends by her acquaintances. She resided in Marysville for a few months recently and it was while she was a resident of this city that she and Mr. **Bristow** began the romance that will have its culmination today at the marriage altar. Mr. **Bristow** is also well known in Yuba county. He has taught school in this county for a number of years and was a member of the faculty of the Marysville grammar school last year. He is now principal of the grammar school at Colfax, at which place he and his bride will reside. He comes from a prominent family and is a young man of sterling worth. The announcement of the wedding will be received with gladness by the many friends of the couple, who had been looking forward to such announcement for some time. It occasioned no surprise, as it was generally known that the wedding would take place in the early fall.

26 Oct 1914 (Monday); Page 5 Col 6; Funeral services over Mrs. Bell **Mitchell** were held Sunday at 10 a.m. from her late home at Ninth and H Streets with Rev. **Rifenbark**, rector of St. John's Episcopal Church officiating. A choir composed of Mrs. Irwin **Salves** and Willard **Roberts** rendered several appropriate vocal selections. Members of Wahnita council, Degree of Pocahontas, of which the

NEWSPAPER INFORMATION

deceased was a member, attended the services in a body. Rev. J. T. **Vineyard** of Smartsville officiated at the cemetery at Smartsville where interment was made under the direction of Kelly Brothers. The following assisted as pall bearers; Milton **Ramsey**, John **Murphy**, John **Beck**, Will **Newbert**, William **McKenzie** and E. E. **Geerv**.

13 Mar 1915 (Saturday); Page 1 Col 1; Morris **Murphy**, a well known former resident of Yuba County, died Friday afternoon in Oroville, following a brief illness. **Murphy** was one of the pioneer residents of the state, coming to California across the plains in the year 1874, when he located in Smartsville in this county, resident there for many years. For the past ? years he has resided in Oroville where he died. He passed the greater part of his boyhood days in this county and is quit well known by the many residents of Smartsville. Deceased was 68 years of age. The remains will arrive in Marysville Sunday evening from Oroville and will be taken in charge by Kelly bros. The funeral services will be on Tuesday morning at 10 o'clock, at the Catholic church at Smartsville and interment will be made in the Catholic cemetery at that place. The following children survive the deceased: Morris **Murphy** of Watsonville, Jno **Murphy** of Folsom **Henry** and Stephen **Murphy** of Oroville, Mrs. **Mary** ? of Paso Rubles, Mrs. Kate **Becker** of Placerville, Mrs. Lillie **Pa?**, Mrs. Florence **Dalsheim** and Miss Rose **Murphy** of Oroville.

6 Apr 1915 (Tuesday); Page 4 Col 2; Calton Wilber **Manasco**, for many years proprietor of the hotel at Mooney Flat, just across the county boundary line, in Nevada County, passed away at the hotel last night. He had been a resident of California for 59 years. The deceased was a native of Alabama, aged 85 years and 28 days. He had been at Mooney Flat for the past twenty-six years. He was a member of Rose Bar Lodge, No. 89, F. & A. M. of Smartsville, and was one of the oldest Masons in California, having received his first degree in Masonry in the early fifties. The Masons will conduct the funeral services, which will be held as arranged later. A wife, Mrs. Amelia **Manasco**, two daughters, Mrs. H. L. **Hite** of Sicard Flat and Miss Grace V. **Manasco** of Oakland, and one son, George V. **Manasco** of Angels Camp, survive. R. E. Bevan & Son of this city (Marysville) have charge of the arrangements.

6 Apr 1915 (Tuesday); Page 4 Col 6; Funeral services over the remains of Mrs. Mary **Mellarkey** were held from the family home at Smartsville yesterday morning at 10 o'clock. Rev. Father Dermody officiated at the services. The interment was at the family plot in the Catholic cemetery at Smartsville. The pallbearers were John **McQuaid**, John **Cramsie**, Al **Wheaton**, Robert **Beatty**, John **Havey** and William **Toland**.

8 Apr 1915 (Thursday); Page 8 Col 4; The funeral of the late C. W. **Manasco** was held from the family home at Mooney Flat this afternoon at 2 o'clock. Services were under the auspices of Rose Bar Lodge, No. 89, F. and A. M., of Smartsville. Burial was in the fraternity cemetery of Smartsville, under the direction of R. E. Bevan & Son of this city.

NEWSPAPER INFORMATION

13 Jun 1915 (Sunday); Page 5 Col 2; Owen **Adkins**, better known as "Jeff" Adkins, died at the home of his brother at Durham Friday afternoon. He was a native of California and was fifty-five years old. He is survived by his mother, Mrs. F. M. **Adkins** of Waldo, three sisters, Mrs. D. T. **Hite** of Auburn, Mrs. F. A. **Colburn** of Loomis, and Mrs. Charles **Murch** of Lincoln, four brothers, J. T. **Adkins** of Los Molines, R. E. **Adkins** of Durham, C. P. **Adkins** of Waldo and B. **Adkins** of Sacramento. Deceased was well known in Marysville and Yuba County, having been a resident of this county for many years. The remains will arrive here this morning on the El Dorado at 6:40 and the interment will be in the I. O. O. F. cemetery at Smartsville. Kelly brothers will conduct the funeral.

8 Jul 1915 (Thursday); Page 4 Col 2; In the death in this city (Marysville) yesterday morning of Daniel **Saunders**, another of the sturdy band of Argonaut pioneers has joined the silent majority of its members, those men and women who braved danger and privation that the great state of California might be what it is today, the Mecca for health and wealth. Deceased, with his wife and child, came to California from New York via Cape Horn and located in 1853 where now stands the town site of Smartsville. His wife at the time being the only white woman in Rose Bar township outside of a few women at the settlement of Roses' Bar. He erected the first house in Smartsville and, in which his wife, at the age of ninety-three, still resides. He was a man of kindly and genial disposition, forgiving and unselfish, very fond of children and loved by them in return. For many years he was a trusted employee of the Blue Gravel Mining company and in turn with the Excelsior Water and Mining company, after its merger with the former. He was noted for his great physical endurance, both in manual labor and walking contests in the early days. He left home early one Sunday morning, walked to Marysville and returned that evening, having carried fifty pounds of flour on the return trip. He was a native of Roscommon, Ireland, aged ninety years, and is survived by his aged wife and two sons, **John** and **Charles** of Smartsville. Well can it be said of him: "Green be the turf above the friend of our better days, None knew thee but to love thee, none named thee but to praise." The funeral services will be held in the Catholic church at Smartsville on Friday morning and the interment will be in the Smartsville cemetery. The remains are at the undertaking parlors of Kelly brothers, who have charge of the funeral.

8 Jul 1915 (Thursday); Page 5 Col 4; Smartsville News – The funeral of the late Mrs. Eva **Sanford** of Woolf was held last Sunday and the interment was made in this place. A large number came from a distance to pay the last tribute of respect. Rev. Wm. **Clark**, Methodist minister of Grass Valley, officiated at the church and grave. The interment was made in the Fraternal conclave.

30 Aug 1915 (Monday); Marysville Evening Democrat; Page 1 & 8 Col 3&4; James **O'Brien**, pioneer miner, farmer and empire-builder of the West, died at his home at Smartsville Sunday morning about 10 o'clock after being closely identified with the development and building up of Northern California for more that half a century. The end followed a long illness due to old age. He was seriously attacked by sickness about a year ago, and though he partly recovered from this spell, his

health for several months had been gradually failing. For the past week or more death was daily expected, relatives having gathered at the bedside, knowing that the inevitable summons was near. He was 83 year old. Funeral services over the remains will be held at Smartsville Tuesday morning at 10 o'clock, with the celebration of a requiem mass in the Catholic church for the repose of the soul. Burial will be made in the Smartsville cemetery, under the direction of Kelly Bros. Friends and acquaintances are invited to attend the services, omitting flowers. Few men in California have had a more interesting career than James **O'Brien**, whose endeavors are linked with many accomplishments that will remain as monuments to his name long after the breaking of mortal ties. And not alone to one community were his efforts of enterprise devoted, for while he spent the greater part of his life in Yuba County, the work he accomplished in the development of this state extended into several counties, including Nevada, Butte and Yolo. With an education gained in the University of the World, James **O'Brien** succeeded in most any enterprise he undertook, and they were many during his long and active life, though he was chiefly engaged with mining, by which industry he was attracted to the West. From a prospector he became an operator and acquired vast holdings in the Smartsville district of Yuba County, which later showed the wisdom of his business foresight. In the days of hydraulic mining in California he was one of the most prominent operators in the state, and took an active part in the fight between the miners and farmers when litigation arose a number of years ago over the choking of the rivers by debris turned loose by monitors. In comparatively recent years, after hydraulic mining was stopped by legislation, he sold a large part of his holdings along the Yuba river to W. P. Hammon and associates for dredging purposes. In addition to his other interests in the mining and power fields, he at one time was part owner in The Democrat. As a man among men James **O'Brien** was greatly admired for his strength of character. The same determination and consistency that brought him success in large undertakings marked his actions in everyday life. Men who worked for him say that there were few more generous to those he liked, though equally as strongly determined against enemies. His activity as a man of enterprise continued almost during his entire life, for it was only when his health and strength began to fail that he sought retirement. Even then he was not content to leave his affairs entirely in the hands of others. His home at Smartsville has been for many years one of the most picturesque places in the foothills. An honorable place among the representative citizens of the Sacramento Valley is accorded James **O'Brien**, who has been identified with the mining interests of California for more than half a century. When he came to the state in 1853, like the great majority of the emigrants of that time, he brought nothing to presage the successful career which has been his, but with the courage of youth, the optimistic nature inherited from his Irish forefathers, and the ability to grasp and make use of the multifold opportunities presented, he has overcome all obstacles and compelled fruitful returns for his efforts. Age had not robbed him of the energy, which distinguished his life, and although over 82 years old, he retained his activity and interest in business affairs, and was counted the leading mining man and rancher of Yuba County, where he has spent the greatest part of his life in the West. A native of County Cork, Ireland, James **O'Brien** was born May 28, 1830, and when 14 years of age was brought to the United States by his parents. They

NEWSPAPER INFORMATION

settled in Westfield, Mass., and in that locality he worked out for a farmer about five years. Resolving to take a part in the stirring success being enacted in California, he left Massachusetts in May, 1853, and came to the Pacific Coast by the Nicaragua route, arriving in San Francisco on July 3 of that year. He made the trip thence to Marysville by steamer, after which he went to Barton's Bar on the Yuba river in Yuba County. He spent the summer in the mines and the winter seasons prospecting. For four years he remained so occupied and in the meantime had become interested in the big irrigation projects of this section and had built the Oroville ditch by contract. In 1858 he built the Boyer ditch a distance of twenty-five miles, extending from Deer Creek to Smartsville; and the following year built the Excelsior ditch from the South Yuba to Smartsville, a distance of about thirty-four miles. In the fall of 1859 he contracted and built the Knight's Landing road to Putah Creek, extending across Yolo County, and in this way was largely identified with the development of this section of the state. In the fall of the next year he located in Smartsville, Yuba County, and bought mining property, and shortly afterward was associated with Prof. William Ashburner, Messrs. Walker, Baker and Hage, of San Francisco, in the building of the Packtolas tunnel at a cost of $80,000. Of the ten shares, Mr. **O'Brien** was the owner of five, and was accordingly made superintendent of the mine, which interests he operated successfully for some years, when they consolidated with the Blue Gravel and Excelsior mines. With the added responsibilities he continued in the position for about four years, when the property was sold to eastern capitalists. About that time (1882) there was considerable litigation over mining operations, and the hydraulic process was stopped, thereby causing Mr. **O'Brien** considerable loss. Withdrawing to a large extent from mining operations, he turned his attention to farming, purchasing a tract of 9200 acres in the Yuba river, he began ranching and stock raising. He was very successful in his work and also fortunate in his choice of land, as later he sold 442 acres to W. P. Hammond for $200 per acre, and 1200 acres to Mr. Cranston for $300 and $350 per acre, the latter being just across the mine from the other. Hammond put in dredgers, taking out $3000 per day and on that location the town of Hammonton sprang up in a couple of years. James **O'Brien** also bonded 4000 acres to a Mr. Hanford, while he rented the remainder of his property for cash on a ten-year lease. He still owns over 7000 acres of land, principally devoted to grazing and the raising of grain. James **O'Brien** was later interested in the promotion of a new enterprise which promises much toward the development of Yuba County when carried to a successful issue. This is the Nevada and Marysville Water and Power Company, which expects to dam the rive rat The Narrows near Smartsville, and there to use the water for power, irrigation of the land north of the Yuba river, and afterward to dredge the river bed, which is rich in gold. This was a great undertaking, especially in consideration of the age of Mr. **O'Brien**, but when completed will stand as one of the best investments in the state. In October 1860, James **O'Brien** was married to Mary **Kirby**, a native of Ireland, whom he had met before immigrating to the west, and who made the trip to California to become his wife. She died November 13, 1894, leaving the following children: **Mary**, wife of J. P. Pierce of Santa Clara; **Katie**, wife of Dr. Holbrook of San Jose; **Josie**, **Helen** and **Agnes** at home; **Isabelle**, who took the veil in the convent at Oakland; **James**, a mining man and the owner of Bunker Hill mine in

NEWSPAPER INFORMATION

Plumas County, and **William**, an assistant to his father in the management of his property. (There is a nice picture of him on the first page of this paper).

3 Dec 1915 (Friday); Page 8 Col 4; Mrs. Katherine **Bushby**, a native of Ireland, aged 77 years, and one of the best-known and oldest residents of eastern Yuba county, died at Smartsville yesterday. She had resided in the county sixty years. Being married to George **Bushby** in this city (Marysville) 54 years ago. The deceased is survived by her husband, four daughters, Mrs. F. W. **Morrill** of Montavo, Ventura county; Mrs. W. M. **Hickeson** of Rose Bar township, Mrs. T. G. **Magonigal** of Smartsville, Miss Sarah **Bushby** of Rose Bar township and two sons, Joseph **Bushby** of Rose Bar township and William **Bushby** of Smartsville. Rev. Father J. J. **Enright** of Smartsville, pastor of the church of the Immaculate Conception, will officiate at funeral services to be announced later. Interment will be under the direction of R. E. Bevan & son.

23 Dec 1915 (Thursday); Page 6 Col 6; Smartsville News - A message reached this place last Monday of the death of Samuel J. **Fisher** in Rough and Ready. The deceased was for several years a highly respected resident of this township. The funeral will take place today and the interment made in the family plot in the fraternal cemetery at 1 o'clock pm.

11 Jan 1916 (Tuesday); Page 4 Col 2; SMARTSVILLE Jan 1916. The body of the late Mrs. Mary **Reid**, pioneer woman of Smartsville, was consigned to a grave in the Smartsville cemetery yesterday afternoon, following funeral services conducted by Rev. George U. **Gammon**, pastor of the First Presbyterian church in this city. The pallbearers, all men the deceased had seen grow from childhood to man's estate during her long residence in the mountain town, were as follows; Robert **Beatty**, William **Peardon**, Leo **Meade**, William **Cramsie**, John **Cramsie** and John **McQuaid**. Kelly Bros. was in charge of the interment.

19 Jan 1916 (Wednesday); Page 1 Col 6; One of the saddest automobile accidents to occur in this vicinity in some time happened about 6 o'clock Tuesday evening and resulted in the death of John J. **Casey**, Yuba County supervisor from the Browns Valley district, and Miss Bessie **Burris**, daughter of Mr. and Mrs. Byron **Burris**, well known residents of Browns Valley and slight injury to Mrs. **Casey** and Mrs. Charles **Taylor**, the other two occupants of the Auto. The accident happened at a point on the Browns Valley grade about five miles north of Marysville, shortly after the automobile party had left this city on their return to their homes in Browns Valley. The automobile, a Ford car, skidded off a six-foot grade, struck a slight rise, and turned over into a small ditch that is being dug in that vicinity. The two occupants of the front seat, **Casey** and Miss **Burris**, were pinned under the auto by the steering gear and the windshield, both their heads being under the foot of water, which the ditch contained and before any assistance could reach them they had, both drowned. Mrs. **Casey** and Mrs. **Taylor**, riding in the rear sear of the auto, escaped with a severe shock and a few minor bruises. Immediately after the accident occurred, Mrs. **Taylor** and Mrs. **Casey**, finding that they could do nothing for the two victims pinned under the car, started out for help. Mrs. **Casey** started

NEWSPAPER INFORMATION

toward Browns Valley and Mrs. **Taylor** started back toward Marysville. After walking about two miles, calling hysterically for help, Mrs. **Casey** came upon Frank Kupser, a well known Browns Valley man, who returned with her to the scene of the wreck, but he found that he was powerless to do anything alone, and he went to telephone for assistance. In the meantime Mrs. **Taylor** located some foreigners but was unable to make them understand and returned to the scene of the wreck. E. S. Austin representative of the A. J. Marx Music House of Sacramento, and L. J. Francis, local representative of the company, came along in an auto, and found Mrs. **Casey** at the scene of the wreck, who implored them to do something but when they attempted to lift the car up they found that more men were needed to accomplish anything, so they brought Mrs. **Casey** and Mrs. **Taylor** to Marysville and notified local authorities and a rescue party was detailed to rush to the scene, which they did in automobiles. When the rescue party from this city reached the scene of the accident, Frank Kupser, Frank Gregory, and Charles Elisalda had succeeded in righting the overturned auto and removed the prostrate forms of **Casey** and Miss **Burris** from underneath. **Casey** was dead when removed but there seemed to be some life in Miss **Burris**, but every effort to resuscitate her proved futile and when Drs. A. L. Miller and Everett Gray arrived on the scene they pronounced the young woman dead. Byron Burris, father of the girl victim, arrived on the scene soon after Kupser, and the other men had removed the victims from beneath the auto. According to the story told by the two women occupants of the car, **Casey** lost control of the car while in a rut, and the road being quite slippery in that part, it swerved to one side and the first thing the occupants of the car knew the machine started down the step grade, and when it struck a slight rise at the bottom of the grade, turned completely over and landed in the ditch. Mrs. **Casey** and Mrs. **Taylor** found the water surging up about them and fearing that they would be drowned, fought desperately for their lives and finally succeeded in tearing the storm curtains from the machine and gaining their freedom. They did not seem to be injured, so they made their way to the top of the grade and started for assistance. Undoubtedly the two victims, **Casey** and Miss **Burris**, were drowned, as the wounds upon the bodies were not sufficient to cause death. Only a slight abrasion was found on the head of Miss **Burris**, while **Casey** sustained a slight bruise on one finger. Immediately upon hearing of the accident, Coroner J. K. Kelly, together with his assistant, rushed to the scene of the accident, and after the victims had both been pronounced dead, Coroner Kelly took charge of the remains which were brought to Marysville, where they are now at the coroner's office. The Ford auto, which **Casey** was driving at the time of the accident, was only slightly damaged, considering the fall over the grade. Aside from a badly damaged windshield and broken top, the car was in fairly good condition when it had been righted. It was removed today and brought to town for repairs. There are many reasons vouchsafed as to the cause of the accident, but the most feasible one is that the driver lost control when he turned around to say something to his wife. The occupants were eating a lunch, which they had with them at the time. When **Casey** turned around to say something he failed to keep the car in the road, and when it struck a sort of rut in the road the steering gear swerved quickly and the car shot down the embankment. **Casey** and his wife and Mrs. **Taylor** and Miss **Burris** had been spending the day in Marysville at the home of Miss Iola **Dunning**, sister of

NEWSPAPER INFORMATION

Mrs. **Casey**, after bidding goodbye to Mrs. **Taylor's** mother, who had departed for San Francisco for a visit with friends. They left Miss **Dunning's** place at about dusk on their return to Browns Valley. Supervisor **Casey** was 52 years of age, and was born and raised in Yuba County, where he had resided practically all his life. He was quite well known, and had many friends, who will regret to hear of the sad affair. He was supervisor of district No. 4 for the past three years, and during 1913 served as chairman of the board of supervisors of Yuba County. He was a prominent member of the Marysville Lodge No. 783, B.P.O.E, Marysville Lodge No. 1, E. Campus Vitas, and Marysville Parlor No. 6, Native Sons of the Golden West. He is survived by a wife, two children, three brothers and two sisters. Miss **Burris**, 23 years of age, was quite a favorite among the younger social set of Browns Valley, and possessed a legion of friends who will mourn her loss. She is the daughter of Mr. and Mrs. Byron **Burris** and has made her home with her parents at Browns Valley from girlhood. Besides her parents she is survived by two sisters and a brother. Funeral services over the remains of the late John J. **Casey**, killed in the automobile accident on the Browns Valley Grade, Tuesday night, will be held from the Elks' home on D street, tonight, at 8 o'clock, when the local lodge of Elks, of which the deceased was a member, will have charge of the services. Friends and acquaintances are respectfully invited to attend. Thursday morning the remains will be taken to Smartsville, where requiem mass will be celebrated at the Church of Immaculate Conception at Smartsville, when Rev. Father Enright will officiate. Interment will be in the Catholic cemetery at Smartsville, under the direction of Kelly Bros., who have charge of the funeral arrangements. The funeral of Miss Bessie **Burris**, the other victim of the fatal automobile accident of Tuesday night, will be held from the late home of her parents, Mr. and Mrs. Byron **Burris**, at Browns Valley, Friday afternoon at 1 o'clock, when Rev. M. Rifenbark will officiate, and interment will be made in the Browns Valley cemetery under the direction of Kelly Bros. Friends and acquaintances are respectfully invited to attend the services.

23 Jan 1916 (Sunday); Page 4 Col 3; Smartville Jan 1916. Frank Wesley **Taylor**, who until a short time previous to his death, was a salesman for the Moran Packing company of this city, and who is well known here, died at this home near Spenceville, Nevada county, yesterday following a brief illness. He was a native of Rough and Ready, California, aged 32 years, 4 months and 17 days. He is survived by a wife and two children, **Audrey** and **Jack**, a mother, Mrs. Sarah **Taylor** of Rough and Ready; two brothers, **Joseph** of Marysville and **Sterling** of Rough and Ready; two sisters, Mrs. Blanche **Maas** of Dunsmuir and Mrs. Maude **Culvert** of Hornitos, California. Taylor was a member of Silver Oak camp No. 185, Woodmen of the World. He enjoyed a large host of friends both in this city and at Spenceville. Funeral services will be held on Monday at 11 a.m. from the home of Mrs. James **Bowman**, near Spenceville, Rev. S. J. **Buck** will officiate. Interment will be in the Smartsville cemetery. Funeral arrangements will be in charge of Kelly Brothers.

8 Feb 1916 (Tuesday); Page 7 Col 2; Mrs. Abigail P. **Casey**, widow of the late J. J. **Casey** was appointed executrix of the estate of the latter in the superior court here

NEWSPAPER INFORMATION

Monday afternoon. The estate, which consists mostly of real property, is valued at $6275. The will was made November 28, 1904, and bequeaths the entire estate to the widow with the exception of $300, which is left to Mrs. Winifred **Casey**, mother of the deceased.

12 May 1916 (Friday); Page 1 Col 4; DOBBINS. May 11 – John Frances **Doyle**, son of the late **James** and Mary **Doyle**, died here today. He was a native of Smartsville and was 40 years of age. He was a brother of Mrs. Henry **Merriam** of Dobbins, Mrs. F. L. **Johnson** of Sacramento; Mrs. J. C. **Baldwin** of Marysville, and Edward **Doyle** of Dobbins. The funeral arrangements have not been made, other than interment will be made under the direction of Kelly Bros., Marysville.

14 Jul 1916 (Friday); Page 4 Col 6; Smartsville Jul 1916 - Mrs. Martha **Toland**, one of the early settlers in Yuba county, died at her home near Smartsville yesterday, at the age of 83 years. Mrs. **Toland** had been a resident of Smartsville since 1855. She left Ireland when she was six years old and arrived in San Francisco in 1855 by way of the Isthmus of Panama. Shortly after arriving in San Francisco she came to Smartsville with her first husband, Robert **McKeague**. Of this marriage there were three children, Mrs. Elizabeth **Green** and Mrs. Margaret A. **McDonald** of Berkeley and Robert A. **McKeague** of Honolulu. In 1864 Robert **McKeague** died and in 1868 she married the late Robert **Toland** in Marysville. Of the second marriage there were four children, **Mary** J., **W**. J., and Hugh **Toland** of Smartsville and John **Toland** of the United States customs service in San Francisco. R. E. Bevan & Son are the funeral directors. The funeral will be held Sunday morning at 10 o'clock from her late home. Rev. **Harrison** of the Methodist Episcopal Church South of Yuba City, will conduct the services. Burial will be made in Fraternal Cemetery at Smartsville.

27 Dec 1916 (Wednesday); Page 4 Col 3; SMARTSVILLE, Dec 26 – Daniel **Fraser**, a resident of this vicinity for forty years died Sunday night at the home of his daughter, Mrs. John **Magonigal**, at Union Ranch near here, at the age of 73 years. He had been ill for months and death was due to old age. He is survived by three daughters – Mrs. Irwin **Hiscox** of Grass Valley, Mrs. John **Magonigal** of Smartsville and Mrs. A. G. **Wheaton** of Smartsville; three sons – James **Fraser** of Smartsville, Edwin **Fraser** of Wheatland and G. W. **Fraser** of Grass Valley. The funeral will be held tomorrow morning from the home of the daughter, at 11 o'clock. Burial will be made in the fraternal cemetery at Smartsville under the direction of R. E. Bevan and Son of Marysville.

27 Dec 1916 (Wednesday); Page 4 Col 3; The funeral of James **Byrne**, pioneer resident of Smartsville, who died in Sacramento Sunday, will be held this morning at 9:30 from the Catholic Church at Smartsville. Burial will be made in the Catholic cemetery under the direction of Kelly Bros. Byrne is survived by four sons; **Robert** of Rio Vista, **James** of Plumas County, **John** of Natomas and Henry **Byrne** of Richmond; three daughters, Mrs. Brady **Compton** of Smartsville, Mrs. Clarence **Owen** of Marysville, and Mrs. J. B. **Wilbur** of Oroville.

NEWSPAPER INFORMATION

8 May 1917 (Tuesday); Page 4 Col 4; Mrs. Katherine **Looney**, a pioneer of Browns Valley, died Sunday morning at the home of her grand daughter, Mrs. Charles **Sperbeck**. She was a native of Ireland, and was 90 years of age. She had made her home near Browns Valley for the last sixty years. She is survived by a daughter, Katherine **Pettit** of Browns' Valley, and a granddaughter, Mrs. Charles **Sperbeck**, and two great Grandchildren, Arlene and Bernice Sperbeck. The funeral services will be held from the Catholic church at Smartsville this morning at 10 o'clock. The Rev. Father Enright will officiate and will be in the Catholic cemetery at Smartsville under the direction of Kelly Bros.

7 Aug 1917 (Tuesday); Page 8 Col 3; Funeral services were held yesterday afternoon for Mrs. Freedom Damon **Crocker**, 87, who died at the home of Rev. John Vineyard near Smartsville Sunday morning. Mrs. **Crocker** was a native of Nova Scotia but had resided in this state for the past 20 years. Rev. Vineyard officiated at the funeral services. Interment took place in the Smartsville Cemetery under the direction Kelly Bros.

22 Sep 1917 (Saturday); Page 1 Col 4; Mrs. Katie Jane **Hickeson**, 48, wife of William **Hickeson**, died at her home near Smartsville yesterday afternoon. Death was not unexpected, as Mrs. **Hickeson** has been failing in health for about a year. She is survived by her husband, William **Hickeson**; one daughter, Miss Minnie **Hickeson**; three sisters, Mrs. Sarah **McGonigal**, Mrs. Ellen **Morrell** and Miss Sarah **Bushby**; two brothers, **Joseph** and William **Bushby**, and her father, George **Bushby**. Deceased was a native of this county. Funeral arrangements are being made by Kelly Bros.

22 Sep 1917 (Saturday); Page 4 Col 1; Smartsville, Sept. 21 – A gloom of sadness extends over this entire community this morning in the extremely sad and untimely death of one of our most beloved young wives and mothers, Mrs. Ada **Peardon**. Mr. and Mrs. **Peardon** and children returned from Marysville at 10:30 o'clock last evening and while **Peardon** had gone downtown for milk Mrs. **Peardon** started to light an alcohol stove to prepare milk for her little girl. In some manner her clothing became ignited and although assistance reached her within a few minutes she was so badly burned that death followed at 4:45 this morning. Dr. W. J. **Guinan** of Marysville was hastily summoned and a graduate nurse was in attendance, everything being done that could possibly save her, but without avail. Mrs. **Peardon** was the daughter of the late Mr. and Mrs. T. **Newbert**, pioneer residents of this place. She was 30 years of age. Besides her grief stricken husband she leaves two small children, **William**, aged 5, and **Helen**, 2 years; three brothers, **Horace, William** and Thomas **Newbert**, and two sisters, Mrs. Henry **Creps**, of Wheatland, Mrs. J. B. **Byrne** of Fairview. The funeral services will be held at the family home at Smartsville, Sunday afternoon at 2 o'clock. Interment will be made in the family plot in the Smartsville cemetery, under the direction of R. E. Bevan & Son.

27 Jan 1918 (Sunday); Page 4 Col 2; While eating his lunch in the Palm café on D street at 2 o'clock yesterday afternoon, William Joseph **Toland** was stricken with heart failure and dropped dead. Deceased was 48 years of age and a native of

NEWSPAPER INFORMATION

Smartsville where he had considerable property interests. He was a member of Rose Bar Masonic lodge no. 89 at Smartsville and Oriental lodge, I. O. O. F., of this city (Marysville). He is survived by three brothers: John **Toland** of San Francisco, **Hugh** and Tandy **Toland** of Smartsville; a sister, Mary J. **Toland** of San Francisco. Coroner J. K. Kelly took charge of the remains and turned them over to R. E. Bevan & Son, where an inquest will be made.

26 Apr 1918 (Friday); page 4 Col 1; OAKLAND, April 25 – A "war wedding," which came as a complete surprise among the smart set here took place in Berkeley today when Miss Elsie **Detrick** became the bride of James Edward **Holbrook**. No formal announcement had been made of the engagement of the couple and the news of their wedding, which was hastened because of government orders received by **Holbrook**, is of great interest to scores of friends. **Holbrook** has been attending the ordinance school on the campus in Berkeley and this week received orders, which will take him away from here. Because of this the couple decided to be quietly married before the young man left for his new post of duty. The wedding took place at the home of Mr. and Mrs. Edington **Detrick**, Jr., on Hillgrass avenue, with Rev. Father Lantry O'Neill of Newman Chapel as the officiating clergyman. Miss Sue **Tuttle** attended as maid of honor at the wedding and Miss Katherine **Bennett** as bridesmaid. The groomsman was Allen **Sproule**, a student at the aviation school at the State University. The bride belongs to one of the best-known families in the bay region and is a niece of Mrs. George Nickel and Mrs. E. Swift Train. She is a graduate of the Berkeley high school and one of the popular members of the younger set in the college town. **Holbrook** formerly attended the University of California, but left college to enter the service of his country. He is a member of the Zeta Psi fraternity and was prominent in student affairs. James **Holbrook** is the son of Mrs. Katherine **Holbrook** of Berkeley. He is a native of Smartville and a nephew of James K. **O'Brien** of that town. Holbrook attended the University of California. He was employed at the Yuba Manufacturing company here for several months.

25 Jul 1918 (Thursday); Page 6 Col 4; Smartville, July 24 – A telephone message Wednesday to J. H. **McQuaid** conveyed the news of the death in Berkeley of Mrs. Julia **Tifft**, a former pioneer resident, which occurred at the home of her daughter, Mrs. J. H. **Harris**. Mrs. **Tifft** was more than 80 years of age, and had been an invalid for more than two years. Deceased was the widow of Dr. R. W. **Tifft**, who was physician here for many years. The family arrived in Yuba County in early days, traveling overland, first locating in the Erie district and later moving to this place. While on route across the plains, their only son, an infant, died, and was buried in an improvised casket made out of a box. A family of five daughters was raised and educated here, only two of who survive their parents. Mrs. J. H. **Harris** of Berkeley and Mrs. John **Hill** of Roseville. Six grandchildren also are left, among whom are R. R. **Beatty**, now attached to the naval reserve in San Francisco. The funeral under the direction of R. E. Bevan & son will take place on Friday, and interment will be in the family plot in the Masonic cemetery.

NEWSPAPER INFORMATION

9 Nov 1918 (Saturday); Page 8 Col 2; SMARTSVILLE, Nov 8. Believed to have been dead for at least two days, the body of James **Campbell**, pioneer miner and prospector of Smartsville, was found in his pretty cottage yesterday by Linn **Havey**, a friend of the dead man who had missed seeing him for several days. **Campbell** was well known in the hills, and leaves several sons who reside in and near New Castle. The body was taken in charge by Coroner J. K. Kelly, who believes Campbell died of infirmities of old age. He was 75 years of age. Coroner Kelly will conduct an investigation, however.

10 Nov 1918 (Sunday); Page 8 Col 3; Funeral services for the late James **Campbell** were held from Kelly Bros. undertaking parlors yesterday afternoon at 4:30 o'clock. Rev. Father Guertin officiated. Interment in Catholic Cemetery.

25 Dec 1918; Page 4 Col 4; Smartsville, Dec 24 – Mary Ann **Bach**, pioneer of California, who has resided at Smartsville since the fifties, died at an early hour yesterday of the infirmities of old age. Mrs. Bach was 83 years of age. John E. **Bach** and Miss Alice **Bach** of Smartsville, and Mrs. John **Toland** of Oakland, children of Mrs. **Bach**, survive. Funeral services will be held Thursday afternoon from the residence, and interment will be in the Smartsville cemetery. Many "old timers" of this mining section will attend. R. E. Bevan has charge of the funeral arrangements.

4 Mar 1919 (Tuesday); Page 1 Col 4; At the advanced age of 89 years, Frances M. **Adkins**, pioneer resident of Yuba County, died at her home in the Earle district, Saturday evening. Mrs. Adkins was a native of Kentucky. The following children survive; Mrs. America **Hite** of Marysville; Mrs. Mary J. **Coburn**, Loomis; Mrs. Frances **Murch**, Lincoln; J. T. **Adkins**, Oregon; Robert E. **Adkins**, Durham; O. P. **Adkins**, Earle and B. F. **Adkins**, Sacramento. The remains are in charge of Kelly Bros.

9 Aug 1919; Page 3 Col 3; Smartsville News - In the death of Captain C. C. **Bitner** of Spencerville, which occurred Tuesday morning suddenly from heart disease. Nevada and Yuba counties suffer the loss of one of their most highly respected pioneer citizens. A man of sterling qualities, loved and honored by all for over half a century, having passed 80 years of a successful life, he leaves to mourn his demise a loving wife and two devoted daughters, Mrs. Ella **Austin** and Mrs. Emma **Anderson** of Spencerville. He was a lifelong member of the Masonic lodge in the east. The funeral will be held Thursday from his home in Spencerville at 10 o'clock. Interment in the fraternal cemetery of Smartsville.

13 Oct 1920 (Wednesday); Marysville Democrat; Page 1 Col 4; Smartsville – Rev. John T. **Vineyard**, pioneer resident of this section, is dead, of double pneumonia aged 84 years. He is survived by two brothers; William **Vineyard** of Lone Tree, and Miles **Vineyard** of San Francisco, and one sister, Mrs. Bertha **Smith**, of Fallon, Nevada. Earl E. **Woodroffe** of Marysville is a stepson of deceased. Mrs. David **Canning** and Mrs. Charles **Schellinger**, both of Marysville, are stepdaughters. Deceased was born in Plantsville, Grant county, Wisconsin, July

NEWSPAPER INFORMATION

16, 1826. He came to California in 1852 and settled first in Marysville. After living here five years he moved to Sucker flat, a suburb of Smartsville, then famous for its gold output. Later he moved to Mooney Flat, remaining there until his death. The funeral will take place Thursday afternoon, following services in Smartsville. The burial will be in Fraternal cemetery. Studies will be dispensed with in Rose Bar School during the obsequies to permit the pupils to pay respect to the memory of deceased, who always took a deep interest in educational matters and in every movement fro the advancement of the commonwealth. During his 68 years residence in California Rev. Vineyard formed a wide circle of friends and acquaintances that now deeply regret his passing. He was of the type of pioneer that founded the great State of California – the type of which only a few now remain.

22 Oct 1920 (Friday); Page 5 Col 4; Smartsville, Oct 21 – The funeral of Mrs. Mary **Bitner**, pioneer resident of Spenceville and widow of the late Captain **Bitner**, was held today from her late home in Spenceville at 2 o'clock. The services were conducted by Rev. R. C. **McAdie**, paster of the Marysville Presbyterian Church. Suitable music was provided at the services by Farwell **Brown** and Mrs. E. E. **Monson** while Mr. Brown also sang at the grave. The interment was in Smartsville cemetery. The pallbearers were L. F. **Pieratt**, A. S. **Walker**, Ed **Jackson**, Frank **Hunt**, Charles **Francis** and H. D. **Williams**. Numerous floral offerings testified to the respect in which the community held the memory of the pioneer woman. The arrangements were made by Kelly Bros. Mrs. Bitner died at her home Monday night after an attach of pneumonia. She was a native of Pennsylvania and was 81 years of age. She is survived by two daughters; Mrs. Ella M. **Austin** and Mrs. Emma **Anderson**, both of Spenceville.

26 Jan 1921 (Wednesday); Page 4 Col 6; The funeral of the late Aden **Wright**, well-known Yuba county pioneer, was held from his home near Waldo yesterday. Rev. C. W. F. **Daniels**, pastor of the Marysville Christian church, officiated, and appropriate vocal numbers were rendered by Farwell **Brown**, with Mrs. E. E. **Monson** as accompanist. Interment was made in the Smartville cemetery by Kelly Bros. The pallbearers were T. D. **Mitchell**, W. R. **Welch**, C. N. **Welch**, C. **Welsh**, R. **Welch**, and J. E. **Beck**.

6 Mar 1921; Page 1 Col 1; William H. **Bowman** passed away yesterday afternoon in this city (Marysville) at the home of his sister, Mrs. Jasper **Hunt**, on F. Street, following an illness of short duration. He was a native of California, and aged 44 years. Elk Grove has been his residence for some time. Besides Mrs. **Hunt**, he is survived by four other sisters, Mrs. O. E. **Swift** of this city (Marysville), Mrs. George **Sheppard** of Salinas; Mrs. Wm. **Barrie** of Smartsville; and Mrs. L. **Rossi** of Earl; two brothers, F. W. **Bowman** of Marysville, and J. M. **Bowman** of San Jose. He was a nephew of Jason **Jones** of this city (Marysville). Funeral arrangements are to be announced by Kelly Bros., who have charge of the remains.

21 Jun 1921 (Tuesday); Page 4 Col 3; SMARTVILLE, June 20 – After an illness, which had confined her to her bed for more than a year Mrs. Alpha Frances

NEWSPAPER INFORMATION

Widener passed away Sunday night at ten o'clock, at the home of her daughter, Mrs. John E. **Bach**. Mrs. **Widener** was a native of Tennessee and was about 69 years old. She had made her home in Smartsville with the exception of a few years, since 1876, and was one of the best-known and beloved pioneers of this section. A sufferer for many years, she endured all with a true Christian patience and fortitude, which endeared her to not only her own immediate family, but to a large circle of friends. She was the widow of the late Russell **Widener** and one son, **William**, also preceded his mother to the grave. She is survived by one son, Joseph E. **Widener** of Pleasant Valley, and one daughter, Mrs. John E. **Bach**; two brothers, McCord **Atchley** of Stockton and Harrison **Atchley** of Los Angeles, and the following grandchildren; **John** and Ethel **Bach** and Lloyd **Royal**, **Ruth** and Hazel **Widener**. The funeral will be held under the direction of R. E. Bevan and Son of Marysville, Tuesday afternoon at 2 o'clock from the home of John E. **Bach**, thence to the M. E. Union church for services. Interment will take place in the family plot in the Fraternal cemetery.

1 Nov 1921 (Tuesday); Page 7 Col 2; SMARTVILLE, Oct 31. A gloom of sorrow was cast over our little village on Sunday afternoon, when the news of the death of William J. **Meade** was received from Oroville. Just a week ago on Sunday, **Meade** was taken to the Oroville hospital for treatment for an abscess in the head from which he had been suffering for some time. Encouraging reports had been received about him, until the last few days, when he commenced to fail rapidly. A native of Smartville, aged 42 years, he spent his entire life here. He was an exemplary young man in every way, one of whom any community might boast, loyal to his family, his friends, his religious duties and his state. He always was found willing and ready to lend aid and means to every worthy cause. Loved and respected by all, he will be sadly missed from the community where he took such an active part in such a manly way. He is survived by his wife, one daughter, **Helen**, and the following brothers and sisters: **John** and **Leo**, Miss Mary **Meade** and Mrs. T. J. **O'Connor**, to whom the sympathy of this entire town and vicinity is extended.

1 Nov 1921 (Tuesday); Page 8 Col 5; William Joseph **Meade**, well known resident of Smartville, died in the hospital at Oroville, Sunday, where he had been a week, but had been in ill health for several months. Deceased was 41 years, and 10 months of age, and a native of Smartville. He was the son of the late Mr. and Mrs. Garrett **Meade**, and is survived by a widow, Agnes Weber **Meade**, and a 2 year old daughter, **Helen**, two sisters, Mrs. Thomas J. **O'Connor** and Miss Mary **Meade**, and two brothers, **John** T. and Leo **Meade**. The funeral services will be held from the Catholic church at Smartville at 10:30 Wednesday morning, with Rev. Father J. **O'Meara** of Grass Valley to officiate at a requiem mass. The cortege will leave the home of W. J. **Dempsey** at 8:30 tomorrow morning. Interment will be made in the Smartville Cemetery under the direction of Kelly Bros.

4 Jan 1922; Page 2 Col 2; Smartville, Jan 3. Michael A. **Wallace**, who came to this section a year ago from Canada on account of filing health, died on his farm near Smartsville yesterday, aged 68 years. Deceased, a native of Nova Scotia, is survived by two brothers, Dr. James I. **Wallace** of Saskatchewan, Canada and Dr.

NEWSPAPER INFORMATION

Walter B. **Wallace** of Oklahoma. Burial will take place Wednesday afternoon in the Odd Fellows' cemetery, Smartsville, under the direction of Kelly Bros. of Marysville.

26 Jan 1922 (Thursday); Page 4 Col 3; A marriage license was issued in Oroville yesterday to Francis Vincent **Johnson**, aged 26, of Smartsville, and Cora Elizabeth **Cramp**, aged 18, of Browns Valley.

20 Apr 1922 (Thursday); Marysville Democrat; Page 5 Col 3; SMARTSVILLE – Word was received here today of the death in Oakland Tuesday of Richard **Beasley**, former resident of Smartsville. He was a brother of William J. **Beasley**, local resident. He was 67 years old and is survived by six children.

25 Apr 1922 (Tuesday); Page 1 Col 4; William G. **Welch**, 70, as old-time resident of Yuba County, died yesterday at the Sutter County hospital after a long illness. Decedent, who was a native of Missouri, came to California 63 years ago and for many years made his home at Smartville. He was a brother of J. B. **Welch** of Los Molinas, James **Welch** and Mrs. M. B. **Wright** of Wheatland. Kelly Bros. will have charge of the funeral, which will be held, at the Smartville cemetery tomorrow afternoon at 2 o'clock.

8 Jun 1922 (Thursday); Page 7 Col 3; OROVILLE – At a nuptial Mass Wednesday morning at 9:30 in St. Thomas Catholic church in this city, Louis Patrick **Havey**, of Smartsville, and Mary Emma **Poor**, of Brown's Valley, were united in marriage by Rev. J. J. **Hynes**, pastor of the local Catholic church. Mr. **Havey** is the son of Mr. and Mrs. John **Havey** of Smartsville, while his bride is the charming and beautiful daughter of Mr. and Mrs. **Poor**, of Brown's Valley. Chester **Havey**, brother of the groom, and his cousin, Miss Irene **Murphey**, of San Francisco, acted as witness to the marriage. After a honeymoon in the Bay cities Mr. and Mrs. **Havey** will make their home in Marigold, Yuba County, where the groom holds a responsible position with the Marigold Dredging Company. Many Oroville friends as well as friends in Smartsville and Brown's Valley will unite in wishing Mr. and Mrs. **Havey** much happiness in the years to come.

9 Jun 1922 (Friday); Page 7 Col 3; SMARTSVILLE – John Henry **Linehan** died at his home here yesterday. He was 59 years of age and a native of California. Deceased is survived by a widow, Anna **Linehan**, his father, Timothy **Linehan**, and two sisters, Mrs. T. J. **McGovern** of Hammonton, and Miss Katherine **Linehan** of Smartsville; and two brothers, T. J. **Linehan** of Ventura and P. F. of San Francisco. Deceased's father is now 100 years old.

14 Jun 1922 (Wednesday); Page 1 Col 4; Francis **Poole**, 90 years old, prominent resident of Yuba county for years, died yesterday afternoon at his home near Marigold. **Poole** was born in England and came to the United States when a boy. He lived in California 74 years, most of that time at Smartville. He is survived by his widow, May Anne **Poole**; two daughters, Mrs. Mary E. **McDonald** of San Francisco and Mrs. John F. Havey of Marigold, and three sons, William C. **Poole**

NEWSPAPER INFORMATION

of Marysville, James F. **Poole** of Kings county, and Frank D. **Poole** of Marysville. He leaves nineteen grandchildren and five great grandchildren. The funeral is to take place from his late home at Marigold Friday morning at 10 o'clock, Rev. A. E. **Butcher**, rector of St. John's Episcopal church, officiating. Interment will be made in the Smartville Fraternal cemetery under the direction Kelly Brothers.

29 Jun 1922; Page 7 Col 3; Smartsville, June 28. The news of the death of John **Fippin**, six-year old son of Mr. and Mrs. Asa **Fippin**, was received here Tuesday morning with deep expressions of sorrow. The little fellow had been ill for one week with an affection of the tonsils but it was not until Saturday when Dr. Barnes of Grass Valley, the attending physician advised removal to the Nevada county Sanitarium, that his illness was considered serious. He was taken to the Sanitarium on Sunday but despite every effort of doctors and of nurses he grew gradually worse until he passed away on Tuesday at 11 o'clock. He was a bright, lovable child, a pupil of the Rose Bar school and his death deeply affect both teachers and pupils. He is survived by is heart broken parents, one brother, **Robert**, and four sisters, **Julia**, **Nellie**, **Eleanor** and **Bernice**. The funeral will take place from the family home on Thursday afternoon at 2 o'clock thence to the M. E. church for services by the Rev. David **Ralston** of Grass Valley. Interment will take place in the fraternal cemetery.

29 Aug 1922 (Tuesday); Marysville Democrat; Page 1 Col 1; Mrs. Veronica Amelia **Manasco**, pioneer resident of Yuba county, 65 years of age died last evening at her home at Mooney Flat, where she was born and has resided ever since. She conducted the Mooney Flat hotel, which was at one time known as the Live Oak hotel, being named after a mammoth live oak tree that formerly grew in front of the building. After the passing of her parents, Mrs. Manasco conducted the hotel. She was also the community nurse and was called on at all times to relieve the suffering or attend to the ills of the old '49 ers. She was noted for her charity and kind deeds throughout the entire community. Of the original children born in Mooney Flat in the early days only one person survives, Mrs. Manasco. Her brother Captain Henry **Schmidt**, passed away in San Francisco in 1921. For the past yea she had been an invalid and spent much of that time with her daughter, Mrs. H. L. **Hite** of Yuba City, where she made many friends. She is survived by three children: George **Manasco** of Hammonton, Mrs. H. L. **Hite** of Yuba City, and Mrs. R. **Beatty** of Oakland and also by a brother William **Schmidt** of Turlock, and a sister, Mrs. Sadie **Gassaway** of San Francisco and six grandchildren. Funeral services will be held at 2 p.m. Wednesday from the Catholic Church in Smartville. Interment will be made in the family plot in the Smartville cemetery under the direction of R. E. Bevan & Son.

10 Dec 1922 (Sunday); Page 6 Col 2; Yuba City, Dec 9 – Patrick K. **Kerrigan**, aged 62 years, passed away here today following a lingering illness extending over many months. The deceased was born in Smartville, Yuba county, and lived practically all of his life in this vicinity where he was well known. He is survived by a brother, Peter **Kerrigan**, of Encino, and two sisters: Mrs. Annie **Nagel**, of San Francisco, and Sister **Mary** Eugene of Notre Dame Convent, San Francisco. A

NEWSPAPER INFORMATION

requiem mass will be said for the repose of his soul in St. Josephs church at 10 o'clock, Monday morning. Interment will be made in the Catholic cemetery under the direction of Kelly brothers.

24 Jan 1923; Page 1 Col 2; Death Summons James **Welch**; is mourned by many. Death last evening claimed James **Welch**, a native of Sutter county, and one of the best known stockmen and ranchers of Sutter and Yuba counties, a man honored and respected, who counted his friends in large numbers. He had been ill about two weeks. Welch died at the family home near Waldo where he was in the stock business. He was born in Sutter County nearly 62 years ago. He was a member of Wheatland Parlor, Native sons of the Golden West. Left to mourn his passing are a widow, Mrs. Lillie **Welch**. He was the father of **James**, **Ira**, **David**, **Clarence** and **Willard** Welch, Mrs. D. F. **Kuster**, Mrs. A. A. **Whiteside**, Mrs. C. **Butler** and Mrs. J. W. **Suffin**, the brother of Joseph **Welch** and Mrs. Mary **Wright**. Funeral arrangements will be announced by Lipp and Sullivan of the Kelley funeral parlor.

10 Apr 1923; Page 8 Col 4; Mrs. Anna **Walsh** is called by Death in Smartsville Home - Mrs. Anna A. **Walsh**, a native of California, and 63 years of age, and a resident of Smartsville for the past 25 years where she was highly respected passed away in Smartsville yesterday. She is survived by her husband, John E. **Walsh**, three children Charles E. **Walsh** of Los Angeles, Walter L. **Walsh** of Marysville, and Mrs. Hollister H. **McKinnon** of San Luis Obispo and a sister, Mrs. Lucy **Pratt** of Alameda. The remains are in charge of Lipp and Sullivan and the funeral arrangements will be announced later.

11 Apr 1923 (Wednesday); Page 8 Col 3; The funeral services for the late Mrs. Anna **Walsh**, who died in Smartsville Monday evening, will be held Thursday afternoon at 2 o'clock at the Smartsville cemetery under direction of Lipp & Sullivan.

1 Aug 1923; Page 1 Col 2; Sherman **Woodroffe** Fatally Hurt on Dredge - Caught in Fall of Gantry; Dies Shortly After - Sherman E. **Woodroffe**, well-known gold dredge man of Hammonton, was fatally injured yesterday afternoon shortly before 4 o'clock when the framework of the stern gantry of a dredge collapsed while he was at the top working, the man being precipitated in the crash 40 feet to the deck below. **Woodroffe** struck on some cables as he pitched headfirst and upon striking the floor was frightfully broken up. Beside a frontal fracture of the skull, both arms were broken and several ribs fractured, and he suffered internal injuries and severe cuts and abrasions. Dr. Ralph Richard Scribner of Hammonton was called and rendered first aid after which preparations were made for the removal of the injured man to the Rideout hospital in this City. He died at about 7:45 o'clock while on the way to the hospital. Coroner Frank Bevan was called on the case. At his request an autopsy examination was made by Dr. P. B. **Hoffman** of this City and Dr. **Scribner**. What caused the stern gantry to collapse is not know. This A frame tower at the stern of a gold dredge. It supports the cables that in turn support the stacker, which carries off the rocks to the stacks behind as the dredge moves forward. When the steel work gave way the cables and the stacker also fell. The

NEWSPAPER INFORMATION

sacker was broken in the crash. **Woodroffe** was employed as an oilier. He was the only man on the gantry at the time of the accident. The deceased was born in California 43 years ago. He was a son of Mrs. J. T. **Vineyard** of Marysville, husband of Janie **Woodroffe** and brother of Mrs. David **Canning** of this City, Mrs. Martin **Kuster** of Wheatland, Mrs. Charles E. **Schellenger** of this City, James and Bruce **Woodroffe** of Chico, and Earl **Woodroffe** of Maysville. He was a member of Court Pride, Nov. 34, foresters of America. The body is at the Bevan funeral home. The inquest and funeral will be announced later.

7 Sep 1923 (Wednesday); Page 8 Col 4; Following an illness during the past eight months, death last evening summoned Mrs. Ida Ellen **Carter**, wife of W. S. **Carter**, at her home at Smartsville. She was 53 years old and had been a resident of Yuba county all her life. Friends will regret to hear of the passing of this good woman, although it was realized that she could not long survive. Funeral arrangements will be announced later by Lipp & Sullivan. Mrs. Carter was the mother of Mrs. Lola Hall **Stafford**, Miss Lilas **Hall**, sister of D. N. **Jones**, W. L. **Jones**, Mrs. Lillie E. **Welch** and Mrs. Dan **Cain**.

20 Sep 1923 (Thursday); Page 8 Col 5; Mrs. Margaret Ellen **Allread**, wife of William **Allread** of Smartsville, passed away yesterday morning at the home of W. J. Dempsey at 411 Seventh street, at the age of 47 years, following a lingering illness. She was born in Smartsville, where she has a host of friends. Deceased was a daughter of John **McQuaid** of Smartsville, and leaves beside her husband two daughters, **Mary** and Helen **Allread**; two brothers, Jack **McQuaid** of Smartsville and Thomas **McQuaid** of San Francisco, and two sister, Mrs. W. J. **Corey** and Miss Mary **McQuaid** of San Francisco. Lipp and Sullivan announce the funeral to be held at 10 a.m. in the Catholic Church at Smartsville.

25 Mar 1924 (Tuesday); Page 5 Col 5; Death yesterday morning ended the useful career of one of this county's best known charitable women when the Grim Reaper took from among the residents of this community, Mrs. Katherine **Barrett**, relict of the late John **Barrett**. She died at the family home, 1117 F street, at the age of 60 years. Surviving deceased are four sons and four daughters, **Dick**, **Dan**, **Jack** and Jim **Barrett**, Mrs. Harold **Gavin**, Mrs. Jack **Young**, Misses **Mamie** and Marjorie **Barrett**; two sisters, Mrs. A. M. **Tufts** of San Francisco, and Mrs. J. F. **Dempsey** of Smartsville, and three brothers, **J. H.**, **J. J.**, and D. F. **Driscoll**. The funeral services will be held Friday morning at 9 o'clock at St. Joseph's Catholic Church in this city and the body will be laid away in the family plot in the Smartsville cemetery by Lipp and Sullivan of Marysville.

12 Jun 1924 (Thursday); Page 5 Col 7; Mrs. Mary Ann **Poole**, well known Yuba county woman, passed away in San Francisco Tuesday. She was a native of England and was 86 years old. Surviving Mrs. **Poole** are the following children: Frank **Poole** of Smartsville, William C. **Poole** of Marysville, Mrs. John **Havey** of Marigold, Mrs. Mary **McDonald** of San Francisco, and James D. **Poole** of Turlock. The remains will arrive here today. Funeral arrangements will be announced by Lipp & Sullivan.

NEWSPAPER INFORMATION

9 Feb 1925 (Monday); Marysville Democrat; Page 3 Col 2; Francis V. **Johnson**, well known former resident of Marysville and Smartsville, died Sunday evening at 9 o'clock in the East Bay hospital following a hemorrhage of the brain suffered earlier in the afternoon. He was born in Santa Cruz, July 18, 1894, and came to Smartsville when a small boy with his parents and he attended the Smartsville schools. He leaves to mourn his passing, besides his wife, Cora **Johnson**, his parents, Mr. and Mrs. W. F. **Johnson** of Smartsville, a sister, Mrs. Frank **Miles** of Marysville, and a brother, Clarence **Johnson** of Smartsville, and two uncles and an aunt, **Clarence** and Dan **Weldon** of Quincy and Miss Lillian **Grove** of Sacramento. Deceased had been in the employ of the Pacific Gas & Electric Company for several years. Three years ago a pole fell with him, pinning him underneath and necessitating amputation of one of his legs, and he had never recovered from the shock. His parents and sister went to Oakland yesterday on receipt of the message. They will accompany the body to this city tonight, arriving here on the Sacramento Northern railway. Funeral services will be held Wednesday morning at 10 o'clock in the Catholic Church at Smartsville and interment will be made in the Smartsville cemetery under the direction of Lipp & Sullivan, who are making the funeral arrangements. Miss Madge **Maynard** will be the soloist.

12 Mar 1925; Page 8 Col 1; Mrs. **Adkins** Dies in Erle District Wheatland – Mar 12 – Mrs. Clara Elizabeth **Adkins**, well known resident of the Lone Tree District, died at the family home. She was aged 51 years, 5 months and 19 days, and was a native of Placer County. Besides her husband, Benjamin **Adkins**, she leaves to mourn her loss, four sisters, Mrs. E. W. **Forbes**, Seattle, Mrs. J. R. **Johnson**, San Francisco, Mrs. W. R. **Grace**, Oroville; Mrs. W. E. **Adkins**, San Francisco, and the following nieces and nephews: Mrs. W. A. **McLaughlin**, Nulaska, Washington; Mrs. Adeline **Vitt**, San Francisco; Mrs. Minnie **Kilbourn**, San Francisco; and Gerald **Forbes**, Seattle; **June**, **Will**, **Robert** and Jack **Grace** of Oroville. Funeral arrangements are being made by Lipp and Sullivan, who announce that services will be held in the Chapel Saturday morning at 11 o'clock. Interment will be made in the family plot in the Smartsville cemetery.

19 May 1925 (Tuesday); Page 4 Col 5; SMARTSVILLE, May 18. Miss Mary **McQuaid**, one of the best known women of Yuba county, who was born in Smartsville 59 years ago, where she has lived all her life time, died here today at the home of William Allread. She leaves a host of friends who regret her passing. Although ill in health for some time she had not been seriously ill until the past week or more. "Aunt Mary" **McQuaid**, as she was affectionately called by those who knew her, was born in Smartsville, April 8, 1866, making her age 59 years, 1 month and 10 days. She was the daughter of Mr. and Mrs. John **McQuaid**, pioneer residents of Smartsville, and was born in the same home where death came to her. She was educated in the schools of Smartsville and began her life work of teaching school at an early age. For eight years she was teacher in the schools of Yuba county. She was teacher of the MacDonald, One Tree and Smartsville schools in succession and many well known residents of Yuba county today first learned their "A B C's" under her guidance. After teaching in this county, she secured a position in the public schools of San Francisco, where for 30 years she was a teacher in the

NEWSPAPER INFORMATION

Bernal Heights school. Two years ago, following the death of her sister, Mrs. William **Allread**, she came to Smartsville to make her home with her brother-in-law. William **Allread**, and her two nieces, **Mary** and Helen **Allread**. To mourn her loss are left two brothers, John H. McQuaid of Smartsville and Thomas E. **McQuaid** of San Francisco. A number of nephews and nieces and other relatives also survive. She was a member of the Marysville Parlor, Native Daughters of the Golden West, and of Marysville Court, Catholic Daughters of America. She was active in lodge affairs and took a keen interest in the affairs of the community. "Aunt Mary" **McQuaid** was a woman of wonderful qualities and exceptional kindness of heart. While actively following her chosen calling for a period of almost 40 years, she was never too much engaged to devote her time to the aid of others, and her devotion to family and friends in time of need has placed in the hearts of hundreds a loving bond the breaking of which will leave a wound that time alone can heal. Funeral services will be held from the Catholic church here Wednesday morning at 10 o'clock. Rev. Father James **O'Meara** of Grass Valley will conduct the mass. Interment will be made in the family plot in the Smartsville Catholic cemetery. Funeral arrangements are in charge of Lipp and Sullivan of Marysville.

20 May 1925 (Wednesday); Marysville Democrat; Page 5 Col 4; SMARTSVILLE, May 20 – One of the largest funeral ever witnessed here was that held this morning for the late Miss Mary **McQuaid**, pioneer school teacher, who died early Monday morning. The services were conducted in the local Catholic church by Father **O'Meary**, of Grass Valley, and **Dermody**, of Lodi. Burial was made in the Smartsville Catholic cemetery. Pall Bearers were: E. V. **Carroll**, Donald **McQuaid**, Frank **McQuaid**, Jack **McQuaid**, Jr., Leo **Carey** and William **Allread**.

19 Jul 1925 (Sunday); Page 1 Col 2; Another old-time pioneer of Yuba county, George **Bushby**, 95 years old, passed away at his home near Hammonton yesterday afternoon. Bushby was a native of England and came to Sutter county in 1861 and has ever since been a resident of this county. Surviving him are two daughters, Mrs. T. G. **Magonigal** of Smartsville, and Mrs. F. W. **Mourill** of Montalba, and one son, William **Bushby** of Hammonton. Funeral services will be held Tuesday at the Smartsville cemetery with Rev. B. F. Butts officiating. The funeral arrangements are under the direction of the Bevan-Bricker funeral home.

14 Oct 1925 (Wednesday); Page 5 Col 2; SMARTSVILLE, Oct. 14 – Word was received here yesterday of the death of Edward J. **McGanney**, prominent New York attorney who was born and raised in this place coming from one of the old pioneer families of Yuba county. The body will be brought here for burial in the family plot beside his late parents, Mr. and Mrs. Daniel **McGanney**, prominent pioneers of the Sucker Flat district. **McGanney** spent his early boyhood in this section attending the Smartsville schools. He later attended St. Mary's College and then went to Columbia University. On his graduation from that institution he took up the practice of law, which he has followed for many years in Astoria, New York. He has a large number of friends in this section. He is survived by two sisters, Mrs. J. E. **Cramsie**, Smartsville, and Mrs. Mary **McLeod**, San Francisco;

NEWSPAPER INFORMATION

and two brothers; Daniel C. **McGanney**, San Francisco and Frank **McGanney**, Salt Lake City.

7 Aug 1928 (Tuesday); Page 1 Col 4; William B. **Burke**, city night watchman of Marysville for 18 years died Tuesday afternoon at his home, 223 Eighth street, following a lingering illness. Ill health forced **Burke** to retire as night watchman several years ago. He was a native of Pennsylvania. Surviving are his widow, **Teresa**, and the following sisters: Mrs. George **Meyers** of Marysville, Mrs. L. A. **Williams**, Mrs. F. C. **Eisen** and Mrs. Ida **O'Banion**, all of Oakland. **Burke** was a member of the Redman and Knights of Pythias lodges. Funeral arrangements will be announced by Lipp and Sullivan.

26 Aug 1929 (Monday); Page 2 Col 3; Robert D. **Wheaton**, 35, rancher of the Smartsville district, died Sunday night at the Sutter hospital in Sacramento. He had been in the hospital for a week and a half during which time he underwent an operation. Born at Smartsville, **Wheaton** had lived all his life at Smartsville. He was a past master of Rose Bar lodge No. 89, F. and A. M., of Smartsville, and a member of Smartsville lodge No. 431, Order of the Eastern Star. Surviving **Wheaton** are his parents, Mr. and Mrs. A. G. **Wheaton**, of Marysville; four sisters, Mrs. Emma **Fippin** of Smartsville, Mrs. Mable McCrea of Rough and Ready, Mrs. A. E. Spencer of Marysville and Mrs. E. L. **Craun** of Marysville, and two brothers, **James** and Allen **Wheaton**, both of Marysville. The body was brought to Marysville, Monday, by Lipp & Sullivan. Funeral services will be held at 2 p.m. Wednesday at the chapel of Lipp and Sullivan. The services at the chapel will be in charge of the Smartsville O. E. S. lodge. At the grave in the family plot in the Smartsville cemetery, the services will be in charge of the Smartsville lodge of F. & A. M.

30 Dec 1929 (Monday); Page 1 Col 2: MARRIAGE LICENSE. Peardon-Blake – In Marysville, Yuba county, Dec 28, 1929, Frank G. **Peardon**, 40, and Dollie **Blake**, 30, both of Marysville.

30 Dec 1929 (Monday); Page 3 Col 3; Miss Dollie **Blake** of Marysville became the bride of Frank G. **Peardon**, Marysville, and businessman, at a quiet wedding performed by Justice of the Peace Henry **Creps** at the Creps home at Wheatland, Saturday night. The couple was attended at the ceremony by the bridegroom's brother and sister-in-law, Mr. and Mrs. W. B. **Peardon**, of Marysville. Following the wedding a supper was served at the bride and bridegroom's home, 717 Ninth street, Marysville. **Peardon** is a member of the grocery firm of Peardon Bros. His bride is a well-known Yuba county girl. She is the daughter of Mrs. Christine **Blake** of District Ten.

14 Oct 1935 (Monday); Page 4 Col 1; SMARTVILLE, Oct 14. Funeral services for Thomas **Flanagan**, who died Saturday in Nevada City, will be held Tuesday. Mass will be said for the response of his soul in the Nevada City Catholic church at 10 a.m. Interment will be in the family plot in the Smartville Catholic cemetery. **Flanagan** was born here 55 years ago and attended the local schools. He was

NEWSPAPER INFORMATION

employed at Graniteville until he became ill with pneumonia. A sister, Mrs. Mary **Gilham** of Nevada City, is the only surviving relative.

11 Dec 1935; Page 1 Col 3; Victim of Sudden Death at Waldo to be buried Thursday. Funeral services for Benjamin F. **Adkins**, found dead at his Waldo district home Monday, are to be held at 1 p.m. Thursday in the Lipp and Sullivan Chapel with Rev. Raymond M. Huston officiating. **Adkins** lifeless body lay at his door several hours before being found by A. Kaiser, Wheatland prospector working on the ranch. Adkins was 67 and spent his entire life in the Waldo district, where he was born. Surviving are a sister and brother, Mrs. Mary **Coburn** of Loomis and Robert **Adkins** of Chico, as well as several nieces and nephews. Interment is to be in the Smartsville cemetery.

15 Jan 1936 (Wednesday); Page 2 Col 2; Funeral services for Mrs. Emma **Poor**, 64, who died at her home at Browns Valley yesterday following a short illness, will be held Thursday afternoon at 2 o'clock from the chapel of Lipp and Sullivan. Interment will be made in Smartsville cemetery. Mrs. **Poor** had been a resident of Browns Valley for the past 20 years being a native of this state. Surviving are her husband, Mark **Poor**, two sons, Ed **Poor** of Browns Valley, and Robert **Poor** of Marysville; one daughter, Mrs. Ethel **Handy** of Browns Valley; one brother, Ed **Bartlett** of Dixon; and 15 grandchildren

7 Feb 1936; Page 4 Col 2; Ethel May **Bach**, 58, died Thursday morning at the Rideout hospital in Marysville following a short illness. She was the widow of the late John E. **Bach** of Smartsville and was a pioneer resident of the Smartsville district. She is survived by three children; **John**, **Ethel** and **Vernon**, and a brother Joe **Widener** of Grass Valley. Funeral services will be held at 2 o'clock Saturday afternoon from the Methodist Episcopal church at Smartsville, with interment in the Smartsville cemetery by W. R. Jefford and Son of Grass Valley. The body will lie in state at her home in Smartsville until time of the funeral. She was born November 15, 1878. Pallbearers who will serve at the funeral services are W. L. **Newbert**, R. R. **Beatty**, J. E. **Cramsie**, W. C. **Huling**, A. B. **Sanford** and A. D. **Fippin**.

18 Feb 1937 (Thursday); Page 5 Col 6; John E. **Cramsie**, one of the most prominent men of the Smartville district of Yuba county, and a native of that town, died in the Rideout hospital in Marysville Wednesday afternoon after a short illness. He was 64 years of age. **Cramsie** had resided in his native community all of his life and had served it for a number of years as justice of the peace. He spent most of his life at farming. Surviving him are his wife, the former Anna **McGanney**, two brothers, W. P. **Cramsie**, superintendent of schools of San Jose, and Frank **Cramsie** of Auburn, and a sister, Mrs. Roy **Van Tiger** of Sutter county. The body was taken to the mortuary of Lipp & Sullivan and is lying in state there until 8 a.m. Saturday, when it will be taken to Smartville, where sacred rites will be held in the Catholic church at 10 a.m. Interment will be made in the family plot in the Catholic cemetery at that place.

NEWSPAPER INFORMATION

24 Feb 1937 (Wednesday); Page 1 Col 5 & 6; Funeral services for W. J. **Dempsey**, long time resident of Marysville and Yuba county, held in St Joseph's Catholic church Wednesday morning, were attended by large concourse of friends and floral tributes represented a great many more. The body was taken to Smartville for interment in the family plot in the Catholic cemetery. Rev. Bernard **McElwee** officiated at both places. Many automobiles conveyed friends to the cemetery, where Lipp & Sullivan conducted the burial. Casket bearers were Robert **Byrne**, T. A. **Gianella**, Richard **Barrett**, B. E. **Bryant**, Leo **Meade** and John J. **Murphy**. **Dempsey**, aged 67, was a native of French Corral, Nevada county, hi s parents, Mr. and Mrs. John **Dempsey** having located there on their arrival in 1852, his father engaging in mining there. Later the family moved to Smartville, where the elder **Dempsey** farmed, his son growing up as a farmer and later operating hay hauling outfits throughout the valley. Afterward he entered into the livery business in Marysville with Amos Lane, continuing until he purchased the John Dobler Saloon at Fourth and C. Streets, which he conducted as such until prohibition days, then engaging in the soft drink business. He purchased the building two years after taking over the business. He is survived by his son, **Kenneth**, who has been living on the old farm at Smartville the past few years, but who has now taken over his father's interests. Mrs. **Dempsey** died in 1913

23 Mar 1937 (Tuesday); Page 4 Col 1; SMARTVILLE, Mar 23. Another old time Yuba county resident died in San Francisco Sunday. He was Daniel C. **McGanney**, clerk of the appellate court for years, and a native of Smartville. He was a brother of Mrs. Annie **Cramsie** of Smartville, Mrs. Mary B. **McLeon** and the late **Edward**, **James** and Frank **McGanney**. Funeral services were held in the bay city Tuesday morning, entombment being made in the Holy Cross cemetery.

21 Jul 1937 (Wednesday); Page 4 Col 2; A friend of the family read prayers for the burial at 10 o'clock Wednesday morning for Mrs. Grace **Bennett** at the graveside in the Smartville cemetery. Mrs. **Bennett**, resident in Smartville for the past 10 years, died Monday morning at the age of 49. Pallbearers were William **Fullcher**, C. **Firebaugh**, D. **Trauger**, J. **Morgan**, H. H. **Thompson** and J. **Moffatt**. Arrangements were by Lipp & Sullivan.

11 Oct 1938 (Tuesday); Page 4 Col 5 & 6; Living all his life in the house where he was born, and granted his wish that he might live to the age of 80, John Henry **Sanders**, for years the village blacksmith of the Empire ranch at Smartville, died Sunday night. He had passed his 80th birthday anniversary by just a week. In all his life, **Sanders** had been away from his home on only two Christmas holidays, the other 78 were spent in the house where he first saw the light of day. **Sanders** was the last of his family, the three other children of the pioneer settlers, **Daniel** and Catherine **Sanders**, having passed on before him. Learning the blacksmith trade during his youth, he conducted the old shop at the Empire ranch for years, giving it up only when the advent of the automobile put the stages, freight wagons and teams out of business and left the blacksmith without business. **Sanders** had been in failing health for some time, and toward the end his wish was that he might keep alive long enough to celebrate his 80th birthday, which he did on his sickbed.

NEWSPAPER INFORMATION

He had enjoyed an eventful and happy life during his younger years and was widely known and highly esteemed throughout the region and in Marysville. For many years he, as a violinist, played for dances throughout the upper part of the county and in Nevada county. Surviving relatives are cousins, James **Beirne** and Annie **O'Brien** of San Francisco. The body will lie in state at his home at Smartsville during Tuesday and funeral services will be held in the Catholic church there Wednesday at 10 a.m., under direction of Hutchison & Merz of Marysville, with Rev. Fr. James **O'Meara** of Grass Valley officiating. Interment will be made in the family plot in the Smartville Catholic cemetery.

24 Oct 1938 (Monday); Page 2 Col 3; Mrs. B. Gassaway Died At Home Of Sister in Hub. Mrs. Bertha **Gassaway** passed away Sunday evening at the home of her sister, Mrs. T. J. **Tyrrell**, 607 Fourth Street, Marysville, after a lingering illness. Funeral services will be held in the Lipp & Sullivan chapel Tuesday at 2 p.m., to be followed by interment in the Smartville Cemetery. Rev. Raymond M. **Huston** will officiate. Mrs. Gassaway was a native of Smartsville, aged 73 years, and had been a resident of Yuba county throughout her life. She is survived by a daughter, Mrs. Ruth **Davey** of Alleghany, Sierra County; a son, Elmer **Colling**, of Marysville; her sister, Mrs. Tyrrell, and two brothers, George **Jeffery** of Chico and W. B. **Jeffery** of Smartville, and by several grandchildren and great grandchildren.

7 Feb 1940 (Wednesday); Page 3 Col 7; Mark E. **Poor**, well known resident of the Sicard Flat district of Yuba county, died in the Rideout hospital Wednesday morning. He had been a patient there about two weeks. **Poor** was visiting his injured daughter, Mrs. J. R. **Handy**, at her home near Browns Valley when he fell in the yard and broke his hip. His advanced age, 78 years, made the injury more serious, and he gradually weakened. Mrs. **Handy** had fallen in her home a few days before, injuring one limb so severely that she was unable to stand. Poor was a native of Mormon island, near Stockton. He and his family moved to Sicard Flat district in 1918, and Mrs. **Poor** died three years ago. Surviving children, in addition to Mrs. **Handy**, are Robert **Poor** of Marysville and Ed **Poor** of Browns Valley. Funeral services will be conducted in the colonial chapel of Hutchison & Merz Friday at 2 p.m., to be followed by interment in the family plot in the Smartville cemetery.

23 Oct 1940 (Wednesday); Page 1 Col 1; Brady F. **Compton**,, resident of Smartville, where he mined, died in a Richmond hospital Tuesday morning following a surgical operation. He had gone to Richmond, where a son resides, less than a week ago to enter the hospital. He was the son of the late **Charles** and Nellie **Compton**, Sicard Flat pioneers. Compton was a native of California, aged 67 years, and last resided in Yuba county for a long time. His survived by his wife, Mary Jane **Compton**, two sons, **Clarence** of Smartville and **Charles** of Richmond, and three daughters, Mrs. Sarah **Link** and Mrs. Dorothy **Taylor**, both of Santa Maria, and Mrs. Marian **Victour** of Pittsburg. There are two bothers, **Gale** of Browns Valley and **Fred** of Oakland and three sisters, Mrs. Ida **Randal**, Mrs. Harry **Randal** and Mrs. Marie **Morris**. Four grandchildren survive him. Mrs. Compton formerly Mary **Byrne**, was visiting with her daughters in Santa Maria

NEWSPAPER INFORMATION

when informed of her husbands death. Funeral services will be held in the Lipp & Sullivan chapel Thursday at 2 p.m., with Rev. Thomas **Tateman** of Nevada City officiating. Interment will be made in the Smartville Masonic cemetery.

7 Jan 1941 (Tuesday); Page 3 Col 2; Miss Mary E. **Mooney**, 80, native of Empire Ranch in Smartville district, where she spent her entire lifetime, died Monday night in Rideout hospital of complications resulting from a fall at her home about a week ago. She had been in failing health. Miss Mooney was a daughter of the late Mr. And Mrs. Tom **Mooney**, Yuba county, pioneer who settled in Smartville section in 1849, Mooney Flat being later named for them. Surviving are a sister, Miss Adelaide **Mooney**, and a brother Thomas **Mooney**, both of Empire Ranch. Funeral services will be held at 10 a.m. Thursday at Empire Ranch, with Rev. D. W. **Thompson** officiating. Interment will be made on Smartville cemetery under direction of Lipp & Sullivan.

26 Apr 1941 (Friday); Page 1 Col 7; Yuba county lost another of its old-time residents Thursday afternoon when Mrs. Fannie **Hapgood** 84, of Timbuctoo died in the home her father erected almost 100 years ago in the historical foothill mining center. Mrs. Hapgood, as native of Yuba county, lived within the immediate area of her birth her entire life, and was a recognized authority on the early history of Yuba county. Mrs. **Hapgood's** parents were among the first settlers, and it is from them she learned facts and background of the community. It is said that several historical writes based most of their writings on information obtained from the Timbuctoo woman. Surviving relatives include her husband, James **Hapgood**; a daughter, Mrs. Lavania **Kay** of Grass Valley; two brothers, George **Marple** of Timbuctoo and Sam **Marple** of Sacramento; a son James L. **Hapgood** of Sacramento; three grandchildren, Doris **Berkeley**, Sacramento; Carol **Hapgood**, Sacramento and James **Hapgood** of the U. S. Navy. Funeral services will be held at 2 p.m. Monday in the Lipp & Sullivan chapel with interment to follow in the Timbuctoo cemetery, a short distance from the Hapgood home.

2 Sep 1941 (Tuesday); Page 2 Col 7; Funeral services for John **Quinkert**, old resident of Smartville, who died in the Rideout hospital in Marysville early Monday morning, will be held in the Smartville Catholic church Wednesday at 10 a.m. and interment will be made in the Smartville cemetery, it was announced Tuesday morning. Rosary service will be held in the Lipp & Sullivan chapel in Marysville Tuesday at 8 p.m. **Quinkert** had been under treatment in the hospital the past five weeks. He was a native of Detroit, Michigan, aged 84 years, and had worked at carpentering during his active years. He had resided at Smartville half a century. His wife, **Clara**, preceded him in death four years ago. Surviving relatives are two sisters, Mrs. Annie **McGanney** and Mrs. Wilhelmina **Plamondon**, both of San Francisco, and the following nieces and nephews: Mrs. William **Burke** of Walnut Creek, **Dan** and James **McGanney** of San Francisco, Sister **Clara** of Notre Dame Convent, Mrs. George **Herboth**, Frank **Conrath** and L. J. **Conrath**.

18 Nov 1941 (Tuesday); Page 4 Col 8; Miss Johanna C. **Sheehan**, past 80 years of age, prominent in educational affairs of Yuba county some 39 years ago, and a

NEWSPAPER INFORMATION

leader in the women's organizations of the Catholic church for many years, died early Tuesday morning at the home of her cousin, Officer Dennis **McAuliffe** of the Marysville police department, and Mrs. **McAuliffe**. She had been ill a long time and bedfast the past two years. Miss **Sheehan** had taught in the Smartville, Dobbins and Brophy schools, and made her home at Smartville until 1921, when she came to Marysville. She was a native of Boston, Mass., but lived here since 15 months of age. She was a member of the Catholic Ladies Relief society, the Catholic Daughters of America and the National Council of Catholic Women. A brother, Daniel **Sheehan**, preceded her in death some years ago. Funeral services will be held in St. Joseph's Catholic church at 9 a.m. Thursday, followed by interment in the Smartville cemetery. Rosary service will be held in the Lipp & Sullivan chapel Wednesday at 8 p.m.

31 Jan 1942 (Saturday); Page 2 Col 5; William L. **Newbert**, 68, native of Timbuctoo and a lifetime Yuba county resident, died Friday night of the county detention home, where he had been acting as custodian. He had been in failing health for some time. Newbert and his wife, Mrs. Nellie Wright **Newbert**, came here from Smartville district about four months ago to take charge of management of the detention home. Surviving Newbert besides his wife are two sisters, Mrs. H. R. **Creps** of Wheatland and Mrs. A. B. **Byrne** of Sacramento, and two brothers, H. C. **Newbert** and P. A. **Newbert**, both of Sacramento. Funeral arrangements are in charge of Lipp & Sullivan and will be announced later.

5 Mar 1942 (Thursday); Page 6 Col 2; Mrs. Hood **Newbert**, the former Mary **McNally** of Smartville, who died in Rio Linda Tuesday, will be buried in the Smartville Catholic cemetery Friday. Funeral services will be held in the Catholic church in North Sacramento at 9 a.m., after which the cortege will leave for Smartville.. Mrs. **Newbert** leaves a son, Louis **Ferrier**, in San Francisco, in addition to her husband. She was a sister of Mrs. John **Poole**, and Mrs. Kate **Brown**, both of San Francisco.

6 Jul 1942 (Monday); Page 4 Col 1; SMARTVILLE, July 6. At the advanced age of 94 years, death Saturday evening removed Mrs. Mary **Murdock**, who had been residing at the home of a daughter, Mrs. A. B. **Sanford**. Mrs. **Murdock** was a native of Nova Scotia, but had resided in California for many years, coming here from Berkeley about three months ago. Funeral services will be held Tuesday at 2 p.m., and interment will be in the Smartville cemetery. Three surviving children of the decedent are Everett **Murdock** of Berkeley, Mrs. Mary **Kelly** of Boston and Mrs. **Sanford**. She also leaves five grandchildren and seven great-grandchildren. She was a sister of Mrs. Louis **Fenton** of Berkeley.

1 Aug 1942 (Saturday); Page 1 Col 7; August **Anderson**, owner of most of what remained of the pioneer town of Spencerville, died there of a heart attack yesterday. Funeral services will be held in the Lipp & Sullivan Chapel here next Tuesday at 2 p.m. **Anderson** was a native of Sweden, aged 77 years. He had spent most of his life at Spencerville, conducting the country store until changing conditions made it no longer profitable. He was the postmaster there for many years. The death of his

NEWSPAPER INFORMATION

wife not long ago left him without kin, excepting a half-brother, L. M. **Larson** of Berkeley, who was interested with him in the property, and a half sister, Mrs. **Austin** of Nevada City. The ghost town of Spencerville, once a prosperous copper mining camp and long important as a stock raising center and a stopping place for teams and stages, is expected to soon disappear, the Camp Beale reservation having included it.

5 Aug 1942 (Wednesday); Page 3 Col 4; Final rites for August **Anderson** were held in the Lipp & Sullivan Chapel Tuesday afternoon, with Rev. D. G. Decherd of the Methodist church officiating. Aaron Burt was the soloist and Miss Vashti Prentiss presided at the organ. Interment was in the Smartville cemetery, where friends acting as casket bearers were Les **Pesano**, Amos **Fitts**, Bud **Kneebone**, Henry **Byer**, Will **Barrie** and John J. **Murphy**.

6 Mar 1943 (Saturday); Page 2 Col 4; NEVADA CITY, Mar. 6. Mrs. Ella M. **Austin**, Nevada county school superintendent for 16 years, died Friday at her home here. Mrs. **Austin**, 85, was the mother of Bert C. **Austin**, San Francisco mining engineer. As a young woman she taught in country schools.

6 Mar 1943 (Saturday); Page 4 Col 2; DIED. Mrs. Ella M. **Austin**. In Nevada City, Mar 4, 1943, former county superintendent of schools, Nevada county. She was born in Pennsylvania, July 21, 1858, and resided in Nevada City for 20 years. Prior to that time she lived in Spencerville. For many years taught school in Sheridan, Pleasant Ridge and Mark well. Married in 1879 to John H. **Austin**, who passed away in 1934. She was the mother of Bert C. **Austin**, mining engineer of San Francisco and Mrs. Hazel **Walker** of New York City. Mother-in-law of Fred **Austin** of Alleghany, sister-in-law of Mrs. Ed **Hall** of Auburn. Mrs. **Austin** was a talented artist and held membership in the Eastern Star Chapter of Wheatland. Funeral service will be held from Trinity Episcopal church in Nevada City, Monday, March 8th at 9:30 a.m. with interment in Smartville cemetery, under direction of Holmes Funeral home.

12 Apr 1943 (Monday); Page 3 Col 7; Nathan B. **Sanford**, member of one of the oldest pioneer families of the Waldo district of Yuba county, now a part of Camp Beale, died Thursday night in Sacramento after an illness of some duration. He had lived there a long time. Funeral services will be held Monday at 11:30 a.m. at the Smartville cemetery, it was announced today by W. J. **Sanford** of Oakland, a brother who came up to make the arrangements. The decedent was a carpenter by trade and followed that calling for many years. He was a native of Waldo district, aged 85 years. It is said that on one occasion when he was employed to shingle the spire of a church steeple in Berkeley, 70 feet above the ground, he hung by his feet in order to reach portions of the job. **Sanford** was a son of **Benjamin** and Euphoria **Sanford**, who settled in Yuba county in 1854 and established one of the early fruit ranches of the district, the product being peddled to the mining camps by the boys of the family. Decedent was a brother of Mrs. Ida E. **Barnes** of Berkeley, A. B. **Sanford** of Smartville and W. J. **Sanford** of Oakland.

NEWSPAPER INFORMATION

26 May 1943 (Wednesday); Page 3 Col 6; Mrs. Ella H. **Jeffery**, well known resident of Smartville, died at her home Tuesday afternoon after a long illness. She was a native of the Erle district and had resided all her 66 years there and at Smartville. Her husband, William B. **Jeffery**, was for many years the village blacksmith, but has been retired for a number of years. Mrs. Jeffrey was a charter member of the Smartville chapter of Order of Eastern Star and had been the secretary the past 13 years. She was a sister of Joseph A. **Perkins** of Marysville, Mrs. Nancy **Whiteside** of Plumes district, Mrs. Hattie **Pittman** of Marigold, Mrs. Louisa **Jones** of Smartville and Mrs. Ida **Creps** of Marysville, Her parents were among the original settlers in the Erle district, locating there in the 1850's. Funeral services will be held in the colonial chapel of Hutchison & Merz Thursday at 2 p.m., with Rev. Carl N. **Tumbling**, rector of St. John's Episcopal Church, as the minister. The body will be laid to rest in the family plot in the Smartville cemetery.

18 May 1944 (Thursday); Page 3 Col 4; John T. ("Jack") **Meade**, age 65 and a native of Smartville, died last night at the home of his sister, Miss Mary **Meade**, 411 Seventh Street. He retired several months ago from farming in the Smartville district. He was the son of the late **Garrett** and Annie **Meade**, pioneers of this community and had lived in Yuba county his entire life. He was the brother of Mrs. Ann **O'Connor**, Miss Mary **Meade** and Leo **Meade** and the uncle of Loretta **O'Connor**, all of Marysville and Thomas C. **O'Connor** of the U. S. Navy. Recitation of the rosary will be at 8:30 Friday evening in the colonial chapel of Hutchison & Merz. Mass will be celebrated Saturday morning at 9:30 in St. Joseph's church. Interment will be in the family plot in the Smartville Catholic cemetery.

16 Nov 1944 (Thursday); Page 4 Col 7; Leo G. **Meade**, 61, a lifetime resident of Yuba county and a member of a pioneer family, died at his home in Marysville Wednesday afternoon after an illness of several months. He was a native of Smartville and is survived by two sisters, Miss Mary **Meade** and Mrs. Ann **O'Connor** of Marysville, and was the uncle of Mrs. Loretta **Matte** of Sacramento, Tommy **O'Connor**, U. S. Navy, Lieut. William **Meade**, U. S. Air Corp, Put, Helen **Meade**, WAC, and Kenneth **Dempsey** of El Cerrito. Recitation of the rosary will be in the Colonial Chapel of Hutchison & Merz Friday evening at 8:30. Mass will be held in St. Joseph's church Saturday morning at 9 o'clock. Interment will be in the Catholic cemetery in Smartsville.

1 Dec 1944 (Friday); Page 1 Col 5; Miss Helen M. **O'Brien**, descendent of a Yuba county pioneer family and former resident of Smartville, died this morning in San Francisco. Miss **O'Brien** was found dead in bed by her sister, Miss Agnes **O'Brien**. Miss **O'Brien** was painfully hurt while visiting here recently when she fell and injured he nose. She was a native of Smartville, the daughter of the late **James** and Mary **O'Brien**. Her father, a representative citizen of the Sacramento valley, was identified with mining interests of California for over half a century. Miss **O'Brien** is survived by brothers and sisters, William A. **O'Brien** of Marysville, James K. **O'Brien**, Miss Agnes **O'Brien** of San Francisco, Mrs. Mamie **Pierce** of New York and Sister **Gertrude** of the College of Holy Names of

NEWSPAPER INFORMATION

Oakland. Funeral services will be held at 10 a.m. Monday in the Catholic church at Smartville under the direction of Lipp & Sullivan.

28 Mar 1945 (Wednesday); Page 6 Col 3; Thomas James **Cann** who has resided in the Smartville district for the past 30 years, died at his home there this morning. **Cann** was born in England on January 24, 1886, and came to California 43 years ago. He is survived by his wife, Maude **Cann**, Smartville, and was father of the following; Phillip **Cann**, Vallejo; Charles **Cann**, Laytonville; James **Cann**, Vallejo; Bert **Cann**, Oakland; Mrs. W. **Orr**, Yuba City; Mrs. Dorothy **Schulze**, San Leandro; a sister, Mrs. Ada **Curry**, Grass Valley. Funeral services will be announced at a later date by the Jones Funeral Home.

29 Mar 1945 (Thursday); Page 2 Col 8; Funeral services for Thomas James **Cann**, long-time Smartville resident, who died Wednesday, will be held Saturday at 2 p.m. in the Jones funeral home at Yuba City, Rev. R. W. Lowry will officiate and the interment will be in the Smartville cemetery.

24 May 1945 (Thursday); Page 2 Col 6; James K. **O'Brien**, a native of Smartville and a leader in the early highway development of this section, died Wednesday in San Francisco, where he has made his home for the past two years. **O'Brien**, about 81, spent, with the exception of the last two years, his entire life in Yuba county where he was active in political and civic affairs. Several times during his life he made trips to Washington, D. C., in behalf of local highway projects. **O'Brien** was active in the organization which fostered the Tahoe-Ukiah highway and also the Garden highway, which was built between Sacramento and Yuba City via Nicolaus. At one time **O'Brien** was strongly urged by a number of Yuba and Sutter county democratic leaders to become a candidate for congress, a honor which he declined. He was a life-long member of the democratic party and was a leader in it locally for a number of years. The **O'Brien** ranch, where James K. **O'Brien** was born, was sold several years ago, and it was at this time that **O'Brien** and his sisters moved to San Francisco to reside. Two of **O'Brien's** sisters died recently, Mary **Pearce** died in New York in April, and Miss Helen **O'Brien**, San Francisco, died last November. **O'Brien** was a graduate of Santa Clara university and was a member of the Grass Valley council, Knights of Columbus. Surviving relatives are Sister **Mary** Gertrude of College of Holy Name in Oakland; Miss Agnes M. **O'Brien**, San Francisco, and a brother, William A. **O'Brien**, ar., of Marysville. A number of nephews and nieces also survive. Rosary will be recited at 8 p.m. Friday at Lipp & Sullivan's chapel with a funeral mass said at 10:30 a.m., Saturday in the Smartville Catholic church. Interment will be in the family plot in the Smartville Catholic cemetery.

9 Jul 1945 (Monday); Page 3 Col 4; Thomas G. **Magonigal** died early Sunday morning at his home near Lincoln. He formally lived in the Smartville district until his property was taken over by the government. He was a native of California, 80 years old, a retired stockman, and had lived in the Lincoln district for three and half years. He leaves a wife, **Hannah**, son **Thomas**, Jr., and daughter Mrs. Lester **Pisani** of Lincoln. The funeral will be held at 2 p.m. Tuesday at the family home

NEWSPAPER INFORMATION

near Lincoln. Interment will be in the Smartville cemetery. The arrangements are in charge of Lipp & Sullivan.

20 Aug 1945 (Monday); Page 3 Col 7; Mary **Cooney**, lifelong resident of Yuba county, a native of Smartville, died in a local hospital Sunday morning. Mrs. **Cooney** was the widow of the late William **Cooney**, well known Smartville ranch operator. Mrs. **Cooney** was 77 years old and had lived in the Smartville district her entire life with the exception of the past several years when she resided in Marysville. Funeral arrangements will be held at the Smartville Catholic church at 10 a.m. Wednesday. Rosary will be recited at 8 p.m. Tuesday at the Lipp & Sullivan chapel.

3 Dec 1945 (Monday); Page 1 Col 6; James H. **Fraser**, life long resident of Smartville district died Sunday afternoon in a Grass Valley hospital where he had been confined for many months. Fraser, a native of Nova Scotia, had lived at Mooney Flat, Nevada county, since he was a small child. Surviving relatives are a sister, Mrs. Emma **Magonigal** of Smartville and a brother, Edward **Fraser**, of Wheatland, and a number of nieces and nephews. Funeral services will be held from the Hooper funeral home in Grass Valley with burial in the Smartville cemetery. The time will be announced later.

22 Dec 1945 (Saturday); Page 4 Col 2; SMARTVILLE, Dec 22. George **Rigby**, who recently returned from the ETO where he served many months with an engineers battalion in England, France and Germany is visiting his parents, Mr. and Mrs. A. **Rigby**.

2 Jan 1946 (Wednesday); Page 2 Col 3; Mrs. Carrie R. **Johnson**, 80, a native of Quincy, Plumas county, died Monday in a Yuba City hospital after an extended illness. Her home was in Smartville where she had lived for 38 years. She was the wife of William F. **Johnson** of Smartville, the mother of Evelyn C. **Miles** and Clarence **Johnson** of Marysville, and the sister of Clarence **Weldon** of Quincy. Mass will be said in the Smartville Catholic church Thursday at 10 o'clock. Recitation of the Rosary will be at 8 o'clock tonight in the colonial chapel of Hutchison & Merz. Interment will be in the Smartville Catholic cemetery.

16 Mar 1946 (Saturday); Page 3 Col 6; Mrs. Caroline **Hermann**, 93, former resident of Smartville and Hammonton, died Thursday in Oakland where she had made her home the past seven years. A native of Germany, she came to this country when she was five years old. Survivors include two sons; **Charles** H. of Oakland, and **Frank** F. of Davis, and two daughters, Mrs. Pauline **Gilbert** of Hammonton, and Mrs. Louisa **Barry** of Davis. Graveside funeral services will be held at 11 a.m. Monday at the Odd Fellows cemetery in Smartville under the direction of Lipp & Sullivan.

26 Apr 1946 (Friday); Page 1 Col 7; Yuba county lost another of its old-time residents Thursday afternoon when Mrs. Fannie **Hapgood**, 84, of Timbuctoo died in the home her father erected almost 100 years ago in the historical foothill mining

NEWSPAPER INFORMATION

center. Mrs. **Hapgood**, a native of Yuba county, lived within the immediate area of her birth her entire life, and was a recognized authority on the early history of Yuba county. Mrs. **Hapgood's** parents were among the first settlers, and it is from them she learned facts and background of the community. It is said that several historical writers based most of their writings on information obtained from the Timbuctoo woman. Surviving relatives include her husband, James **Hapgood**; a daughter, Mrs. Lavania **Ray** of Grass Valley; two brothers, George **Marple** of Timbuctoo and Sam **Marple** of Sacramento; a son, James L. **Hapgood** of Sacramento; three grandchildren, Doris **Berkeley**, Sacramento; Carol **Hapgood**, Sacramento, and James **Hapgood** of the U. S. Navy. Funeral services will be held at 2 p.m. Monday in the Lipp & Sullivan chapel with interment to follow in the Timbuctoo cemetery, a short distance from the Hapgood home.

11 Jan 1947 (Saturday); Page 5 Col 5; SMARTVILLE, Jan. 11.- Joe **French**, a resident of the Smartville vicinity since 1902, was buried in the local catholic cemetery Friday. He died in a Marysville hospital, where he had been confined about a week. **French** was born in Detroit, Mich., 91 years ago and moved to Chicago at an early age. He at one time owned a restaurant there but sold it in the time of the Colorado gold rush to establish mining claims. Unsuccessful in mining, he packed his belongings on two burros and began a trip from Leadville, Colo., which lasted 33 years and took him many places and finally to Smartville, where he built a cabin on the Yuba River Bar. He read the government river gauge below the narrows. He lead a solitary life until about two years ago, when friends persuaded him to move nearer town because of his advanced years. Father James **O'Rielly** of Grass Valley conducted graveside services. Friends acted as pallbearers.

21 Jan 1947 (Tuesday); Page 1 Col 2; Death today claimed Mrs. Mary Josephine Ryan **Byrne**, 65, a native of Smartville and a lifelong resident of this Community. She is survived by her husband, Robert E. **Byrne**. She died at her home on D Street following a short illness. Mrs. **Byrne** was a graduate nurse of St. Mary's hospital and was a graduate of the convent of Notre Dame, the alumni of which was one of her constant activities. She also was very active in Catholic Ladies societies. Her parents were pioneers of the early west, settling in the Smartville area. Rosary will be recited at 8 p.m. Wednesday in the Lipp & Sullivan Chapel. Mass will be held at St. Joseph's Catholic Church in Marysville Thursday at 10 a.m. and interment will be in the family plot in the Smartville cemetery.

14 Jul 1948 (Wednesday); Page 3 Col 3; Frank D. **Poole**, a retired rancher of Smartville, died at 1:45 a.m. today in a Yuba City hospital. He was 80 years old. **Poole** is survived by his wife, Mrs. Daisie **Poole** of Smartville; two sons, **Clarence** of Yuba City and **Harry** of Roseville; a sister, Mrs. J. F. **Havey** of Richmond; five grandchildren and several nieces and nephews. Funeral services will be at 10 a.m. Friday from the Jarvis funeral home and interment will be in Smartville cemetery.

5 Aug 1949; Page 7 Col 1; SMARTVILLE, August 5 – Funeral services for the late Samuel **Marple**, who died Tuesday in Sacramento were held today in Hooper – Weaver chapel in Grass Valley. Interment was in Timbuctoo cemetery not far from

NEWSPAPER INFORMATION

the home where he was born 70 years ago. Surviving relatives are a brother, **George**, the last of a family of 16 children; a niece, Mrs. Charles **Ray** of Grass Valley; a nephew Lester **Hapgood** of Sacramento and a brother-in-law, James Hapgood of Timbuctoo.

17 Jan 1950 (Monday); Page 8 Col 5; Harriett Amelia **Call**, 85, died yesterday in Yuba City, where she had been living for the past month. Mrs. **Call** was born in Timbuctoo and had made her residence in Smartville where the late Salmon M. **Call** was engaged in farming in the Camp Beale area. She is survived by one son, George C. **Call** of Smartville. Funeral services will be held in Hutchison's Colonial chapel at 2 p.m. Thursday with Rev. Bernard W. Lowry or the First Methodist church of Yuba City officiating. Interment will be in Smartville cemetery

6 July 1950 (Thursday); Page 12 Col 4; OBITUARIES.– Mrs. Chloe Mae **Parkison**, 62, a native of Iowa and a resident of Smartville for the past 10 years, died last night in a Marysville hospital. Survivors include her husband, **John**; four daughters, Mrs. Helen **Laughlin** of Smartville, Mrs. Grace **Stering** of Meridian, Mrs. Beulah **Cottrell** of Sacramento and Mrs. Bessie **Robinson** of Washington; four sons, **Bert** of Washington, **John** of Auburn, **Robert** of Smartville and **Bernard** of Montana; three sisters, Mrs. Grace **Van Winkle**, Mrs. Jessie **Beghtel** and Mrs. Martha **Roberts**, and one brother Herman **Parrish**, all of Iowa; 19 grandchildren and two great-grandchildren. Funeral services will be conducted at 2 p.m. tomorrow in Hutchison's Colonial chapel. Interment will be in Smartville cemetery.

18 Oct 1950 (Wednesday); Page 10 Col 8; George Washington **Marple**, 79, a native of Yuba county, died in his home in the Golden West hotel, Marysville, yesterday. He formerly worked for the Southern Pacific Company in their Sacramento shops but had recently made his home in Smartville and Marysville. Marple is survived by a niece, Mrs. Lavinia E. **Ray** of Grass Valley, and a nephew, James L. **Hapgood** of Sacramento, J. L. **Hapgood**, a brother-in-law of Smartsville, also survives. Marple was affiliated with the Moose Lodge of Chico. Funeral arrangements will be announced later by Lipp & Sullivan chapel, Marysville.

27 Nov 1950 (Monday); Page 10 Col 7; OBITUARIES. A Smartville resident, Mrs. Alice E. **Whitmore**, 41, died Saturday night in a local hospital. She was a native of Iowa and had lived in Yuba county four years. Survivors include her husband, Jess F. **Whitmore**; two sons, **Elvin** Lee and Donald **Whitmore**, both of Smartville; a daughter, Virginia **Whitmore**, Smartville; a brother Robert **Montgomery**, Nevada; a sister, Mrs. Virginia **Brott**, Petaluma, and her parents, Mr. and Mrs. Horace **Montgomery** of Los Angeles. Services are pending at Lipp & Sullivan chapel.

25 Jan 1951 (Thursday); Page 7 Col 7; James Alexander **McClure**, 71, was found about 4 p.m. yesterday alongside Hwy. 20, between Smartville and Timbuctoo, Yuba County Sheriff John R. Dower reported today. Cecil Stanley, 1060 Packard

NEWSPAPER INFORMATION

Ave., Linda, who was driving along the highway, found the body, the sheriff said. According to Yuba County Coroner Warren Hutchings, **McClure** apparently died of natural causes. An autopsy will be performed however, to determine the exact cause of death. **McClure** apparently was walking along the highway when he was stricken, Hutchings said. The dead man was a native of Ireland, but had lived in Smartville since 1900. He was a retired miner. Surviving are his brother, Joseph Patrick **McClure** of Marysville, and a sister in Boston, Mass. Funeral services are pending at Hutchison's Colonial Chapel.

29 Jan 1951 (Monday); Page 9 Col 6; Rosary will be recited at 8 p.m. Sunday in Hutchison's Colonial Chapel for James Alexander **McClure**. The mass will be said at 10 a.m. Wednesday in the Smartville Catholic Church and burial will be in the Smartville Catholic Cemetery.

12 Feb 1951 (Monday); Page 8 Col 4; Matthew Henry **McCarty**, 82, a lifelong resident of Timbuctoo, died in his home there Friday evening. He was a native of the one-time thriving mining community. He was born on New Year's Day, 1869. For many years he had been employed by the Marigold Dredging Co., now known as the Yuba Consolidated Gold Fields, Inc. Surviving are two sisters, Mrs. G. A. **Forges** of Del Paso Heights and Mrs. H. P. **Calligan** of Marysville; a brother, Robert E. **McCarty** of Marysville; three nieces, Mrs. Alice **Herboth** and Mrs. Ruth **Michel** of Marysville and Mrs. Marguerite **Hofer** of Berkeley; and three nephews, **Andrew**, **Clarence** and George **Galligan**, all of Marysville. Rosary will be recited in Hutchison's Colonial Chapel, Marysville at 9 p.m. today, and mass will be celebrated at 10 a.m. tomorrow in the Smartville Catholic Church. Interment will be in the Smartville Catholic Cemetery.

3 Mar 1951 (Saturday); Page 8 Col 4; Funeral services for Charles **Martien**, who died in an Auburn Hospital Tuesday after an extended illness, were conducted Thursday. Mass was celebrated at 10:30 o'clock in the Church of the Immaculate Conception, Smartville, by Father James Sullivan of Grass Valley. Interment was in the Catholic Cemetery in Smartville. **Martien**, a native of Pennsylvania, was a resident of Marysville for many years. He is survived by his wife, Elizabeth Daugherty **Martien** of Smartville.

4 Apr 1951 (Wednesday); Page 9 Col 1; Funeral services for Nora **Bushby** who died Friday in Auburn, were held at 1 p.m. Tuesday in the Lipp & Sullivan Chapel. Rev. Carl Tamblyn officiated and Mrs. Raymond Huston was the soloist with Lowell McDaniel playing the accompaniment. Marvin **Pitmar**, Earl **Bushby**, George **Magonigal**, Frank **Countryman**, Tom **McDivitt** and Les **Passni** acted as pallbearers for the interment which was in the Smartville Cemetery.

9 Apr 1951 (Monday); Page 6 Col 2; Ida **Rigby** died yesterday in her home at Smartville after a short illness. She was born in Sucker Flat May 16, 1871, the daughter of the late Mr. and Mrs. Daniel **Daugherty** and lived her entire life in this area. Her husband was the late Albert **Rigby**. Surviving her are two sons, **Joseph** and George **Rigby** and a sister, Mrs. Lizzie **Martein**, all of Smartville. Funeral

NEWSPAPER INFORMATION

services will be at 10 a.m. Wednesday when mass will be celebrated in the Church of the Immaculate Conception at Smartville with interment in the family plot in the Catholic Cemetery. Hooper – Weaver Mortuary of Grass Valley will be in charge.

6 Dec 1951 (Thursday); Page 11 Col 6; Following a long illness, Joseph **McClure**, 65, died yesterday in a local hospital. A native of Massachusetts, **McClure** had lived in this community for 17 years and formerly had been in Smartville. He has no known relatives. Lipp & Sullivan will be in charge of Graveside services in the Catholic cemetery at Smartville at 10 a.m. Saturday.

8 Sep 1952 (Monday); Page 7 Col 4; William F. **Hacker**, 71, longtime resident of Wheatland died yesterday in the Sutter Hospital, Sacramento. Hacker was a native of California and resided in Loomis, Placer county for the past 10 years. Surviving are his wife, **Cora**, and the following sons and daughters: Marele E. **Flowers**, Del Paso heights; Arlene M. **Welsh**, Sacramento; Edna B. **Peske**, Loomis; Francis W. Hacker, Del Paso heights; and Harvey J. **Hacker**, Loomis. In addition, there are seven grandchildren, and seven great grandchildren surviving. Funeral services will be held Wednesday, September 10, at 2 p.m. in the Lipp & Sullivan chapel, Marysville. The service will be under the auspices of Roses Bar Lodge No. 89, F. & A.M. Interment will be in the Smartville Masonic cemetery.

17 Oct 1952 (Friday); Page 7 Col 1; Mrs. Alvina **Gunning**, 87, widow of S. O. **Gunning**, former auditor and recorder of Yuba County, died at her San Francisco home yesterday. Mrs. **Gunning** is survived by three sons, **Samuel** and Thomas **Gunning** of San Francisco and Robert **Gunning** of Bakersfield, and three daughters, Mrs. Alvina **Geraghty** of San Francisco, Mrs. Ella **Young** of Daly City and Mrs. Margaret **McCay** of Bakersfield. **Gunning** was Yuba county auditor and recorder from 1894 to 1914. Funeral services will be conducted tomorrow morning in Marysville.

22 Jan 1953 (Thursday); Page 22 Col 2; OBITUARIES. Mrs. Emily Murdock **Sanford** longtime Yuba county resident, died unexpectedly at her home in Smartville Wednesday evening. A native of Nova Scotia, she would have been 83 years old on Tuesday, Jan 17. She was married to Alfred Bruce **Sanford** on May 10, 1901, and had made her home in Smartville since that time. The couple celebrated their golden wedding anniversary in 1951 when the community honored them. In addition to her husband, a sister, Mrs. George **Kelly** of Lynn, Mass. And a brother, Everett **Murdock** of Berkeley, survive her. Funeral arrangements are being made under direction of Hooper-Weaver Mortuary of Grass Valley and will be announced later.

26 Jan 1953 (Monday); Page 4 Col 4; Funeral services will be held Tuesday at 1:30 p.m. in Lipp & Sullivan Chapel, Marysville, for James Chester **Welch**, 66, a retired farmer who died at his Pleasanton Rd. home near Lincoln, Saturday afternoon. **Welch** was a veteran of the first World War and had lived in Lincoln for 12 years up to the time of his death. He is survived by two sisters, Mrs. A. A. **Whiteside** of Yuba City and Mrs. Chris **Butler** of Grass Valley; four brother's, I. L. **Welch** of

NEWSPAPER INFORMATION

Lincoln, Nevens **Welch** of Lincoln, Ray **Welch** of Browns Valley and Willard **Welch** of Sacramento. Interment will be in Smartville Cemetery.

5 Feb 1953 (Thursday); Page 13 Col 1; Funeral services are pending in Lipp & Sullivan Chapel for Mrs. Georgie Elizabeth **McGanney**, 73, a native of Marysville, who died yesterday in a local hospital following a long illness. Mrs. **McGanney** had resided at 712 F St. for the past nine years and lived in Salt Lake City for 35 years prior to her return here. She is survived by a daughter, Mrs. Noreen **Facer** of Marysville; a son Frank J. **McGanney** of Tacoma, Wash.; two sisters, Mrs. Carrie G. **Glyce** of Mill Valley and Mrs. Irene **Watkins** of Glendale; one grandson and one granddaughter.

4 Apr 1953 (Saturday); Page 11 Col 5; Horace **Newbert**, a native of Timbuctoo, died yesterday at his home in Rio Linda following an illness of two weeks duration. He was a former dredgerman. Surviving are a son, **Louis**, a brother, **Tom**, and two grandchildren, all of Sacramento, and a number of nieces and nephews. Interment will be Tuesday in Smartville Catholic Cemetery at a time to be announced later.

14 Jul 1953 (Tuesday); Page 10 Col 5; Adelaide **Mooney**, 88 year old life-long resident of the Empire ranch near Smartville, died last night in a local hospital following a long illness. Miss Mooney was the daughter of Mr. & Mrs. Thomas **Mooney**, early settlers in this area. Funeral services will be conducted at Lipp & Sullivan chapel Thursday at 10 a.m. Interment will be in the Smartsville cemetery.

15 Jul 1953 (Wednesday); Page 8 Col 2; William Bernard **Jeffery**, 84 year old miner-blacksmith despondent over ill health, committed suicide by shooting himself in the chest with a shotgun in his Smartville home yesterday. Yuba county sheriffs' deputies reported today. Jeffrey was discovered at 4:40 p.m. by a neighbor, Mrs. Frances Rigby, sprawled in the kitchen doorway of the house where he lived alone. A 12-gauge shotgun lay near his feet. Both barrels had been discharged and deputies theorized the triggers had been released by the victim with the end of a kitchen poker. An unsigned note lay on a nearby table. I complained of ill health "and I think this is the best way out." Jeffrey was born in Smartville and had resided in that area all his life. In his early days he was a miner and later a blacksmith. He was the son of the late **George** and Maria **Jeffery**, pioneer settlers of that area. He is survived by nephews, Elmer **Colling** of Smartville, **George** and Justin **Jeffery** of Chico; three nieces, Mrs. Ruth **Davey** of Alleghany, Mrs. Veca **Bacus** of Chico. Funeral services will be conducted at 2 p.m. Friday in Hitchinson's Colonial chapel. Interment will be in the Smartville Masonic cemetery.

31 Jul 1953 (Friday); Page 7 Col 3; Robert Emmett **McCarty**, 72, a native and lifelong resident of Yuba County, died today in his home at 1306 H St., Marysville. **McCarty**, a retired dredgerman for the Yuba Consolidated Gold Fields, was born at Timbuctoo. Survivors include his wife, **Ella P.**; two sisters, Mrs. H. P. **Galligan** of Marysville and Mrs. Rose **Forbes** of Del Paso Heights; and two stepchildren,

NEWSPAPER INFORMATION

Leroy **Hay** of Sacramento and Ruth H. **Puttman** of San Francisco. Funeral services will be announced later by Lipp & Sullivan chapel, Marysville.

1 Aug 1953 (Saturday); Page 12 Col 2; Rosary for Robert E. **McCarty** will be recited at 8 p.m. Sunday in Lipp & Sullivan Chapel. Funeral mass will be said at 9 a.m. Monday in St. Joseph's Catholic Church, Marysville and interment will be in the Catholic Cemetery at Smartville.

26 Aug 1953 (Wednesday); Page 16 Col 7; Requiem mass will be celebrated in St. Joseph's Catholic Church in Marysville at 9 a.m. Friday for Peter F. **Linehan**, 89, Smartville, who was found dead near his Timbuctoo Rd. home yesterday. Rosary will be recited in Hutchison's Colonial Chapel, Marysville, at 8 p.m. tomorrow. Interment will be in the Smartville Catholic Cemetery. **Linehan** was born in San Francisco but lived the greater part of his life in the Smartville area. He had made his home with his sister, Mrs. Mary A. **McGovern**. He also is survived by Miss Katherine **Linehan**, a sister, Miss Helen **McGovern**, a niece, and Jack **McGovern**, a nephew, all of Smartville.

9 Apr 1954 (Friday); Page 18 Col 2; Robert E. **Byrne** of Marysville, 75 year old dredgerman, died last night in an Oroville hospital following a long illness. **Byrne** was born in Smartville and lived most of his life in this area. He is survived by a brother, **Henry** of Richmond, and a sister, Mrs. B. E. **Compton** of Santa Maria. Funeral services are pending at Lipp & Sullivan Chapel, Marysville.
10 Apr 1954 (Saturday); Page 12 Col 4; Rosary will be recited Sunday at 8 p.m. in Lipp & Sullivan chapel for Robert E. **Byrne**. Requiem mass will be celebrated at 10 a.m. Monday in St. Joseph's Church. Interment will follow in the Smartville Catholic Cemetery;.

17 Sep 1954 (Friday); Page 8 Col 2; Mrs. Grace **Beatty**, a former Smartville resident, died unexpectedly yesterday in Livermore. She was the daughter of the late Mr. and Mrs. C. **Manasco** of Mooney Flat, Nevada County. She attended Smartville and Mooney Flat schools and was a registered nurse. She made her home in Smartville for many years, leaving there in 1914 for Livermore. Besides her husband, Richard Ray **Beatty**, she is survived by three daughters, Mrs. Frances **Henke** and Bobbie **Beatty** of Livermore and Margaret **Beatty** of Los Angeles; a grandson, Freddy **Lienke** of Livermore, and a sister, Mrs. Ida **Hite** of Yuba City. Rosary will be recited Sunday evening at 8 o'clock in Lipp & Sullivan Chapel, Marysville and Mass will be celebrated at 10 a.m. Monday in the Church of the Immaculate Conception, Smartville, followed by interment in the Fraternal Cemetery.

15 Dec 1954 (Wednesday); Page 16 Col 1; Rosary will be recited in Lipp & Sullivan Chapel at 8 p.m. tomorrow for Miss Agnes M. **O'Brien** who died Monday in San Francisco. Requiem mass will be celebrated at 10:30 a.m. Friday in the Immaculate Conception Church in Smartville. A native of Smartville, Miss **O'Brien** was the daughter of the late **James** and Mary **O'Brien**. She was a sister of the late **James** K., **William**, **Josephine**, **Helen** and Isabel **O'Brien**, Mrs.

NEWSPAPER INFORMATION

Kathleen **Holbrook** and Mrs. Richard **Pierce**. Interment will be in the Smartville Catholic Cemetery.

3 Jan 1955 (Monday); Page 14 Col 7; Committal services for Hugh Thomas **Dykes** of Berkeley were held at 2 p.m. today in Smartville Fraternal Cemetery. **Dykes**, a Berkeley real estate and insurance broker, who died after a lengthy illness, was well known in Smartville, having visited often at the former home of his wife, who was Grace **Black**. His mother and a brother also survive him. Memorial services were held yesterday in Berkeley. Hooper – Weaver Mortuary of Grass Valley made the local arrangements.

22 Apr 1955 (Friday); Page 4 Col 6; Rosary will be recited for Alois **Hug** (), who died April 19, in Lipp & Sullivan Chapel Sunday at 8 p.m. Requiem mass will be celebrated Monday in the Smartville Catholic Church at 10:00 a.m. Interment will be in Smartville Catholic Cemetery.

12 May 1955 (Thursday); Page 19 Col 3 & 4; William **Bushby**, 89, a native of Yuba County, died early today in an Auburn hospital. He was a retired farmer and lived at 510 Fourth St., Marysville. **Bushby** was a member of Foresters of America. He is survived by his wife, **Fannie**; a son, Earl of Yuba City; a daughter, Blanche **Foreman**, Marysville; two sisters, Hannah **Magonigal** and Sarah **Bushby**, both of Loma Rica; three grandchildren, and four great grandchildren. Funeral services are pending at Lipp & Sullivan Chapel.

13 May 1955 (Friday); Page 6 Col 2; Funeral services for William **Bushby** will be conducted at 2 p.m. Saturday in Lipp & Sullivan Chapel, Marysville, with Rev. Lawrence Wells, pastor of the Marysville First Christian Church, officiating. Interment will be in Smartville Cemetery.

1 Aug 1955 (Monday); Page 6 Col 7; Clarence Albert **Compton**, 56, died yesterday afternoon in Marysville Hospital following a illness of two years. He was a native of Smartville and had lived there all of his life. He had been employed by the P.G. & E. Co. for many years. **Compton** is survived by two sons, **William** of Marysville and **Farrell**, now serving in the USAF and stationed in North Africa; his mother, Mrs. Mary Jane **Compton** of Santa Maria; one brother, **Charles** of Richmond; three sisters, Mrs. Marian **Vitor** of Concord, Mrs. Dorothy **Taylor** and Mrs. Sally **Sink**, both of Santa Maria; two grandsons also survive. Funeral services are pending at Hutchison's Colonial Chapel.

2 Aug 1955 (Tuesday); Page 14 Col 2; Requiem mass for Clarence Albert **Compton** will be celebrated tomorrow at 10 a.m. in the Smartville Catholic Church. Interment will follow in the Smartville Catholic Cemetery. Rosary will be recited Tuesday evening at 8 p.m. in Hutchison's Colonial Chapel. **Compton** was 56 and a life long resident of Smartville. He had been an employee of the P.G. & E. Co. for many years.

NEWSPAPER INFORMATION

8 Dec 1955 (Thursday); Page 14 Col 4; Miss Sarah **Bushby**, 81, native of Yuba County and lifelong resident in this area, died this morning in her Loma Rica home. She had long been ill. Miss **Bushby** is survived by a sister, Mrs. Hannah **Magonigal** of Loma Rica. Rosary will be recited in the Lipp & Sullivan Chapel, Marysville, at 8 p.m. tomorrow. Requiem mass will be solemnized at 10 a.m. Saturday in St. Joseph's Catholic Church. Interment will be in the Smartville cemetery.

17 Nov 1956 (Saturday); Page 5 Col 2; Funeral services will be conducted Monday at 2 p.m. in Hooper – Weaver Mortuary, Grass Valley, for Alfred Bruce **Sanford**, who died Friday in an Auburn hospital. Born in Smartville, Nov. 29, 1869, he was the son of **Benjamin** and Euphemia **Sanford**, who located there in the early 1850's. He is the last member of a large family. He was married May 11, 1901 to Emily **Murdock**, who died several years ago. Surviving relatives include a number of nieces and nephews, several of whom he raised after the death of their mother. Interment will be in the family plot in Smartville Fraternal Cemetery.

9 Apr 1957 (Tuesday); Page 4 Col 3; Mrs. Mary **McGovern**, 81, of Smartville died this morning while visiting friends in San Francisco. Born in Smartville, she had lived there most of her life. Before her marriage, she had been to nurse's training in San Francisco and had worked there as a nurse for several years. She was a member of the Catholic Daughters of America, Marysville Chapter. Her husband, Thomas **McGovern**, drowned near Hammonton more than 30 years ago. She is survived by a daughter, Helen **McGovern**, and a son, Jack **McGovern**, both of Smartville. Funeral arrangements are pending at Lipp & Sullivan Chapel in Marysville.

11 Apr 1957 (Thursday); Page 18 Col 5; Rosary will be recited tonight at 8 o'clock in Hutchison's Colonial Chapel in Marysville for Mrs. Mary **McGovern**, 82 of Smartville who died Tuesday while visiting in San Francisco. Mass is planned for tomorrow morning at 10 o'clock in the Smartville Catholic Church. Burial will follow in the Smartville Cemetery.

15 Apr 1957 (Monday); Page 9 Col 1; Funeral services were held this afternoon at the Smartville Fraternal Cemetery for Edwin (Edward) **Davis** of Oakland, who died there Friday after a long illness. Survivors include his wife, the former Mae **Black** who lived with her grandparents, the Benjamin **Sanford's**, as a girl, and a daughter and a son, all of the East Bay area. **Davis**, who formerly operated a garage in Oakland, was a frequent visitor here. Interment was in the family plot.

23 Apr 1957 (Tuesday); Page 8 Col 2; Mrs. Mary Byrne **Compton**, 80, a native of Smartville, died Monday in Santa Maria following a long illness. She is survived by three daughters, Mrs. Dorothy **Taylor** and Mrs. Sarah **Sink** of Santa Maria and Mrs. Marian **Vetour** of Richmond; a brother, Henry **Byrne** of Live Oak; and four grandchildren. Her husband Brady **Compton**, and two sons, **Clarence** and **Charles**, predeceased her. Interment will take place in Smartville at a time to be announced later.

NEWSPAPER INFORMATION

24 Apr 1957 (Wednesday); Page 4 Col 4; Rosary will be recited tomorrow night at 8 o'clock in the Chapel of the Twin Cities for Mary B. **Compton**, 80, a native of Smartville who died Monday in Santa Maria following a long illness. Requiem mass is planned for 10 a.m. Friday in the Smartville Catholic Church. Burial will follow in the Smartville Cemetery.

14 May 1957 (Tuesday); Page 6 Col 5; Mrs. Mary Elizabeth **Galligan**, 1019 F St., Marysville, who would have celebrated her 93rd birthday anniversary today, died yesterday. Mrs. **Galligan** was a lifelong resident of Yuba County, having been born May 14, 1864 in Timbuctoo, the early-day mining town. She was one of seven children born to **Andrew** and Susan (Flanigan) **McCarty**, both natives of Ireland who came to the United States in the Gold Rush days. **McCarty** settled in Rose Bar in 1853 and his wife joined him in California in 1960, after making the journey by way of the Isthmus of Panama. Mrs. **Galligan** was the widow of Hugh P. **Galligan**, of Marysville, who came to this state from Iowa in 1864 by the overland route. Two of her sons who preceded her in death, **Andrew** L. and Clarence F. **Galligan**, also were born in Timbuctoo. A daughter, Mrs. Alice **Herboth** also, pre-deceased her. She is survived by one son, **George**, of Marysville; a daughter, Mrs. Ruth **Michel**, who had been residing with her in the old family home, Marysville; seven grandchildren, and 10 great grandchildren, several of whom reside in this area. In the past few years, Mrs. **Galligan** had become enfeebled by failing health. She was a member of the Altar Society of St. Joseph's Catholic Church, Marysville. Rosary will be recited at 8 p.m. tomorrow in the Lipp & Sullivan Funeral Chapel. Requiem mass will be at 10 a.m. Thursday in St. Joseph's Church, and interment will be in the Smartville Catholic Cemetery.

16 Nov 1957 (Saturday); Page 5 Col 3; Mrs. Sophia **Peckham**, 96, native and longtime resident of the Waldo area north of Wheatland, died early today in Orangevale. She had moved to Orangevale in 1942, when the family home became part of Beale Air Force Base. She made her home there with a daughter, Mrs. Ernest **Pittman**. Two other daughters, Mrs. Bertha **Olson** and Mrs. Mable **Anderson**, live in Wheatland. Funeral arrangements are pending at Lipp and Sullivan Chapel.

18 Nov 1957 (Monday); Page 14 Col 3; Funeral services will be conducted tomorrow in Marysville for Mrs. Sophronia **Peckham**, 96, native and longtime resident of Yuba-Sutter area, who died Saturday in Orangevale. Born in the Nicolaus area, she was the daughter of Mr. and Mrs. J. L. **Wallis**, who came to California from Illinois in a covered wagon. She was married at the age of 20 to T. W. **Peckham**, a stock man, and she made her home for 51 years on a ranch near Smartville, where the couple raised sheep and cattle, on more than 1500 acres. **Peckham** died in 1935. In 1943, when the ranch became part of Beale Air Force Base, Mrs. **Peckham** moved to the home of her daughter, Mrs. Ernest L. **Pittman**, in Orangevale. She had been in failing health since she fractured her hip about three years ago. She was a charter member of Smartville Chapter, Order of Eastern Star, which was formed in 1924. In addition to Mrs. **Pittman**, she is survived by five other daughters, Mrs. Mable **Anderson** and Mrs. Bertha **Olson**, both of

NEWSPAPER INFORMATION

Wheatland, Mrs. Cora **Hacker** of Loomis, Mrs. Pearl **McCormick** of San Francisco and Mrs. Dean **Woodworth** of Yucipa; a son, W. T. **Peckham** of Rio Oso; 12 grandchildren, 20 great grandchildren and 10 great great-grandchildren. Services will be conducted tomorrow morning a 11 o'clock in Lipp & Sullivan Chapel,l with Rev. Victor Hatfield, rector of St. John's Episcopal Church in Marysville, officiating. Burial will be in the Smartville Masonic Cemetery.

16 Dec 1957 (Monday); Page 4 Col 2; Theresa Jane **Niles**, 74, died last night in Rideout Memorial Hospital following a short illness. She was a native of Dublin, Ireland and came to San Francisco at the age of six months. She lived there until five years ago when she moved to Smartville. She and her husband, **George** celebrated their golden wedding anniversary six years ago. Mrs. **Niles** was a member of the Third Order of St. Francis. Besides her husband, she is survived by four daughters, Mrs. Margaret **Juardo** and Mrs. Mary Ellen **Rouch** of San Francisco, Mrs. Josephine **Allen** of Smartville and Mrs. Frances **Yarboraugh** of Grass Valley; two sons, **William**, now in the Army and stationed overseas, and Anthony of Smartville. Funeral services are pending at Hutchison's Colonial Chapel in Marysville.

17 Dec 1957 (Tuesday); Page 15 Col 4; Rosary will be recited tomorrow night at 8 o'clock in Hutchison's Colonial Chapel for Mrs. Theresa Jane **Niles**, 74, of Smartville, who died Sunday night in Rideout Memorial Hospital. Mass is planned for Thursday morning at 10 o'clock in Smartville Catholic Church. Interment will be in Smartville Catholic Cemetery.

3 Mar 1958 (Monday); Page 10 Col 5; A requiem mass will be celebrated at 11 o'clock tomorrow in Church of the Immaculate Conception, Smartville for Ellen Callaghan **Gaffney**, 90, a former resident. The daughter of Mr. and Mrs. Patrick **Callaghan**, she was born in New Jersey and raised in Mooney Flat but had been a resident of San Francisco for many years. Surviving her are a sister, Mrs. Annie **O'Brien** of Marysville; four nephews and two nieces. Interment will be in the family plot in Smartville Catholic Cemetery under the direction of a Bay area mortuary.

4 Oct 1958 (Saturday); Page 9 Col 1; Funeral services for Jesse Franklin **Whitmore**, 67, of Smartville, who died Thursday in a local hospital will be held at 9 a.m. Monday in Lipp & Sullivan Chapel, Marysville. Interment will be in the Smartville Masonic Cemetery. **Whitmore**, who was a native of Texas and had lived in this community for 20 years, was an employee at the McClellan Air Force Base, Sacramento. He was a widower, and is survived by two sons, Elvin **Whitmore** of Yuba City and Donald **Whitmore** of Marysville, and a daughter, Virginia **Whitmore** of Marysville.

12 Nov 1958 (Wednesday); Page 4 Col 3; Funeral services for Charles Vance **Ray** of Grass Valley were conducted today by Rose's Bar Lodge No. 89, Free and Accepted Masons at 2 o'clock in Hooper – Weaver Mortuary, Grass Valley. **Ray**, 69, was a long-time resident of Yuba and Nevada Counties. He was a former

NEWSPAPER INFORMATION

superintendent of Bonanza Ranch, when it was operated by Interstate Land Holding Company and later owned a service station on Auburn St. in Grass Valley, for many years before ill health caused his retirement. While in poor health for several years, he only entered the hospital Sunday morning and died that evening. He is survived by his wife, **Lavina**. Interment was in the Fraternal Cemetery at Smartville.

19 Jan 1959 (Monday); Page 7 Col 5; Funeral services will be conducted tomorrow in Marysville for Mrs. Bertha E. **Pisani**, 55, of Loma Rica. She died Saturday afternoon, collapsing suddenly at her home on Loop Rd., in the Loma Rica area of Yuba County. Death was reported due to a heart attack. She was a native of Smartville. Survivors include her husband, **Leslie**, a daughter, **Lorraine**, and her mother, Mrs. Hannah **Magonigal**, all of Loma Rica; and a brother, George **Magonigal** of Lincoln. Services will be conducted tomorrow at 2 p.m. at Lipp & Sullivan Chapel, with Rev. Floyd W. Thomas of the Bethel Temple, Assembly of God, officiating. Burial will be in the Smartville Cemetery.

6 Apr 1959 (Monday); Page 10 Col 1; James Mortimer **Hapgood**, 87, Timbuctoo's last pioneer resident, died Saturday, in Jones Memorial Hospital, Grass Valley, where he was hospitalized briefly after being ill at the home of his daughter, Mrs. Charles **Ray**, for the past few months. He was one of the few persons who could recall Timbuctoo's busier days. He was born March 28, 1872 at Parks Bar. He moved as a youth to Timbuctoo and spent his entire life there. He was a carpenter. He was a past master of Rose Bar Lodge No. 89 F & AM and had held an office most of the years since his affiliation in 1916. His wife, Fannie Marple **Hapgood**, also a native of Timbuctoo, and a daughter of a '49er, died several years ago. Surviving besides his daughter are a son, **James** Lester of Sacramento; a sister, Mrs. Tessie **McGinnis** of Oakland; three grandchildren and six great grandchildren. Masonic services will be held tomorrow at 2 o'clock in Hooper - Weaver Mortuary, Grass Valley, with interment in Timbuctoo Cemetery.

13 Jun 1959 (Saturday); Page 3 Col 2; Funeral services will be conducted Monday in Marysville for Leslie Pomp **Pisani**, 55, who died yesterday at his home on Loop Rd., Loma Rica, following a long illness. A native of Nevada City, he had been a lifetime resident of the Yuba-Nevada County area. He was a farmer. His wife, **Bertha**, died unexpectedly of a heart attack in January. Survivors include a daughter, **Lorraine** of Loma Rica; two brothers, **George** of Nevada City and **Joseph** of Stockton; and two sisters, Mrs. Mary **Kneebone** of Newcastle and Mrs. Amelia **Fisk** of Fallon, Nev. Funeral services will be conducted Monday at 2 p.m. in Lipp & Sullivan Chapel, with Rev. Floyd W. Thomas of Bethel Temple, Assemblies of God Church, officiating. Burial will be in the Smartville Masonic Cemetery.

23 Jan 1960 (Saturday); The Appeal-Democrat Centennial Edition; Page F-14; Col 4; SMARTVILLE NAMED FOR 1856 MAN. The "Gem of the Foothills" was organized in 1856 by William **Smart** who built a hotel; and for 53 years the town, one of the important centers in the pioneer era of Yuba County, was known as

NEWSPAPER INFORMATION

Smartsville. Then someone in the Post Office Department in Washington, D.C. decided to eliminate the middle "s" and by a edict dated April 7, 1909, the place officially became Smartville. Many an old-timer refused to accept the change, and maps and legal records still show the original spelling of the town's name. In fact, a modern-day attempt has been made to resume it. Miners In The Hills. Smartville was a close neighbor to Sucker Flat (first called Gatesville) and Timbuctoo, both started in 1850. Located in the foothills these towns were not far above the original Rose's Bar, which by the spring of 1849 had so many miners crowding along the Yuba riverbank that it was necessary to limit the size of claims to prevent trouble. In 1850, there were 2,000 said to be working at that bar alone – although there were camps along the river every two or three miles. When flood conditions late in 1850 drove the miners out of the riverbed, they climbed higher up in the hills. They found gold near Sucker Flat, which geologically was an extension of Rose's Bar. The flat first was called Gatesville, after an early arrival in the diggings. After he left it became "Sucker Flat" in obvious reference to his home state of Illinois. Sucker flat (One mile from Smartville) was considered an important area when hydraulic mining began. In 1860 there was notable mining there, with J. P. **Pierce** the principal owner of the famed Blue Gravel Mining Co., which then employed close to 60 men. By 1879 the Sucker Flat-Rose Bar district claimed 400 population. In that year Smartville was reputed to have the most extensive gravel mines in the nation. Companies operated under the names of Golden Gate, Smartsville, Blue Point, Blue Gravel, Pittsburgh, Yuba Mining Co. and Excelsior Water & Mining Co. The latter concern, with Daniel **McGanney** as superintendent, had a giant nozzle for water-mining constructed at the Empire Foundry in Marysville. Rich Rewards. On May 15, 1871, the Smartsville Hydraulic Mining Co. delivered in Marysville amalgam valued at $54,000. In January, 1908, a small amount of platinum was found at the Blue Point mine on supposedly worthless dumps. Among other pioneer citizens of the Smartville-Sucker Flat section were Thomas **Conlin** livery and stage; George **Jeffrey**, blacksmith; C. **Slattery**, builder; William **Cramsie**, water agent; M. M. **McConnell**, druggist; Will **Chamberlain**, saloon – billiards; John H. **McQuaid**, road master; Robert **Beatty**, postmaster; John **Peardon**, hotel keeper; Benjamin **Sanford**, farmer; L. **Crary**, J. P. **Pierce**, Patrick **Campbell** and James **O'Brien**. **O'Brien** "grew up" with the country, having come from his native Ireland in 1853 when he was but 21 years of age, and mined for four years on the Yuba river. In 1859 he was interested in the Smartville Irrigation projects. **O'Brien** bought mining properties and with San Francisco capitalists backing him, built the $80,000 Pactolas tunnel, which later consolidated with the Blue Gravel and Excelsior Mines. But when hydrolyzing stopped in 1882, **O'Brien** lost heavily and then went into farming. At one time he tried without success to have a narrow guage real road built through Smartville to link Marysville and Grass Valley. He died Aug 31, 1915 in Smartville. John H. **McQuaid** Jr. was a deputy under Postmaster **Beatty** for 24 years, beginning in 1898, and then succeeded to the position. He continued another 14 years as postmaster before removing in 1928 to Marysville. The Smartville hotel then was scene of a costume party to honor Mr. and Mrs. **McQuaid**. Mrs. Aurelia **Colling** (Carroll) succeeded **McQuaid**, and when she left the town Mrs. Nell **Compton** (Anderson) took the position, followed by Mrs. Lola

NEWSPAPER INFORMATION

Olive **Colling**, who died unexpectedly Dec. 31, 1957. At present Mrs. Vera **Woods** is acting postmaster. As Smartville continued to grow, in 1863 a Protestant union church was built by Episcopal, Presbyterian and Methodist members. The church burned, was reconstructed again was destroyed by fire and at present is in a remodeled Beale building. First Catholic services in the area took place at Rose's Bar in 1851 and the first church "St. Rose" put up through the efforts of Rev. Peter **Magganotta** who also built the original church for St. Joseph's parish in Marysville. The Smartsville church burned in 1870 and after being rebuilt was named the Church of the Immaculate Conception. It still is in use, one of the interesting pioneer church structures in the county. Priest Drowned. The accidental drowning of Rev. Andrew **Twomey**, 36, was one of the most tragic occurrences in Smartsville's history. The Irish-born priest, ordained in June, 1888, at Dublin, came two months later to California and had served the Smartville church for 14 years. Through his efforts a church was built also at Dobbins and early in 1902 he was planning one for Rackerby. But as he tried to drive his buggy over flood – swollen Dry Creek, the morning of March 8 1902, Father **Twomey** was caught in the swirling waters. The horses were swept off their feet and the buggy overturned to trap the priest underneath near Cartwright Crossing. The body of the priest, who had been on his way to Peoria schoolhouse to hold services, was recovered some distance downstream. His funeral was one of the largest-attended ever held in the foothill region. On March 11, he was buried in the Smartville Catholic cemetery. This is one of two pioneer burying grounds near the old village still in use. The one-teacher Rose Bar Elementary school at Smartville now serves that town, Timbuctoo and Sucker Flat, with an attendance of about 20 pupils.

27 Oct 1960 (Thursday); Page 23 Col 6; Fannie **Bushby**, a resident of Marysville for the past 80 years, died today in the Neibert Nursing Home in Paradise, following a long illness. She was 87 years old. A native of Missouri, she came to the Marysville area in 1880. Before her illness, she made her home at 510 Tenth St., Marysville. She is survived by a son, **Earl** of Yuba City; a daughter, Mrs. Blanche **Foreman** of Marysville; a brother, Edward **Wimberly** of Albuquerque, N. M.; and by three grandchildren, four great grandchildren and one great great-grandchild. Funeral services are scheduled Saturday at 10 a.m. at Lipp & Sullivan Chapel in Marysville. Interment will be in Smartville Masonic Cemetery.

10 Apr 1961 (Monday); Page 18 Col 1; Elizabeth Jane **Martien**, 90, lifelong resident of Smartville, died Saturday in a Marysville Hospital following an illness of several years. She was a native of Smartville and had lived there all her life. She celebrated her 90th birthday April 2. She is survived by three nephews, George **Rigby** of Smartville, Joseph **Rigby** of Marysville, and Jack **Keegan** of Grass Valley. A rosary will be recited tonight at Hutchison's Colonial Chapel at 9 o'clock and a requiem mass will be celebrated tomorrow at 10 a.m. at Smartville Catholic Church. Interment will be in Smartville Catholic Cemetery.

3 Jul 1961 (Monday); Page 9 Col 4; William Franklin **Johnson**, 96, who had lived in Yuba County since 1908, died yesterday morning in a Marysville hospital after a long illness. A resident of Smartville until 1946, when he moved to Marysville.

NEWSPAPER INFORMATION

He made his home at 1018 F St. at the time of his death. He had retired after working for 38 years for the Pacific Gas & Electric Co. **Johnson** was a native of Frankfurt Hill, Utica, New York, and was the husband of the late Carrie **Johnson**. Survivors include a daughter, Mrs. Evelyn **Miles** of Marysville; and a grandson, Donald **Johnson** of San Jose. Rosary will be recited tomorrow at 7 p.m. at Hutchison's Colonial Chapel. Requiem mass will be celebrated Wednesday at 10 a.m. at St. Joseph's Catholic Church. Interment will follow at Smartville Catholic Cemetery.

15 Nov 1961 (Wednesday); Page 26 Col 7; Mrs. Emma E. **Magonigal**, a resident of Smartville since 1896, died today at a Yuba City nursing home. She was 91. Mrs. **Magonigal** came to Smartville from Nova Scotia, Canada, with her parents, Mr. and Mrs. Daniel **Fraser**. She was the widow of John **Magonigal**, also a pioneer resident of the area. She is survived by three sons, **William**, **John**, and **Henry**, all of Smartville; two daughters, Mrs. Clarence **Poole**, and Mrs. Olive **Sweetland**, both of Yuba City; seven grandchildren and 10 great grandchildren. Funeral services will be conducted Saturday at the Ullrey Memorial Chapel at 10 a.m., with the Rev. Edgar Nelson, of the Yuba City First Methodist Church, officiating. Interment will be in Smartville Cemetery.

6 Mar 1962 (Tuesday); Page 5 Col 6; Funeral services will be conducted at 1 p.m. Thursday in Marysville for Cora Elizabeth **Hacker**, 80, a native of Wheatland who died yesterday in Sacramento. The daughter of the late Mr. and Mrs. Thomas W. **Peckham**, who settled in the Lone Tree District of Yuba County 100 years ago, Mrs. **Hacker** had spent most of her life in the Wheatland area. She had in recent years made her home at 111 Callison St. in Loomis. She was a charter member and past matron of Smartville Chapter, Order of Eastern Star. Survivors include three daughters, Mrs. Arlene M. **Welch** of Sacramento, Mrs. Murle **Flowers** of Wilton and Mrs. Edna **Teske** of Citrus Heights; and two sons, Francis W. **Hacker** and Harvey J. **Hacker**, both of Alturas; five sisters, Mrs. Pearl **McCormick** of San Francisco, Mrs. Mable **Anderson** and Mrs. Bertha **Olson**, both of Wheatland, Mrs. Myrtle **Pittman** of Orangevale and Mrs. Anita **Woodworth** of Yucalpa; a brother, William T. **Peckham** of Rio Oso; 11 grandchildren and 12 great grandchildren. Services will be conducted Thursday at 2 p.m. at Lipp & Sullivan Chapel. Burial will be in Smartville Cemetery

18 Jun 1962 (Monday); Page 9 Col 4; Funeral services for Earl **Woodroffe**, 69, a life long resident of Yuba County, will be conducted tomorrow at 10 a.m. at Lipp & Sullivan Chapel. **Woodroffe**, who lived at 905 G. St., Marysville, was born at Mooney Flat, Aug. 23, 1892. He died yesterday at Fremont Hospital. **Woodroffe** is survived by three nieces and a nephew. He was a member of the Marysville Elks Club, and officers of the club will officiate at his funeral. Interment will be in Smartville Cemetery.

23 Aug 1962 (Thursday); Page 24 Col 1; Funeral services were held today for Mae Black **Davis**, 80, a former Smartville resident, who died Monday in Albany. Her death followed that of her brother, Alfred B. **Black** of Grass Valley, who died on

NEWSPAPER INFORMATION

Saturday. Her husband died several years ago. She is survived by a daughter, Mrs. Lucille **Verdier** and a son Elvon **Davis**, both of Oakland; a brother, William **Black** of Grass Valley; and two sisters, Grace **Dykes** and Effie **Black** of Berkeley. Services were today at 2 o'clock at Hooper – Weaver Mortuary in Grass Valley followed by interment in the family plot in Smartville Fraternal Cemetery.

14 May 1963 (Tuesday); Page 12 Col 3; John Joseph **Murphy**, 77, former mayor of Marysville, died this morning in a local hospital following a long illness. **Murphy**, who made his home at 1115 I St., served two terms on the Marysville City Council from 1946 to 1954. He was constable of the Marysville Judicial District from 1935 through 1951. Born in Fort McDermott, Nev., he had lived in the Yuba-Sutter area for 71 years. His childhood was spent in Smartville. Hi is survived by his wife, **Ida**, of Marysville; a daughter, Ruth M. **McDowell** of Marysville; a sister, Nellie **Livingston** of Stockton; and three grandchildren. Funeral services will be conducted Thursday at 2 p.m. at the Lipp & Sullivan Chapel. Officiant will be Rev. Hugh McCallum of the Marysville First Christian Church. Burial will follow in the Smartville Masonic Cemetery.

31 May 1963 (Friday); Page 9 Col 5; Frank Gilbert **Foreman**, 67, a native of Smartville, was pronounced dead yesterday upon arrival at a Marysville hospital. **Foreman** collapsed at his home yesterday. He died after a heart attack, according to a Yuba County coroner's report. A lifelong resident of Smartville and Marysville, he lived at 5352 Feather River Boulevard. He was retired from the Burns Trucking Co. in Yuba City, where he had worked for over 30 years. Prior to that he had driven freight for Wells Fargo between Marysville and Smartville. Survivors, in addition to his wife, **Frieda** include three daughters, Mrs. Louise **Hutchison** and Mrs. Matilda **Colletti**, both of Sacramento and Mrs. Dorothy **Cole**, who is living out of the state; a son, Raymond **Hanson** of Sacramento; and 13 grandchildren. Funeral services will be conducted tomorrow at 10:30 a.m. at Hutchison's Colonial Chapel by Rev. Walter Singer, pastor of the Faith Lutheran Church. Burial will be at Smartville Masonic Cemetery.

5 Jul 1963 (Friday); Page 4 Col 5; Elise D. **Holbrook** of Berkeley died yesterday. She had been ill for some time. A native of San Francisco, she was the daughter of the late **Eddington** and Albertine **Detrick**. She made her home at 17 Plaza Drive, Berkeley. She is survived by her husband, **James** of Berkeley; a son **James** Jr. of Orinda; and four grandchildren. Rosary will be recited tonight at 8 o'clock at Berkeley Hills Chapel, 1600 Shattuck Ave., Berkeley, and requiem mass will be celebrated tomorrow at 9 a.m. at St. Augustine's Catholic Church, Alcatraz Avenue and Dana Street, Oakland. Interment will be Monday at 11 a.m. at Smartville Catholic Cemetery. Memorials to the Alameda County Heart Association have been suggested.

3 Dec 1963 (Tuesday); Page 14 Col 4; Nellie Wright **Newbert**, 76, a native of Yuba County, died yesterday in Rideout Memorial Hospital. She made her home at 1115 I St., Marysville. Widow of the late W. L. **Newbert** of Timbuctoo, she was a member of Marysville Parlor 162, Native Daughters of the Golden West, and

NEWSPAPER INFORMATION

Grass Valley Chapter, Neighbors of Woodcraft. Survivors include two sisters, Ida E. **Murphy** of Marysville and Edith **McKinsey** of Yuba City, and a brother, Aden J. **Wright** of Hagerman Valley, Idaho. Funeral services will be conducted Friday at 2 p.m. at Lipp & Sullivan Chapel. Burial will be in Smartville Masonic Cemetery.

4 Jan 1964 (Saturday); Page 14 Col 8; Oscar John **Rose**, 91, descendant of the family which settled historic Rose's Bar in the Yuba County foothills, died last night in a Marysville hospital. A native of Smartville and lifelong resident of Yuba County, the retired farmer made his home at the Burris Ranch in Browns Valley. He was the son of the late Mr. and Mrs. John **Rose** who settled Rose's Bar 115 years ago. Survivors include a sister, Mrs. Ethel **Burris** of Browns Valley. Funeral services will be conducted Tuesday at 10 a.m. at Lipp & Sullivan Chapel. Cremation will follow at Sierra View Memorial Park.

18 Feb 1964 (Tuesday); Page 13 Col 5; Patrick J. **Quilty**, 77, of Smartville, died Sunday in Grass Valley. A native of Canada, he had lived in Smartville since 1938 and had been ditch agent for the Nevada Irrigation District until his retirement. **Quilty** was a diamond driller by trade, employed extensively on bridge construction throughout the western states and Canada. He had worked on the Golden Gate, Oakland Bay and San Rafael Bridges, among others. Survivors include his wife, **Thelma**, of Smartville; a brother, Walter **Quilty**, and two nephews and a niece, all of Canada. Rosary will be recited tonight at 8 o'clock at the Hooper - Weaver Mortuary, 246 South Church St., Grass Valley. Funeral services will be held tomorrow at 10 a.m. at the Smartsville Catholic Church. Interment will be at the Catholic Cemetery, Smartsville. Memorials to a favorite charity have been suggested.

7 Apr 1964 (Tuesday); Page 8 Col 4; George **Herboth**, 74 year old native of Marysville, died last night in Rideout Hospital shortly after being stricken in his home at 605 Eighth Street, Marysville. He had been in ill health for several years. **Herboth** was retired. He was the owner of Herboth's Machine Shop in Marysville. He was a member of Marysville Lodge of Elks, Marysville Council of Knights of Columbus, and the Independent Order of Foresters. The family home was at 603 Eight Street. Survivors in addition to his wife, **Elizabeth**, include a sister, Louise **Herboth** of San Francisco; a brother, David **Herboth** of Vallejo; and three nephews, Louise **Conrath** of Marysville, Byron **Conrath** of Lansford, N. D., and Joseph **Herboth** of Yuba City. Rosary for Knights of Columbus members will be recited at 7:30 p.m. tomorrow in Lipp & Sullivan Chapel. The public rosary will be recited at 8 p.m. Requiem mass will be said at 10 a.m. Thursday in St. Joseph's Catholic Church. Interment will be in Smartville Catholic Cemetery.

26 Jun 1964 (Friday); Page 11 Col 4 & 5; Winifred Helen **Welch**, 50, died about 4:15 p.m. yesterday in the Yuba County Hospital, according to the Yuba County Sheriff-Coroner's Office. She had been living with her daughter, Mrs. Irene Ellen **Harryman** of 5516 South Gledhill Ave., Linda. Funeral arrangements are pending at Lipp & Sullivan Chapel.

NEWSPAPER INFORMATION

27 Jun 1964 (Saturday); Page 5 Col 7; Last rites for Winifred Helen **Welch**, 50, of Loma Rica, will be conducted Monday at 2 p.m. at Lipp & Sullivan Funeral Chapel. Burial will be at the Smartville Masonic Cemetery. Mrs. **Welch** died Thursday in a Marysville hospital following a long illness. She was a native of Quincy and had lived in the Yuba County area since infancy. Survivors include her husband Clarence R. **Welch** Sr., of Loma Rica; six sons, James **Welch** and Joe **Welch** of Browns Valley, and Clarence **Welch** Jr., Neal **Welch**, Tom **Welch** and Donald **Welch** of Loma Rica; and three daughters, June **Kibbe** of Browns Valley, Irene **Harryman** of Linda and Mary **Welch** of Loma Rica.

15 Feb 1965 (Monday); Page 12 Col 6; Funeral services for Noble **Hinkle**, 54, of Smartville will be conducted Wednesday morning at 10 o'clock by Rev. Jeremiah Boland of St. Patrick's Church in Hooper – Weaver Mortuary, Grass Valley. Rosary will be recited at 8 o'clock Tuesday evening. Interment will be in Fraternal Cemetery at Smartville. **Hinkle**, an electrician, died unexpectedly although he had been in poor health for several years since an accident. He was born at Prosser, Washington, June 1, 1911 and served with the Army in World War II. He is survived by his sister, Mrs. Thelma **Quilty**, Smartville; his mother, Mrs. Maud **Curl**, Grass Valley; two sons, Pat **Hinkle** of Citrus Heights and Gene **Hinkle** of Sacramento; and three grandchildren. The family has requested that those wishing to do so contribute to their favorite charities in lieu of flowers.

26 Dec 1966 (Monday); Page 4 Col 7; Funeral services will be conducted tomorrow in Marysville for Robert Brush **Williamson**, 53, who was found dead Friday, victim of what the Yuba County Sheriff's Office termed a self – inflicted bullet wound. A native of Labontan, Nev., he was a University of California graduate in chemistry. He was a veteran of World War II. He had lived in this area for 24 years. He was a laboratory technician at Fremont Hospital and made his home at 5318 Feather River Blvd., Marysville. Survivors include a brother, **Donald** of Oakland; and a sister, Mrs. Ruth **Garson** of Castro Valley. Private services will be conducted tomorrow at 10 a.m. at Hutchison's Colonial Chapel. Burial will be in Smartville Cemetery.

6 Jun 1966 (Monday); Page 21 Col 5; Funeral services are scheduled tomorrow for Dorothy Ann **Denney**, 44, of Smartville, who died Friday at Sutter General Hospital in Sacramento. A native of Eureka, she had lived in Smartville most of her life. She was a graduate of Marysville Union High School of Nursing in San Leandro. She was office manager for Ross Scott Physical Therapy Center in Marysville. Survivors, in addition to her husband, Albert T. **Denney** of Smartville, include a son, **Dean**, with the Navy; her mother, Mrs. Maud **Cann** of Smartville; a sister, Winona **Stukey** of Yuba City; three brothers, Charles **Cann** of Pittsburgh, James **Cann** of Yuba City and Phillip **Cann** of Napa. Funeral services will be conducted at 9 a.m. tomorrow at Hutchison's Colonial Chapel by Rev. Edgar Nelson of Yuba City First Methodist Church. Interment will be in Smartville Masonic Cemetery.

NEWSPAPER INFORMATION

5 Jul 1967 (Wednesday); Page 16 Col 2; Funeral services for James Monroe **Fitzhugh**, 87, of Smartville, who died of a heart attack yesterday morning at his home on Blue Gravel Road, will be conducted Saturday at 10 a.m. at the Chapel of the Twin Cities. A retired cattleman and large land holder in Modoc County, he was a native of Alturas and had lived there all his life before coming to Smartville in 1953. He was a member of Modoc Lodge 278 of Odd Fellows for 61 years and member of the Northern Rebekah Lodge 208 in Alturas for 51 years. Survivors include his wife, **Grace** of Smartville; a daughter, Mrs. Adelaide **Addington** of Marysville; three grandchildren; 10 great grandchildren. He was the last of a family of 14 brothers and sisters. His father was the first white man to settle in the **Fitzhugh** Creek area of Modoc County. Rev. Thomas J. Tweedle of the First Presbyterian Church in Marysville will officiate at the services. Burial will be in the Smartville Masonic Cemetery. The family has suggested memorials to the Heart Fund.

18 Dec 1967 (Monday); Page 4 Col 5; Rosary for Cecilia Ethelyn **LeBoeuf**, 60, of Smartville, will be recited tomorrow at 8 p.m. at Lipp & Sullivan Funeral Chapel. Miss **LeBoeuf**, a native of Portland, Ore., died yesterday at Rideout Hospital. She had lived in Smartville for one month. Survivors include a brother, Merrill **LeBoeuf**, of Marysville; and a sister, Bertha **Collins** of Smartville. A Requiem Mass will be said Wednesday at 10 a.m. at St. Joseph's Catholic Church in Marysville. Burial will be at the Smartville Catholic Cemetery. Memorials to the American Cancer Society have been suggested.

26 Dec 1967 (Tuesday); Page 4 Col 1 & 2; Funeral services for Clarence Ray **Welch** Jr., 34, of Alleghany will be conducted tomorrow at 2 p.m. at the Lipp & Sullivan Funeral Chapel. **Welch**, a native of Marysville, died Friday due to injuries suffered in an auto accident. He had lived in Yuba County until three years ago. He was a miner employed by the Dickie Exploration Mine Co. Survivors include his wife, Georgia L. **Welch** of Alleghany; three stepdaughters, Joanne **Taylor**, Donna **Hartman** and Carolyn **Vierria** of Marysville; a stepson, Roy **Taylor** of Marysville; four brothers, James R. **Welch** and Joseph C. **Welch** of Browns Valley, Neil C. **Welch** of Loma Rica and Donald **Welch** of Fort Lewis, Wash.; three sisters, Mary **Escobedo** of Marysville, June **Kibbee** of Penn Valley and Irene **Harryman** of Browns Valley. Rev. Ortiz Weniger of Yuba City First Baptist Church will officiate at the last rites. Burial will be at the Smartville Masonic Cemetery.

14 May 1968 (Tuesday); Page 4 Col 4; Frederick Charles **Wirth**, 76, of Smartville, died yesterday at the Martinez Veterans Hospital after a long illness. A native of Avoca, Penn., he had lived in the Smartville area for the past 32 years. A veteran of World Wars I and II, he was a member of the Naval Fleet Reserve Association. He was a retired Naval Chief Gunners Mate. Survivors include his wife, **Maybel** R., of Smartville; and a sister, Nellie **Dehes** of Bridgehampton, N. Y. Funeral services will be conducted Thursday at 10 a.m. at Lipp & Sullivan Funeral Chapel. Beale Air Force Base personnel will officiate. Burial will be in the Masonic Cemetery in Smartville.

NEWSPAPER INFORMATION

25 Nov 1968 (Monday); Page 4 Col 2; Bert Benjamin **Davis**, 78, died this morning at a Marysville convalescent hospital following a long illness. A native of Helena, Mont., **Davis** had lived in Smartville for the past 15 years. He was a retired forest ranger and a veteran of World War I. He was a member of the Masonic Lodge in Libby, Mont. Survivors include his wife, **Adeline**, of Smartville. Funeral services are pending at Hutchison's Colonial Chapel.

26 Nov 1968 (Tuesday); Page 4 Col 2; Bert **Davis**, 78, of Smartville died yesterday in a Marysville convalescent home. A native of Montana, he had been a resident here for the past 15 years. He is survived by his wife, **Adeline**. Graveside services are scheduled tomorrow at 10 a.m. at Smartville Cemetery with Rev. Frank Von Christianson of Marysville First Presbyterian Church officiating. Hutchison's Colonial Chapel is in charge of arrangements. Memorials to the American Cancer Society have been suggested.

12 Dec 1968 (Thursday); Page 8 Col 3; Asa David **Fippin**, 83, a retired gold miner who was born in Rough and Ready in 1885, was declared dead yesterday at Rideout Hospital of an apparent heart attack. He lived on Blue Gravel Road in Smartville. Survivors include his wife, Emma Annabelle **Fippin**; and nine children. Funeral arrangements are pending at the Hooper & Weaver Mortuary in Grass Valley.

13 Dec 1968 (Friday); Page 4 Col 4; Funeral services are scheduled tomorrow in Grass Valley for Asa David **Fippin**, 83, of Smartville, who was pronounced dead on Wednesday at Rideout Hospital in Marysville of an apparent heart attack. A native of Rough and Ready, he was a retired gold miner and made his home on Blue Gravel Road, Smartville. He was a past master and 30 year secretary of Rose's Bar Lodge, F & AM, Smartville and a member of the Order of Eastern Star of Smartville and of the Smartville Fire Department. In addition to his wife, **Emma**, survivors include three sons, **Jess** of Smartville, **Sidney** of Olivehurst and **Robert** of Auburn; six daughters, Julia **Tremewan** of Grass Valley, Nellie **Driggs** of Washington, Jean **Brown** of San Lorenzo, Bernice **Yore** of Idaho, Verna **Belveal** of Bakersfield and Winifred **Daley** of Citrus Heights; a brother, **Jess** of Sacramento; 28 grandchildren and 18 great grandchildren. Services are set for 2 p.m. tomorrow at Hooper & Weaver Mortuary at Grass Valley with officers of Rose's Bar Lodge, F & AM, officiating. Interment will be in Smartville Fraternal Cemetery.

30 Oct 1969 (Thursday); Page 10, Col 3; Maud **Cann**, 85, of Blue Gravel Road, Smartville, died yesterday at her home. Born in Indian Springs, she lived here for 35 years. She is survived by a daughter, Mrs. George **Stukey** of Yuba City; three sons, James of Yuba City, **Phillip** of Napa and **Charles** of Pittsburgh; two sisters, Mrs. Carrie **Duggins** of Bangor and Mrs. Kate **Murchie** of Modesto; seven grandchildren and 15 great grandchildren. Graveside services will be conducted Saturday at 4 p.m. at Smartville Masonic Cemetery by Rev. Harry Vise of the Marysville First Methodist Church. Services are under the direction of Hutchison's Colonial Chapel.

NEWSPAPER INFORMATION

1 Nov 1969 (Saturday); Page 18 Col 3; Funeral services will be conducted Tuesday in Yuba City for Grace Olive **Fitzhugh**, 80, of Blue Gravel Road, Smartville, who died yesterday in Miner's Hospital, Nevada City. A native of North Clove, N.Y., she had lived in Smartville since 1953. Before that she lived in Modoc County. She was a member of North Rebekah Lodge 208 of Alturas. Survivors include a daughter, Mrs. Adelaide **Addington** of Pilot Hill; three grandchildren and 10 great grandchildren. Funeral services will be Tuesday at 10 a.m. in the Chapel of the Twin Cities, with Rev. William Wallace Morgan of the Marysville First Presbyterian Church officiating. Burial will be in the Smartville Masonic Cemetery.

10 Nov 1969 (Monday); Page 22 Col 5; Funeral services will be conducted Monday for Edmund J. **Puff** Sr., 64, who died Saturday in Yuba General Hospital. A native of Hazelton, Iowa, **Puff** was a retired painter. He had lived in this area 12 years and made his home on Hwy. 20, near Smartville. He was a member of the Painters Local No. 146. Survivors include his wife, **Ethel**, and a son, **Edmund** Jr., both of Smartville; a sister, Evelyn **Squires** of Palo Alto; and a brother, **Leo**, of Iowa. Funeral services will be conducted at 10 a.m. Wednesday at the Loma Rica Church of Jesus Christ of Latter-day Saints, with Jack Littlefield officiating. Donations to the Heart Fund or a monument fund have been suggested. Funeral arrangements are under the direction of Hutchison's Colonial Chapel.

3 Jun 1970 (Wednesday); Page 4 Col 3; Clarence Ray **Welch**, 74, a lifetime resident of Yuba County and well known member of the Loma Rica community, died yesterday in Rideout Hospital. **Welch**, who made his home at 11414 Loma Rica Road, was a retired truck driver for the Yuba County Road Department. Survivors include five sons, James R. **Welch**, Joseph G. **Welch**, Thomas W. **Welch**, and Neil C. **Welch**, all of Browns Valley, and Donald E. **Welch** of Loma Rica; three daughters, Irene **Harryman** of Browns Valley, Mary E. **Escovedo** of Marysville, and Arelea Rae **Gerth** of San Jose; two brothers, Bill **Welch** of Sacramento and Nevin **Welch** of Lincoln; a sister, Ruth **Whiteside** of Tierra Buena; and 33 grandchildren. Funeral services will be conducted Saturday at 10 a.m. in Lipp & Sullivan Chapel with Rev. John R. Moy of the Loma Rica Community Church officiating. Interment will be in the Smartville Masonic Cemetery.

8 Oct 1970 (Thursday); Page 4 Col 3; Hannah **Magonigal**, 98, a native and life long resident of this area, died yesterday at the Sierra Convalescent Hospital in Roseville. A native of Smartville, she was the daughter of Mr. and Mrs. George **Bushby**, pioneer Yuba County residents. She most recently lived at Lincoln. Survivors include a son, George **Magonigal** of Lincoln; four grandchildren and five great grandchildren. Funeral services will be Saturday at 10 a.m. at Lipp and Sullivan Chapel with Rev. William Wallace Morgan, pastor of the First Presbyterian Church in Marysville, officiating. Burial will be in Smartville Masonic Cemetery.

NEWSPAPER INFORMATION

7 Nov 1970 (Saturday); Page 11 Col 1; Rosary will be recited tomorrow for Joseph C. **Naglee**, 53, of Marysville, who died at Rideout Hospital yesterday after suffering a heart attack Thursday. A native of Elmira Heights, N. Y., **Naglee** had been a resident of the Yuba-Sutter area for the past 12 years. His residence was at 1402 D St., Marysville, although the family formerly lived at Smartville. **Naglee** was a life insurance underwriter for New York Life and belonged to the Marysville Rotary Club, the Life Underwriters Association, St. Joseph's Catholic Church and his activities included work with the Boy Scouts and the 4-H Club. Survivors include his wife, **Elizabeth**; three daughters, Mrs. Karen **Holmes** of Santa Monica, Mary **Naglee** and Barbara **Naglee**, both of Marysville; three sons, Bruce **Naglee** of Smartville, Peter **Naglee** and Brian **Naglee**, both of Marysville; two sisters, Mrs. Nora **Johnston** of Sayre, Pa. and Mrs. Eloise **Councilman** of Horseheads, N.Y.; and a brother, Arthur **Naglee** of Elmira Heights, N. Y. Rosary will be recited at 8 p.m. tomorrow at Hutchison's Colonial Chapel. Mass will be said at St. Joseph's Catholic Church at 10 a.m. Monday. Interment will be in Smartville Catholic Cemetery. Arrangements were under the direction of Hutchison's Colonial Chapel.

21 Nov 1970 (Saturday); Page 12 Col 2; Funeral services will be Monday for Vern **Shields**, 66, of Smartville who died yesterday at Fremont Hospital following a long illness. A native of Canton, Ill., he was a retired carpenter who had lived in this area for 29 years. He was a member of the Carpenter's Union Local 1750 of Yuba City for 30 years and made his home on Highway 20, Smartville. Survivors, in addition to his wife, **Myrna**, include a daughter, Mrs. Pat **Collier** of Smartville; a son, Ted **Shields** of Etna; three brothers, Paul **Shields** of Montana, Donovan **Shields** of Fullerton and Joe **Shields** of Tucson, Ariz.; six grandchildren and three great grandchildren. Graveside services will be at 2 p.m. Monday at Smartville Masonic Cemetery with members of the carpenter's union officiating. Services are under the direction of Hutchison's Colonial Chapel. The family suggests memorials to the City of Hope at Duarte.

2 Dec 1970 (Wednesday); Page 4 Col 3; Funeral services were scheduled for today at 2 p.m. at the Hooper-Weaver Mortuary Chapel in Grass Valley for Richard **Carney**, 23, grandson of Mr. and Mrs. W. R. **Carney** of Smartville. **Carney**, son of Mr. and Mrs. William R. **Carney** of Grass Valley, was pronounced dead on arrival at 10 p.m. Saturday at Sierra Nevada Memorial Hospital, where he was taken by ambulance. The former Navy man reportedly collapsed while trying to start a vehicle stuck in new fallen snow at an American Hill cabin, a few miles north of Alleghany, according to reports. **Carney** sat at the wheel of the car while an unidentified man tried to push the idling car from the snow. When the men traded places, **Carney** reportedly collapsed. Although final reports are not expected until later in the week, Nevada County Deputy Coroner William Mullis said **Carney** apparently died from a combination of over-exertion and carbon monoxide poisoning. **Carney** served in Vietnam as an electrician and was discharged in April. He then joined the Merchant Marines. He was a graduate of Nevada Union High School and attended Sierra College, where he was a member of the Antique Car Club. **Carney** also was a life long member of the Order of DeMolay for Boys. In addition to his parents and grandparents, he is survived by a

NEWSPAPER INFORMATION

sister, Charlene **Brown** of Grass Valley; and a brother, James **Carney** of Sacramento. Rev. Haven Martin of the United Methodist Church of Nevada City will officiate at the funeral services. Burial will be at the Smartville Masonic Cemetery with full military rites by the Affiliated Veterans Council of Nevada County. The family has requested memorials to the Fred Finch Children's Home, 3800 Coolidge Ave., Oakland.

21 Dec 1971 (Tuesday); Page 4 Col 8; Graveside services were held today for Kim Loring **Carlson**, 20, of Mooney Flat Road, Smartville, who was fatally injured in a three – vehicle crash on Sunday. He was pronounced dead on arrival at Rideout Hospital after the accident. A native of Aberdeen, Wash., **Carlson** had lived in the Yuba-Sutter area for 17 years. He was a student at Yuba College. Survivors include his parents, James **Carlson** of Colusa and Reva **Rouse** of Linda; a sister, Judy **Carlson** of Sacramento; and a brother, Ric **Carlson** of Timbuctoo. Graveside services were held today at 11 a.m. at the Timbuctoo Cemetery at Smartville under the direction of Hutchison's Colonial Chapel.

24 Dec 1971 (Friday); Page 4 Col 1; An 81 year old Smartville man died at his home yesterday from a self-inflicted gunshot wound in the head, according to the Yuba County Sheriff-Coroner's Office. Walter Nelson **Brett** left a note to his wife, **Rose**, expressing his despondency over ill health, according to a coroner's report. He shot himself in the head with a .32 caliber pistol, deputies reported. Mrs. **Brett** found his body on a service porch at their home about 5:30 p.m., according to the sheriff's office. **Brett** was born in Santa Barbara. Funeral arrangements are pending at Hutchison's Colonial Chapel.

27 Dec 1971 (Monday); Page 4 Col 5; Funeral services will be conducted Wednesday at 10 a.m. at Smartville Community Church for Walter Nelson **Brett**, 80, with Jerry Russell of the Smartville Community Church Officiating. **Brett** died Thursday at his Smartville home from a self-inflicted gunshot wound, according to the Yuba County Sheriff-Coroner's Office. Born in Santa Barbara, he had lived here six years. He is survived by his wife, **Rose**. Burial will be at Smartville Masonic Cemetery. Memorials to the Smartville Community Church have been suggested by the family. Arrangements are by Hutchison's Colonial Chapel.

1 Jun 1972 (Thursday); Page 6 Col 3; Funeral services will be conducted at 10 a.m. Saturday at Lipp & Sullivan Funeral Chapel for Ernest **Le Bourveau**, 94, who died Tuesday at Marysville Convalescent Hospital. Funeral services will be conducted by officers of the Rose Bar Masonic Lodge of Smartville. Burial will be in the Smartville Masonic Cemetery.

14 Jul 1972 (Friday); Page 6 Col 3; Funeral services are scheduled Monday for Elizabeth E. **Herboth**, 82, a life-long resident of Yuba County, who died yesterday in Rideout Memorial Hospital following a long illness. A native of Yuba County, she had made her home at 605 8th Street, Marysville. She was the daughter of Mr. and Mrs. Louis **Conrath**, pioneer settlers in the Yuba County area and was the widow of George **Herboth**, a prominent Marysville businessman who died in 1964.

NEWSPAPER INFORMATION

A member of St. Joseph's Catholic Church Altar Society, she was also a member of the Notre Dame Alumnae Association. She is survived by two nephews, Louis **Conrath** of Marysville and Byron **Conrath** of Lansing, N. D. A Rosary will be recited Sunday at 8 p.m. in the Lipp & Sullivan Chapel in Marysville. Monday at 9 a.m. a requiem mass will be celebrated at St. Joseph's Catholic Church. Burial will follow in the Smartville Catholic Cemetery.

21 Mar 1973 (Wednesday); Page 6 Col 2; A longtime resident of Yuba County and native of Smartville, Brownie Frank **Peardon**, 84, of Los Gatos, died yesterday afternoon at his home. **Peardon**, a retired grocer, was a son of the Peardon family that ran the Smartville Hotel for a number of years. He later owned, along with his brother **William**, the **Peardon** Grocery Store on D Street in Marysville. He had lived in Los Gatos for the past eight years. Survivors include a brother, **James** of Marysville, and several nieces and nephews. His brother **William** preceded him in death. Funeral services will be conducted at 10 a.m. Saturday at Hutchison's Colonial Chapel with Rev. Harry Vise, pastor of the Marysville First United Methodist Church, officiating. Burial will be in the Smartville Masonic Cemetery.

18 Mar 1974 (Monday); Page A-6 Col 5; Funeral services will be conducted tomorrow for Hazel Claire **Greever**, 78, of 931 Elinor Ave., Linda. She died yesterday at Fremont Hospital. A native of Ames, Okla., she had lived in the Yuba-Sutter area for the past 16 years. She was a housewife. Survivors include her husband, **Kyle** of Linda; two sons, **Aldon** of Marysville and **Kyle** of Nevada City; a daughter, Billie **Singleton** of Marysville; a brother, Earl L. **Cramer** of Carmen, Okla.; 10 grandchildren and 13 great grandchildren. Rev. Mike Lamb will conduct the services at 10 a.m. tomorrow at Lipp & Sullivan Chapel. Burial will be in Smartville Masonic Cemetery.

30 Sep 1974 (Monday); Page A-10 Col 4; Funeral services will be held Thursday for James Radford **Welch**, 42, who died yesterday at Rideout Hospital following a long illness. A native of Nevada County, Welch was a self-employed farmer who made his home at 17 Spartan Lane, Marysville. He was a veteran of the Korean War. Survivors include three daughters, **Sally**, **Susie** and Sari **Welch**, all of Loma Rica; four brothers, **Joe** of Browns Valley, **Neil** of Marysville, **Tom** of Browns Valley and **Donald** of Marysville; and three sisters, June **Kibbee** of Browns Valley, Mary **Brislin** of Browns Valley, and Irene **Harryman** of Oregon House. Services Thursday will be at 2 p.m. at Lipp & Sullivan. Burial will be in the Masonic Cemetery in Smartville.

5 Nov 1974 (Tuesday); Page A-8 Col 4; Funeral services for Charles H. **Hermann** of Gridley, 88, who died Nov. 1 in Memorial Hospital at Gridley following a long illness, are scheduled tomorrow at 10 a.m. in Lipp & Sullivan Chapel, Marysville. Burial will be in the Masonic Cemetery at Smartville. **Hermann**, a native of Herman, Mo., was a retired machinist for Glidden Paint Co. at Oakland. He was employed there from 1926 until his retirement in 1967. He was a resident of Marysville from 1967 until recent months when he moved to Gridley due to his illness. Survivors include two sisters, Mrs. Pauline **Gilbert** of Marysville and Mrs.

NEWSPAPER INFORMATION

Louise **Barry** of Carmichael; a brother, Frank **Hermann** of Davis; and a niece and nephew. Dr. Wayne Long of Gridley Methodist Church will officiate at the services and members of Odd Fellows Lodge 169 of Davis will conduct graveside services. Visitation will be in the Gridley Funeral Chapel until tonight and then at Lipp & Sullivan until the time of services.

3 Dec 1974 (Tuesday); Page A-4 Col 5; Ethel **Puff**, 67, of 11463 Hill Road, Loma Rica, died early today at Yuba General Hospital following a long illness. A native of England, she had lived in the Yuba-Sutter area for 13 years. She was a retired insurance salesman. Survivors include a son Edmund **Puff** Jr. of Loma Rica. Funeral services are pending at Hutchison's Colonial Chapel.

5 Dec 1974 (Thursday); Page A-8 Col 4; Funeral services will be held at 11 a.m. Saturday for Ethel Wotherspoon **Puff**, 67, of 11463-B Hill Road, Loma Rica, who died early Tuesday at Yuba General Hospital. A native of Middles borough, England, Mrs. **Puff** lived 13 years in the Yuba-Sutter area. She was a retired insurance agent. Survivors include a son, Edmund **Puff** Jr. of Loma Rica; and two brothers, Al **Wotherspoon** of Grass Valley and Kenneth **Wotherspoon** of Utah. Services will be held at the Church of Jesus Christ of Latter-day Saints in Loma Rica, under the direction of Hutchison's Colonial Chapel. Burial will be in Smartville Masonic Cemetery.

17 Dec 1974 (Tuesday); Page A-6 Col 5; Funeral services will be conducted tomorrow for William Ray **James**, 86, of French Corral Road, Smartville, who died yesterday at Rideout Hospital. A native of Tomah, Wis., he had lived in the Yuba-Sutter area for 38 years. He was a retired employee of the Pacific Gas and Electric Co. Survivors include four sons, **Bud** of Marysville, **Chester** of Redding, **Donald** of Fall River Mills and **Larry** of Keno, Ore.; two daughters, Thelma **Vetkos** of Klamath Falls, **Ore**. and Winifred **Vetkos** of Grants Pass, Ore.; a sister, Crystal **Coffield** of Vancouver, Wash.; 29 grandchildren and 39 great grandchildren. Funeral services will be conducted at 2 p.m. tomorrow at Hutchison's Colonial Chapel. Burial will be in the Smartville Masonic Cemetery.

12 Jun 1975 (Thursday); Page A-4 Col 4; Smartville native James Edward **Holbrook**, 79, of Berkeley died yesterday at a Berkeley hospital. He moved to Berkeley in 1907 and was the retired vice-president of the Pabco Corp. He was past president of the San Francisco Rotary Club and a member of the Zeta Phi Fraternity, having graduated from the University of California in 1919. Survivors include a son, **James** Jr. of Lafayette, and a cousin, William **O'Brien** of Marysville. Funeral services will be conducted at 9 a.m. Saturday at the McNary & Morgan Chapel, 3030 Telegraph Ave., Berkeley. A Mass of Christian Burial will be said at 9:30 a.m. Saturday at St. Augustine's Catholic Church, 490 Alcatraz Ave., Oakland. Burial will be in Smartville Cemetery.

6 Aug 1975 (Wednesday); Page A-6 Col 4; Pauline Ella **Gilbert**, 85, of 6147 Griffith Ave., Linda, died yesterday at her home. She was born in Herman, Mo.

NEWSPAPER INFORMATION

Her survivors include her son, Carl A. **Gilbert**, of Linda. Funeral arrangements are pending at Hutchison's Colonial Chapel.

7 Aug 1975 (Thursday); Page B-2 Col 4; Funeral services are scheduled Saturday for Pauline E. **Gilbert**, 89, of Marysville, who died Tuesday at her home at 6147 Griffith Ave. A native of Hermann, Mo., she had lived in the Yuba-Sutter area 57 years. Survivors include her husband, **Absalom** of Marysville; a daughter, Mrs. Bernice **Rada** of San Rafael; a son, **Carl** of Marysville; a brother, Frank **Hermann** of Davis; three grandchildren and two great grandchildren. Funeral services are scheduled at 2 p.m. Saturday at Hutchison's Colonial Chapel. Burial will be in the Smartville Cemetery.

26 Aug 1975 (Tuesday); Page B-2 Col 4; Rosary will be recited at 7:30 p.m. tomorrow at Lipp & Sullivan Chapel for Walter L. **Walsh**, 85, of Marysville, who died Sunday following surgery at Sunnyvale Hospital. He was a native of Smartville and a lifetime resident of Yuba County. A retired electrician, he made his home at 817 11th St., Marysville. Survivors include his wife, **Sadie**, of Marysville; two daughters, Margaret **Bennion** of Sunnyvale and Patricia **Gunderson** of Seattle, Wash.; a son, **Francis** of Los Altos; and five grandchildren. A Mass of Christian Burial is planned for 9 a.m. Thursday at St. Joseph's Catholic Church. Burial will be in the Smartville Masonic Cemetery.

15 Apr 1976 (Thursday); Page A-4 Col 4; Rosary will be recited at 7 p.m. Saturday at Nazareth House in San Rafael for Alvina Gunning **Geraghty**, 90, who died there Tuesday. A native of Marysville, she was the daughter of the late S. O. **Gunning**, who served several terms as Yuba County Recorder. She had lived in the San Rafael area since 1963. Survivors include two brothers, Samuel **Gunning** of Smartville and Robert **Gunning** of Bakersfield; a sister, Mrs. Robert **Young** of Daly City; a daughter, Mrs. William **Breen** of San Rafael; five grandchildren, and one great grandchild. Funeral mass will be celebrated at 9 a.m. Monday at the Nazareth House. Burial will be in the Smartville Cemetery.

17 Apr 1976 (Saturday); Page A-6 Col 3; Funeral services will be conducted Monday for Nellie Gladys **Bartlett**, 73, of Smartville, who died Thursday at Rideout Hospital. A native of Coos Bay, Ore., she had lived in the Yuba-Sutter area for 50 years. She was a member of Smartville Chapter No. 431 Order of Eastern Star, Order of the White Shrine of Yuba City, Sutter Buttes Court No. 149 Order of the Amaranth, Ladies Auxiliary of the Veterans of Foreign Wars in Grass Valley, the American Legion Auxiliary of Marysville, the Smartville Dinner Club and the Cootiette Club of Smartville. Survivors include a stepson, John **Bartlett** of Reno, Nev.; a brother, William **Fisher** of Medford, Ore.; a sister, Mary **Miller** of Medford, Ore.; and one grandchild. Funeral services will be conducted at 2 p.m. Monday at Lipp & Sullivan Chapel with officers of the Smartville Chapter Order of Eastern Star officiating. Burial will be in the Smartville Masonic Cemetery. Memorials to the Smartville Order of Eastern Star Memorial Fund or the J. Clifford Lee Cancer Fund have been suggested by the family.

NEWSPAPER INFORMATION

27 Aug 1976 (Friday); Page A-6 Col 4; Ric **Carlson**, 30, pastor of a Morning Star Ministries group in South Lake Tahoe the past two years, died yesterday at Rideout Hospital following a lengthy illness. A native of Oakland, he had lived in the Marysville-Yuba City area for 15 years before moving to Tahoe. He was awarded the Air Medal and Army Commendation Medal as a member of a U.S. Army Black Beret unit in Vietnam. Survivors include his wife **Angela**; a son, **Andrew**; and two daughters, **Serene** and **Rebekah**, all of South Lake Tahoe; his father, Jim **Carlson** of Colusa; his mother, Reva **Rouse** of Marysville; and a sister, Judi Louise **Carlson** of Sacramento. Graveside services are scheduled tomorrow at 2 p.m. at Timbuctoo Cemetery with Jerry Russell of the Community Church of Marysville officiating. Lipp & Sullivan Chapel is in charge of arrangements. Memorials to the Ric Carlson Memorial Fund at the Marysville Community Church, 17th and Elm Streets, have been suggested.

12 Apr 1977 (Tuesday); Page A-4 Col 4; Last rites will be conducted tomorrow for Selma Frances Esthelda **Byrne**, 92, of Paradise, who died April 8 at a Paradise hospital following a long illness. A native of Nimshew, she was a lifetime resident of Butte County. She was a retired nurse. She was a member of the St. Thomas More Catholic Church of Paradise and a 50-year lifetime member of the Altar Society. She was also a charter member of the Honey Run Covered Bridge Society and the Oroville museum. Survivors include a nephew, Raymond **Crandall** of Santa Cruz and a niece, Suzanne **Hopkins** of Paradise and several grandnieces and grandnephews. Rosary will be recited at 7 o'clock tonight at the Rose Chapel in Paradise. Mass will be celebrated at the St. Thomas More Catholic Church at 11 a.m. tomorrow with Rev. Raymond Renwald and Rev. Paul Dagman officiating. Graveside services will be conducted at 3 p.m. tomorrow at the Catholic Cemetery in Smartville.

21 Oct 1977 (Friday); Page A-4 Col 4; Funeral services will be conducted Monday for Ida May **Hite**, 94, of 2915 Monroe Road, Yuba City, who died yesterday at the Driftwood Convalescent Hospital. A native of Challis, Idaho, she had lived in the Yuba-Sutter area for 62 years. She was a member of Fidelia Chapter, Order of Eastern Star. Survivors include two daughters, Naida E. **Bahling** and Serena **Hickey**, both of Sacramento; two grandchildren and nine great grandchildren. Funeral services will be conducted at 2 p.m. Monday at Ullrey Memorial Chapel with Rev. Harry Vise of the First United Methodist Church of Marysville and officers of the Fidelia Chapter Order of Eastern Star officiating. Burial will be in the Smartville Cemetery.

26 Jan 1978 (Thursday); Page A-6 Col 5; Funeral services are planned Saturday for Billie Pauline **Singleton**, 64, of 931 Elinor Ave., Marysville. She died yesterday at Fremont Hospital. She was a native of Kiowa Kan. Survivors include her husband, Pete T. **Singleton** of Marysville; her father, Kyle A. **Greever** of Marysville; and two brothers, Kyle W. **Greever** of Nevada City and Aldon **Greever** of Marysville; and numerous nieces and nephews. Services will be conducted Saturday at 10 a.m. at Lipp & Sullivan Chapel, with burial in Smartville Cemetery.

NEWSPAPER INFORMATION

1 Oct 1979 (Monday); Page A-4 Col 2; Private funeral services will be conducted tomorrow for LaVerne J. **Hudson**, 70, of Smartville, who died Saturday at Rideout Hospital. A native of Salt Lake City, Utah, she lived in the Smartville area for many years with her husband, Ralph **Hudson**, on Main Street. Other survivors include two sons, Richard **Hudson** of Sacramento and Robert **Hudson** of Cotati; a brother, Ralph **Hansen** of Tahoe; and five grandchildren. Graveside services will be at the Smartville Masonic Cemetery tomorrow. Friends may call at Lipp & Sullivan Funeral Home from 8 a.m. to noon tomorrow.

6 Jan 1981 (Tuesday); Page A-4 Col 6; John **Parkison**, 96, of 9300 Jones Road, Marysville, died Sunday in Rideout Hospital. A farmer, he was born in Glendon, Iowa, and had lived in this area since 1939. Survivors include four sons, John **Parkison** of Marysville, **Bert** and Bernard **Parkison** of Montana and Bob **Parkison** of Weed; four daughters, Helen **Laughlin** of Marysville, Gracie **Stering** of Chico, Bessie **Robinson** of Washington and Beulah **Cottrell** of Berry Creek; a sister, Grace **Ardizone** of Illinois; a brother, Charlie **Parkison** of Coulton; 79 grandchildren and great great-grandchildren. Graveside services will be conducted at 2 p.m. tomorrow in Smartville Cemetery. Rev. Wayne Vincent, Loma Rica Community Church, will officiate. Services are under the direction of Hutchison & Carnes Colonial Chapel.

6 Mar 1981 (Friday); Page A-4 Col 4; Mable Ruth **Wirth**, 85, of 500 Pomona Ave., Oroville, died Wednesday in Oroville Medical Center Hospital. Born in West Deerfield, Mass., she had lived in the Smartville area 40 years. Survivors include a daughter, Ruth **Johnson** of West Deerfield; two nieces, Delores **Martinez** of Oroville and Elizabeth **Fabela** of Yuba City; a nephew, Tirso A. **Ramos** of Yuba City; 25 grandchildren and numerous great-grandchildren. Funeral services will be conducted at 2 p.m. Monday in Lipp & Sullivan Chapel. Burial will be in Smartville Masonic Cemetery. Friends may call at the chapel until 8 p.m. tomorrow, from 9 a.m. to 8 p.m. Sunday and until 2 p.m. Monday.

21 Mar 1981 (Saturday); Page A-6 Col 6; A memorial reception will be held from noon to 5 p.m. tomorrow at 7252 Star Rt. 3 in Smartville for Patricia Ethyl **Collier**, 54, of Smartville. She died Thursday at Mercy Hospital in Sacramento of natural causes. A native of Absorkee, Mont., she was a homemaker and had been a Smartville resident since 1945. Survivors include her husband, John A. **Collier** of Smartville; two sons, John V. **Collier** of Marysville and Bruce **Collier** of Chico; a daughter, Marian **Vargas** of District 10; a brother, Ted **Shields** of Sacramento; and seven grandchildren. The family has suggested contributions be sent to the American Cancer Society, Yuba Sutter Unit, P. O. Box 106, Marysville. Burial will be in the Smartville Masonic Cemetery. Area arrangements are under the direction of the Lipp & Sullivan Funeral Chapel.

20 Jul 1981 (Monday); Page A-4 Col 6; William Raymond **Carney**, 94, a resident of Smartville area since 1930, died yesterday in Grass Valley. A native of Olean, Mo., he was born Nov. 16, 1886. He spent his early life in Missouri and in 1905 moved to San Francisco. He was there during the great San Francisco earthquake.

NEWSPAPER INFORMATION

He attended schools in the Bay Area and during all his working career was a consultant and engineer with the U. S. Bureau of Mines. He traveled worldwide as a mining consultant. He is survived by his wife, **Cora**, of Smartville; a son, William **Carney** Jr. of Grass Valley; a sister, Nellie **Reynolds** of Kansas City, Mo.; two grandchildren and seven great grandchildren. Funeral services will be conducted at 2 p.m. Wednesday in Hooper – Weaver Mortuary at Grass Valley. Burial will be in Smartville Cemetery.

28 Sep 1981 (Monday); Page A-6 Col 5 & 6; Kyle Aldon **Greever**, 88, of 931 Elinor Ave., Linda, died at Marysville Convalescent Hospital Saturday. The 25 year Yuba Sutter resident was a retired self-employed baker. The Kiowa, Kan., native was a member of Ostrom Grange No. 751 and had belonged to Wiley's Cove Lodge No. 524 Free and Accepted Masons of Leslie, Ark., since 1916. Survivors include two sons, Aldon Lee **Greever** of Marysville and Kyle Warren **Greever**, of Nevada City; 10 grandchildren, 19 great grandchildren and two great great-grandchildren. Funeral services are scheduled for 10 a.m. Wednesday at Lipp & Sullivan Chapel in Marysville with Rev. Ray Wisner of the Marysville Church of the Nazarene officiating. Burial will be in Smartville Masonic Cemetery. Friends may call at the chapel until 8 p.m. today, from 9 a.m. to 8 p.m. tomorrow and until the service Wednesday.

19 Oct 1981 (Monday); Page A4 Col 5; A memorial reception will be held next Sunday for John Arthur **Collier**, 58, of Smartville, who died Saturday of an illness on his houseboat at Englebright Boat Marina. Born in Idaho, Collier was a carpenter. He had graduated from Sutter High School and lived in the Yuba-Sutter area most of his life. He served in Europe in the Army during World War II. He was a member of Carpenters Local 1147 of Roseville. His survivors include his wife, **Virginia**, of Smartville; two daughters, Marian **Vargas** of Marysville and Janet **Ceresa** of Sebastopol; three sons, John **Collier** of Marysville, Bruce **Collier** of Chico and Jeffrey **Ceresa** of Sebastopol; his mother, Clara **Rach** of Folsom; three brothers, Leo **Collier** of Oregon, Gene **Collier** of Sacramento and Marvin **Collier** of Folsom; two sisters, Wilma **Chapman** of Pennsylvania and Colleen **Frouke** of San Jose, and by nine grandchildren. Funeral arrangements and cremation services are being handled by the Hooper and Weaver Memorial Chapel in Grass Valley. A memorial reception will be held at the family home from 2 –5 p.m. Oct. 25. Memorials to favorite charity have been suggested by the family.

14 Nov 1981 (Saturday); Page A-6 Col 3 & 4; Services for Frank Earl **Smith**, 68, a resident of Smartville since 1970, will be held 2 p.m. Monday at Lipp & Sullivan Chapel. He died yesterday at Rideout Hospital. A native of Richmond, he was a supervisor for Shell Oil Co. for 38 years until he retired. He was a veteran of World War II, in which he served in the Navy, and a member of the Veterans of Foreign Wars, Banner Mountain Post No. 2655, of Grass Valley. He last lived on McGanny Lane in Smartville. He is survived by his wife, **Catherine**, of Smartville; three brothers, **John**, of Oakland, **Floyd**, of Grass Valley, and **Bill**, of Martinez; and a sister, Gladys **Young**, of Martinez. Friends may call at Lipp & Sullivan Chapel from 9 a.m. to 8 p.m. Sunday and from 9 a.m. until the service on

NEWSPAPER INFORMATION

Monday. Chaplain Brian K. Hunter of Beale Air Force Base will officiate over the services. Burial will be in Smartville Cemetery. Memorials to the Yuba Sutter Chapter of the American Cancer Society are suggested by the family.

17 Nov 1981 (Tuesday); Page A-4 Col 4 & 5; Funeral services are planned for 11 a.m. Friday in the chapel of Hooper & Weaver Mortuary at Grass Valley for Emma Wheaton **Fippin**, 88, of Smartville, who died Sunday in Grass Valley Medical Facility following a long illness. Rev. Haven Martin will officiate, assisted by members of Smartville Chapter 431, Order of the Eastern Star. Burial will be in Smartville Masonic Cemetery. A resident of Smartville, Mrs. **Fippin** was born at Mooney Flat in Nevada County, May 4, 1893. She attended schools at Mooney Flat, Smartville and Marysville. She was the daughter of the late **Allen** and Mary **Wheaton**. Her husband, Asa **Fippin**, died in 1968. She was a charter member of Smartville Chapter, OES. Survivors include three sons, Jess **Fippin** of White City, Ore., Sidney **Fippin** of Smartville and Robert **Fippin** of Auburn; six daughters, Julia **Tremewan** and Nellie **Driggs**, both of Santa Clara, Jean **Brown** of San Lorenzo, Bernice **Your** of Gooding, Idaho, Berna **Baker** of Marysville and Winifred **Daley** of Citrus Heights; two sisters, Viola **Craun** of Bishop and Martha **Spencer** of Marysville, 27 grandchildren and 52 great grandchildren.

26 Dec 1981 (Saturday); Page A-6 Col 4 & 5; Services for Rose Mary **Brett**, 81, of Wheatland, a resident of this area since 1965, will be held at 10 a.m. Monday at Hutchison & Carnes Colonial Chapel. She died Thursday in Fremont Hospital. A native of Wisconsin, she was a housewife and last lived at 121 C St., Wheatland. She has no known survivors. Rev. Kenneth Baser of the Wheatland Assembly of God Church will officiate at the services, and burial will be in Smartville Masonic Cemetery. Friends may visit until 8 p.m. today and from 10 a.m. to 8 p.m. tomorrow at Hutchison & Carnes.

7 May 1982 (Friday); Page A-4 Col 4 & 5; Samuel O. **Gunning**, 91, a native of Yuba County and a resident of Smartville for most of his life, died early today in Sierra Nevada Memorial Hospital in Grass Valley. He had been under treatment for cancer for the past several years. Funeral arrangements are pending at Hooper & Weaver Mortuary in Grass Valley.

10 May 1982 (Monday); Page A-4 Col 4 & 5; Rosary will be recited tomorrow at 8 p.m. at Lipp & Sullivan Chapel for Samuel O. **Gunning**. A native and lifelong resident of Yuba County, **Gunning** died Thursday at Sierra Nevada Memorial Hospital in Grass Valley at the age of 91. He was born in his family's home on G Street in Marysville and lived for the past 40 years in Timbuctoo near Smartville. A retired U. S. government machinist, he served in the Navy in World War I and was a member of the Veterans of Foreign Wars Post in Marysville. His father, Samuel O. **Gunning**, Sr., was Yuba County Recorder from 1895 to 1910. He is survived by one brother, Robert E. **Gunning** of Bakersfield, and many nieces and nephews. Mass of Christian Burial will be said Wednesday at 10 a.m. at St. Joseph's Catholic Church, the Rev. John J. Moore officiating. Burial will be in Smartville Catholic Cemetery.

NEWSPAPER INFORMATION

25 Sep 1982 (Saturday); Page A-6 Col 1 & 2; Funeral services are planned Monday for Dorothy Elizabeth **Manford**, 70, of Smartville, who died Thursday at the Grass Valley Convalescent Hospital. The Springfield, Mo., native lived in this area for 15 years. She had worked as a secretary for the United States government for 10 years. Survivors include her husband, James B. **Manford** of Smartville; a brother, Howard **Sheward** of Antioch; and four nephews. Rev. Richard Markle of the First Presbyterian Church of Marysville will conduct services at the Smartville Masonic Cemetery at 3 p.m. Monday. Services are under the direction of Hutchison & Carnes Colonial Chapel in Marysville.

14 Dec 1982 (Thursday); Page A-11 Col 4; Memorial services for Vera Lee **Magonigal**, 76, of Smartville will be conducted at 10 a.m. Saturday in Smartville Community Church on Main Street, with Daniel Poole of the church officiating. Mrs. **Magonigal** died Tuesday in Sierra Memorial Hospital at Grass Valley. Her husband, John **Magonigal**, died several years ago. Survivors include a sister, Fern **Taylor**, of Portland, Ore.; a sister who lives on the East coast, brothers-in-law and several nieces and nephews.

13 Apr 1983 (Wednesday); Page A-9 Col 5; Funeral services will be held tomorrow for Betty "Billie Lee" **Lessley**, 41, of 412 East 16th St., Marysville, who died Saturday in a fire that destroyed a camper in the backyard of her home. She was pronounced dead at the scene by Yuba County Sheriff-Coroner's deputies after firemen extinguished the flames about 1:30 p.m. Saturday. The cause of the fire is under investigation. A native of Elk City, Okla., she had lived here for the past 10 years. Survivors include three sons, **Franky**, **Tony** and Danny **Martin**, all of Marysville; three daughters, Mary **Martin** of Los Angeles, Jenny **Martin** of Yuba City and Suzy **Martin** of Marysville; her parents, **Aldon** Lee and Maxine **Greever**, both of Marysville; four brothers, **Jack**, **Joe** and Jim **Greever**, all of Marysville; and Butch **Griffith**, of Illinois; and two sisters, Ruby **Brown** of Marysville and Delores **Rose** of Los Angeles. Rosary will be recited at 8 o'clock tonight at Lipp & Sullivan Chapel. Mass will be at 11 a.m. tomorrow at St. Joseph's Church with Rev. John J. Moore officiating. Burial will be private.

24 Jan 1984 (Tuesday); Page A-11 Col 3 & 4; Funeral services will be scheduled tomorrow at 2 p.m. at Lipp & Sullivan Funeral Chapel for Absalom Andrew **Gilbert**, 94, of Linda, who died Sunday at Yuba City Convalescent Hospital. A Yuba Sutter area resident since 1918, he was a retired winch man for Yuba Consolidated Gold Fields. He was a native of California. He was a member of Operating Engineers Local No. 3 of Marysville. He is survived by a son, Carl A. **Gilbert** of Marysville; a daughter, Bernice **Rada** of Marysville; five grandchildren; and nine great grandchildren. Visitations will be until 8 o'clock tonight and from 9 a.m. until 2 p.m. tomorrow at the funeral home. Burial will be at Smartville Cemetery.

20 Feb 1984 (Monday); Page A-13 Col 5; John Thomas **Grace**, 89, of Timbuctoo Road, Smartville, died Sunday at Rideout Hospital. He was born in Brownsville,

NEWSPAPER INFORMATION

Pa. His survivors include a sister, Rose E. **Grace** of Smartville. Funeral arrangements are pending at Lipp & Sullivan Chapel.

21 Feb 1984 (Tuesday); Page A-11 Col 6; Funeral services will be held at 2 p.m. Thursday at Lipp & Sullivan Chapel for John Thomas **Grace**, 89, of Smartville. He died Sunday at Rideout Hospital. A native of Brownsville, Penn., he had been a resident of the Yuba Sutter area since 1929 and had worked as a self-employed mechanic and a gold miner for 50 years. He was a veteran of World War I. Survivors include a sister, Rose Ellen **Grace** of Smartville. Rev. Nicolaus Phelan of St. Patrick's Catholic Church in Grass Valley will officiate at the services. Burial will be in the Smartville Cemetery. Friends may call from 1 to 8 p.m. tomorrow and from 9 a.m. until services Thursday at Lipp & Sullivan Chapel.

9 Apr 1984 (Monday); Page A-13 Col 7 & 8; Charles William **Bebout**, 62, of Smartville died at his home Sunday. A native of Milton, Ky., he had lived in Smartville 10 years. He served with the Air Force 32 years, with assignments in Korea during World War II and in Vietnam. He retired as a lieutenant-colonel. Survivors include his wife, Hilary (Jo) **Bebout** of Smartville; two sons, Charles W. **Bebout** II of Yuba City and James W. **Bebout** of Marysville; a brother, Robert **Bebout** of Austin, Ind.; three sisters, Joan **Madden** of Deputy, Ind., Charlotte **DeVaudrevil** of Lisbon Falls, Maine and Dorcas **Malm** of Milton, Ky.; his mother, Carrie **Grimes** of Austin, Ind. and five grandchildren. Funeral services will be conducted at 10 a.m. Wednesday in Lipp & Sullivan Chapel, with burial in Smartville Cemetery. The family suggests contributions to the Yuba Sutter Chapter of the American Cancer Society, P. O. Box 106, Marysville. Friends may call at the chapel from 9 a.m. to 8 p.m. Tuesday.

29 May 1984 (Tuesday); Page A-13 Col 5; Sadie Marie **Walsh**, 90, of Puyallup, Wash., formerly of Marysville, died Saturday in Puyallup. A native of Ireland, she lived in Marysville for 60 years. She was a member of the Catholic Ladies Relief Society and Catholic Daughters of America in St. Joseph's parish. She moved to Puyallup three years ago. Her husband, **Walter**, who died in 1965, worked for Pacific Gas & Electric Co. for many years and at the same time ranched in Wheatland and then Browns Valley. She is survived by a son, Joe **Walsh** of Los Altos; two daughters, Peggy **Bennion** of Nevada City and Patricia **Gunderson** of Yelm, Wash.; and four grandchildren. Rosary will be recited at 7 p.m. Thursday in Lipp & Sullivan Chapel. Mass will be said at 10 a.m. Friday in St. Joseph's Catholic Church, with Rev. John J. Moore officiating. Burial will be in Smartville Cemetery. The family suggests contributions to the American Diabetes Association, 255 Hugo St., San Francisco, 94122. Friends may call at the chapel from 9 a.m. until the rosary Thursday.

15 Jan 1985 (Friday); Page A-15 Col 3; Services will be Monday for James Bunch **Manford**, 80, of Smartville, who died Tuesday at Rideout Hospital. Born in Yuba City, he returned to this area to live in 1965. He worked as a plumber for 30 years for the U.S. government. He was a Navy veteran and a member of Rose Bar Masonic Lodge No. 89 of Smartville. Survivors are a sister, Lurline **Jones** of

NEWSPAPER INFORMATION

Marysville; and several nieces and nephews. Graveside services will be held at 11 a.m. Monday at the Smartville Masonic Cemetery under the direction of Hutchison & Carnes Funeral Home. Members of Rose Bar Lodge will officiate. Visitation will be Sunday from 9 a.m. to 8 p.m. at the Hutchison & Carnes Funeral Home.

31 Aug 1985 (Saturday); Page A-11 Col 1; Graveside services are planned at 11 a.m. Tuesday at Smartville Cemetery for Georgina Louise **Welch**, 60, of Marysville, who died Wednesday at Rideout Hospital. A native of San Leandro, she had lived in the Yuba Sutter area since 1949. She had been a clerk for the Salvation Army. Survivors include three daughters, Donna **Hammerschmidt** of Marysville, Carolyn **Hassler** of Dobbins and Joanne **Henry** of Bakersfield; a brother, Ed **Gnadig** of Sparks, Nev.; two sisters, Clara **Hampton** of San Diego and Merna **Totten** of Watsonville; 10 grandchildren and three great grandchildren. Services are under the direction of Hutchison & Carnes Colonial Chapel.

17 Jun 1986 (Tuesday); Page A-7 Col 5; Funeral services will be held Thursday for Henry E. **Magonigal**, of Penn Valley, who died June 15, 1986, in Grass Valley. He was 80. A native of Nevada County, he lived in Yuba and Nevada Counties most of his life. He was a cattle rancher. He was one of the original founders of the 4-H in Nevada County and was awarded a 35-year leader's pin. He also founded the Future Farmers of America in Nevada County. He was a retired member of the Sierra Nevada Memorial Hospital board of directors; a retired member of the Nevada County Soil Conservation District; and a 22-year member of the Nevada County Fair Board. He chaired the board for 20 years. He was one of the founders of the Beef Breeders Association of Nevada County, which was organized for the sole purpose of the sale of animals at the fair. He was a 25-year member of the Pleasant Valley school board, serving as president and secretary of that board. He was cattle range man on Camp Beale and Beale Air Force Base property for many years. When the property was taken over by the military during World War II, what wasn't used for military purposes was leased back to local cattlemen for grazing purposes. He kept track of how many head of beef were grazing the land. He was also a member of the state Cattlemen's Association and served as chairman for one year. He was director of the Four Counties Cattlemen's Association for Placer, Nevada, Yuba and Sutter. He was also a member of the Kentucky Flat Farm Bureau. He is survived by his wife, **Alice**, of Penn Valley; two sons, **Clayton** of Elk Grove and **Jerry** of Davis; a daughter, Lois **Miles** of Oregon; two sisters, Olive **Page** of Yuba City and Hazel **Poole** of Smartville; and six grandchildren. Funeral services will be at 10 a.m. in the Nevada City Chapel of the Hooper & Weaver Mortuary, followed by internment in the Smartville Masonic Cemetery. Contributions to the Sierra Nevada Memorial Hospital memorial fund are preferred.

7 Oct 1986 (Tuesday); Page A-7 Col 3; Graveside funeral services were scheduled this afternoon at 2 p.m. at Smartville Masonic Cemetery for Alma Lou **Likens**, 76, who died Oct 3, 1986, at her Smartville home. A native of Texas, she had lived in the Yuba Sutter area since 1959. She is survived by her husband, H. B. **Likens** of Smartville; three sons, R. C. **Likens** and Donald **Likens**, both of Smartville, and Ben **Likens** of Marysville; five sisters, Essie **Proyer**, Jessie **Flatt**, Alpha **Russell**,

NEWSPAPER INFORMATION

Viola **Higenbotham** and Elsie **Day**, all of Texas; 11 grandchildren and six great-grandchildren. The Rev. David Dunkinson of the Olivehurst Seventh-day Adventist Church was to officiate at the services. Arrangements were under the direction of Hooper & Weaver Mortuary of Grass Valley. Memorials to the American Cancer Society have been suggested by the family.

8 Oct 1986 (Wednesday); Page A-9 Col 5; Services are scheduled Saturday for Ida Elsie **Murphy**, 96, a lifelong area resident, who died Oct. 7, 1986, at her Marysville home. Born in the Waldo area of Yuba County, Mrs. **Murphy** was the daughter of early Yuba County ranchers **Aden** and Mary **Wright**. She was the widow of the late John J. **Murphy**, a former Marysville mayor and councilman and Yuba County constable. She was a civilian employee at Camp Beale and for many years worked as a clerk for Hughes Variety in Marysville. Mrs. **Murphy** was a member and past president of the Native Daughters of the Golden West, Marysville Parlor 162; a member of the Mary Aaron Museum, and a life member of the Neighbors of Woodraft. Survivors include a daughter, Ruth M. **McDowell** of Marysville; a brother, Aden J. **Wright** of Chico; three grandchildren, Donald F. **Johnson** of Marysville, Elizabeth A. **Tomlinson** of Yuba City and Eileen **Hoogland** of Placentia; and four great grandchildren and one great great-grandchild. Services Saturday are scheduled for 10 a.m. at Lipp & Sullivan Chapel, with the Rev. David Pummill of the First Christian Church of Marysville and the Native Daughters of the Golden West officiating. Burial will be in the Smartville Masonic Cemetery. Visitation will be Friday from 10 a.m. to 8 p.m.

13 Nov 1986 (Thursday); Page A-11 Col 3; Graveside services were conducted at Smartville Masonic Cemetery today for Hance B. **Likens**, 74, of Smartville, who died Nov. 10, 1986, at Rideout Hospital. A native of Winters, Tex., he moved to Marysville in 1959. He moved to Smartville 15 years ago. He is survived by three sons, R. C. **Likens** and Donald **Likens** both of Smartville and Ben **Likens** of Marysville; 11 grandchildren and nine great grandchildren. The Rev. David Dunkinson officiated at the services. Funeral arrangements were under the direction of Hooper & Weaver Mortuary in Grass Valley.

3 Oct 1985 (Thursday); Page A-11 Col 1; Funeral services are planned at 11 a.m. Saturday at Lipp & Sullivan Chapel for Homer Clyde **Mills**, 61, who died Monday at his Timbuctoo home. A native of Colbert, Okla., he had been an area resident for 20 years. He was a self-employed mechanic for the past 40 years. He is survived by his wife, Maria Isabel **Mills** of Timbuctoo; three sons, Louis **Mills** of Sacramento, Steve **Mills** of Modesto and Jack **Mills** of Auburn; four daughters, Johanna **Mills** of San Diego, Claudia **Mills** of Modesto, Lorena **Mills** of Timbuctoo and Brenda **Mills** of Smartsville; three brothers, J. C. **Mills**, Billy **Mills** and Harry **Mills**, all of Las Cruces, N. M.; three sisters, Lucille **Dixon** of Las Cruces, N. M., Mildred **Young** of Campbell, Tex., and Ruby Jean **Walden** of Konowa, Okla.; and four grandchildren. Pastor Joseph Rice will officiate at the services. Burial will be in Smartville Cemetery. Friends may call at the chapel today from noon to 8 p.m. and Friday from 9 a.m. to 8 p.m.

NEWSPAPER INFORMATION

1 Apr 1986 (Tuesday); Page A-5 Col 3; A Mass of Christian Burial is planned Thursday in Grass Valley for Thelma **Quilty**, 81, a former Marysville resident. A native of Prosser, Wash., she died March 26 in Sacramento. She had traveled extensively in the western states with her husband, who was a diamond driller. They moved from Grass Valley to Smartville in 1939 and lived there until his death in 1964. She was the housekeeper at Sacred Heart Rectory in Maxwell for 12 years, then moved to Marysville until health problems required that she move to Sacramento. She is survived by a nephew, Gene **Hinkle** of Sacramento, and numerous grand nieces and nephews, all of Sacramento. Rosary will be recited tomorrow at 7:30 p.m. at Hooper & Weaver Mortuary in Grass Valley. The funeral mass will be Thursday at 10 a.m. in St. Patrick's Catholic Church, and burial will be in the Smartville Catholic Cemetery.

22 Apr 1987 (Wednesday); Page A-7 Col 3; A Smartville man shot to death as he charged Highway Patrol officers waving a machete was carrying small amounts of Cocaine and marijuana, Vallejo police reported Tuesday. George C. **Miller**, 56, also was carrying a stone pipe and other paraphernalia that police Detective George Bowart said may have been used to take drugs before Sunday's fatal confrontation. Toxicology tests will be performed to determine if Miller was indeed under the influence of drugs at the time. The drugs, and an unloaded .22-caliber pistol, were found when officers searched **Miller's** possessions after the shootout on the Carquinez Bridge. The shooting forced closure of the span and created a 10 mile traffic jam on Interstate 80. **Miller** stopped on the bridge after being chased by officers from a rest stop. Citizens at the rest stop had reported a man had threatened people with a machete. Police said **Miller**, wearing a rubber skull mask and military style fatigues, told officers he had been sent by God and was going to slash the officers and make their cars disappear. "He claimed he was the grim reaper and was sent by God to clean up the place," Highway Patrolman Bill Mulcrevy said shortly after the shooting. Officers attempted to negotiate with him, and finally turned a fire hose on him. **Miller** ran through the high-pressure stream and was shot by the officers.

23 Apr 1987 (Thursday); Page A-11 Col 6; Funeral services are scheduled Friday for George Charles **Miller**, 56, of Smartville, who was killed by California Highway Patrol officers during a tense standoff Sunday on the Carquinez Bridge in Vallejo. **Miller**, armed with a machete and wearing a rubber skull mask and military fatigues, had taken a swipe at a Solano County sheriff's deputy with the machete and was charging other law enforcement officers at the scene when he was fatally shot, the CHP, Vallejo police and Solano County sheriff's deputies reported. A native of Missouri, **Miller** had lived in the Yuba Sutter area four years. He belonged to the Elks Lodge of Montana, the Fraternal Order of Eagles, American Legion and the Veterans of Foreign Wars. He was a building contractor. He is survived by two sons, Mark A. **Miller** of Smartville and Anthony **Miller** of Wyoming; two brothers, Alfred **Miller** of Montana and Leo **Miller** of Missouri; four sisters, Mildred **Miller** of Pacifica, Donna **Fox** of Emeryville, Bernice **Victoria** of San Ramon and Vivian **Williams** of Colorado; and two grandchildren. Services are scheduled at 3 p.m. Friday at Hutchison & Carnes Colonial Chapel

NEWSPAPER INFORMATION

with burial in Smartville Masonic Cemetery. Visitations are scheduled until 8 o'clock tonight at the funeral home.

11 May 1987 (Monday); Page A-7 Col 3; Funeral services are scheduled Tuesday in Marysville for Eleanor Germain **Magonigal** of Smartville who died May 8, 1987, in Grass Valley. She was 84. A San Francisco native, she is survived by her sister, Mary **McQuaid** of San Francisco; two nieces, Mary F. **Scafire** of San Mateo and Dolores **Angelisch** of Tiburon; and a nephew, John **Keating** of Santa Rosa. Services are scheduled at 11 a.m. at St. Joseph's Catholic Church with the Rev. Thomas Marshall officiating. Burial will follow in Smartville Catholic Cemetery. A rosary will be said today at Hutchison & Carnes Colonial Chapel beginning at 7 p.m.

29 Dec 1987 (Tuesday); Page A-7 Col 5; Graveside services are scheduled at 2 p.m. Thursday at the Masonic Cemetery in Smartville for Cora May **Carney**, 97, of Smartville, who died Dec. 27, 1987, in Rideout Hospital. A Smartville resident for 50 years, **Carney** was a housewife. She was born Dec. 7, 1890, in Hayfork. Survivors include a stepson, William **Carney** of Grass Valley. The Rev. Joseph Rice will officiate over the services. Arrangements are being handled by Hooper & Weaver Mortuary of Grass Valley.

23 May 1989 (Tuesday); Page A-5 Col 5; Services are pending at Lipp & Sullivan Funeral Chapel for William Glenn **Gilliam**, 75, of Smartville, who died May 22, 1989, at Rideout Hospital.

26 May 1989 (Friday); Page A-11 Col 3; A Mass of Christian Burial is scheduled Tuesday at 10 a.m. at St. Joseph's Catholic Church for John J. **Barrett** Jr., 64, of Marysville, who died May 24, 1989, at Stanford University Hospital. A native of Marysville, he had lived in the Yuba Sutter area his entire life. He was a retired accountant. He was preceded in death by his father, former Marysville Assistant Fire Chief John **Barrett** Sr. and his mother, Maud **Barrett**. He is survived by six cousins, Mary **Barrett**, James H. **Barrett** Jr. and J. William **Young**, all of Marysville, Sister Ann **Stubbe** of Pacific Grove, Dick **Stubbe** of South San Francisco and Marjorie **Ranberg** of Benicia. A rosary will be recited at 7 p.m. Monday at Hutchison & Carnes Colonial Chapel. The Rev. Thomas Marshall will officiate at the services. Visitation will be from noon to 8 p.m. Monday at the Chapel. Burial will be in Smartville Catholic Cemetery.

27 May 1989 (Saturday); Page A-11 Col 3; Memorial services are scheduled Tuesday at 11 a.m. at Penn Valley Community Church for William "Glenn" **Gilliam** of Smartville who died May 22, 1989, at Rideout Hospital. He was 75. A native of Louisville, Ky., he had lived in the Yuba Sutter area 15 years. He had been in the road construction business for 35 years and was a heavy equipment operator. He was a member of Operating Engineer's Local No. 3. He was a member of Masonic Lodge F&AM of Yuba City. He is survived by his wife, Dorothy **Gilliam** of Smartville; two daughters, Emma Jean **Collins** of Smartville and Donna **Alexander** of Rio Linda; two brothers, Leslie **Gilliam** of Santa Rosa

NEWSPAPER INFORMATION

and V. B. **Gilliam** of Rancho Murrieta; a grandchild and four great grandchildren. Dr. Burt Hall of the Penn Valley Community Church will officiate at the services. Inurnment will be in Smartville. Memorials to the George Ohsawa Foundation of Macrobiotic Cancer Research, 1511 Robinson St., Oroville, Ca. 95965, have been suggested by the family. Arrangements are under the direction of Lipp & Sullivan Funeral Directors.

14 Jan 1991 (Monday); Page A-7 Col 5; Services are scheduled Tuesday for Charles Vernon **Clayton**, 82, of Smartville, who died Jan. 11, 1991, at Fremont Hospital. A native of Yreka, he had lived in the Yuba-Sutter area 25 years. He was a superintendent of construction for Teichert Construction Co. for 25 years. He was a member of the National Rifle Association, was a Master Shooter, a member of the San Francisco Police and Revolver Club, and the San Mateo Rifle Club. He was a member of Operating Engineers Union Local No. 3. He is survived by his wife, Esther **Clayton** of Smartville; a daughter, Jeanne **Wilkinson** of Smartville; a sister, Alma **Jones** of Anderson; nine grandchildren and three great grandchildren. Services will be at 2 p.m. Tuesday at Lipp & Sullivan Funeral Chapel with the Rev. Norman Nelson of Penn Valley Community Church officiating. Memorial contributions to the American Cancer Society or a charity of choice are suggested by the family.

25 Apr 1991 (Thursday); Page A-9 Col 6; Services are pending at Lipp & Sullivan Funeral Directors for William **Gruber**, 90, of Smartville, who died April 24, 1991, at Rideout Hospital.

29 Apr 1991 (Monday); Page A-5 Col 5; Graveside services are scheduled at 11 a.m. Tuesday at Smartville Cemetery for Alan M. **Campbell**, 46, of Portland, Ore. He died April 24, 1991, in an automobile accident near Port Orchard, Wash. A native of Washington, D.C., he was a former Smartville resident. He was self-employed as a consultant. Survivors include his wife, Brenda **Campbell** of Portland, Ore; three sons, Michael **Campbell** and Christopher **Campbell**, both of Portland Ore., and Brent **Campbell** of Binghamton, N.Y.; his parents, **George** and Mildred **Campbell** of Washington, D.C.; and a brother, Donald **Campbell** of Washington, D.C. Arrangements are under the direction of Lipp & Sullivan Funeral Directors.

30 Apr 1991 (Tuesday); Page A-7 Col 3; A Mass of Christian Burial will be said Friday at 10 a.m. at St. Joseph's Catholic Church for William Henry **Gruber**, 90, of Timbuctoo, who died April 24, 1991, at Rideout Hospital. A native of Hannibal, Mo., he had lived in the Yuba-Sutter area 50 years and had been a self-employed prospector for 30 years. He is survived by a nephew, Gene **Gruber** of El Paso, Texas; and a sister-in-law, Eveline **Gruber** of Scranton, N.D. The Rev. Patrick Lee will officiate at the services. Visitation will be Thursday from 2 to 8 p.m. at Lipp & Sullivan Chapel. Burial will be in Smartville Cemetery.

NEWSPAPER INFORMATION

27 Jan 1992 (Monday); Page A-7 Col 6; Arrangements are pending at Lipp & Sullivan Funeral Directors for George Albert **Rigby**, 82, of Smartville, who died Jan. 26, 1992, at Rideout Hospital.

15 Sep 1992 (Tuesday); Page A-5 Col 5; Graveside services are scheduled Wednesday at 2 p.m. at the Smartville Cemetery for Kyle Lawrence **Reese-Turner** who died Sept. 12, 1992, at Sutter Memorial Hospital in Sacramento. Born in Sacramento, he was 12 days old. He is survived by his parents, Kevin **Reese** of Smartville and Kristi **Turner** of Sacramento; a sister, Mikala **Reese** of Sacramento; his grandparents, **Richard** and Ramona **Reese** of Smartville, and Thomas **Turner** of Arkansas; and his step-grandfather and grandmother, **Richard** and Dianne **Ellsworth** of Sacramento. The Rev. Wayne Vincent of the Loma Rica Community Church will officiate at the services. Arrangements are under the direction of Hutchison & Carnes Colonial Chapel.

22 Sep 1992 (Tuesday); Page A-7 Col 5; Funeral services are scheduled for 10 a.m. Thursday at Hooper & Weaver Mortuary Chapel in Nevada City for Sidney Wallace **Fippin**, 58, of Smartville, who died Sept. 18, 1992, in Byron. A native of Smartville, he had lived there his entire life. He was a heavy equipment operator in the construction business and was a member of Operating Engineers Local No. 3. He had served as a member and past-president of the Smartville Volunteer Fire Department for 27 years. He was a member of the Cemetery board and had helped with the restoration of the Little Red Schoolhouse in Smartville. He is survived by his wife, Betty **Fippin** of Smartville, a daughter, two brothers and five sisters. Visitation will be from 10 a.m. to 7 p.m. Wednesday at the Nevada City Chapel. Burial will be in Smartville Cemetery. The family suggests donations to the Smartville Volunteer Fire Department.

7 Jan 1993 (Thursday); Page A-9 Col 3; Graveside services are scheduled for 2 p.m. Monday at Smartville Cemetery for Grass Valley resident Christopher Scott **Zeisloft**, 21, who died Jan. 2, 1993, at University of California-Davis Medical Center in Sacramento. A native of New Jersey, he grew up in Smartville and attended Marysville Schools. He is survived by his wife, Dianna M. **Zeisloft** of Grass Valley; a daughter, Tiffany **Zeisloft** of Grass Valley; two sisters, Stephanie **Zeisloft** of Smartville and Shannon **Zeisloft** of Grass Valley; and his grandparents, **Marvin** and Martha **Zeisloft** of Colfax. Arrangements are under the direction of Hooper & Weaver Mortuary in Nevada City. The family suggests memorials to the U.C. Davis Medical Center Life Flight.

6 Oct 1993 (Wednesday); Page A-2 Col 3; Private inurnment has been scheduled for Ralph Rule **Hudson**, 87, of Smartville, who died Oct 4, 1993, at Rideout Hospital. A native of Wyoming, he lived in the Yuba-Sutter area 30 years. He was a retired carpenter. He is survived by two sons, Robert **Hudson** of Cotati, and Richard **Hudson** of Smartville; a sister, Ruth **Sutherland** of Washington; and four grandchildren. Inurnment will be in Smartville Cemetery. Arrangements are under the direction of Hutchison & Carnes Colonial Chapel, Marysville.

NEWSPAPER INFORMATION

19 May 1994 (Thursday); Page C-2 Col 3; Hazel Mae **Poole**, 86, of Smartville, died May 17, 1994, at Fremont Medical Center. Born in Pleasant Valley, she had lived in the Yuba-Sutter area 65 years. She was a cattle rancher and a homemaker, and she was a member of a pioneer Yuba County family that settled in Smartville in 1863. Survivors include her husband of 68 years, Clarence **Poole** of Smartville; two sons, Bill **Poole** and Dan **Poole**, both of Smartville; a daughter, Barbara **Tuttle** of Rodeo; a sister, Olive **Page** of Yuba City; eight grandchildren and 11 great grandchildren. The family suggests donations to the Smartville Cemetery District, P. O. Box 198, Smartville, Calif. 95977. A service is scheduled for 11 a.m. Saturday at Ullrey Memorial Chapel. Visitations will be from noon to 8 p.m. Friday at the funeral chapel. Burial will be in Smartville Cemetery.

1 Jun 1994 (Wednesday); Page C-2 Col 5; Frances Mary **Rigby**, 76, of Smartville, died May 30, 1994, at her residence. Born in Illinois, she had lived in Smartville since 1947. She was a waitress for 20 years. Survivors include a daughter, Nora **Compton** of Marysville; a brother, Patrick **Collins** of Marysville; and a sister, Katherine **Ferguson** of Marysville. She was preceded in death by her husband, George **Rigby** in 1992, and a son, Robert **Rigby** in 1956. The family suggests memorials to the American Heart Association. Memorial services are scheduled for 2 p.m. Friday at Lipp & Sullivan Chapel. The Rev. Patrick Lee of St. Joseph Catholic Church will officiate.

25 Aug 1994 (Thursday); Page C-2 Col 2; Gladys D. **Gann**, 77, of Smartville, died Aug. 23, 1994, at her residence. Born in Sweetwater, Tenn., she lived in the Yuba - Sutter area for 20 years. She was a registered nurse for 15 years at Long Beach Memorial Hospital and also worked as the town nurse with the children of the community. She also served on the Election Board for many years and helped the Smartville Gleaners. She was the secretary-treasurer for the Smartville Women's Auxiliary. Survivors include her husband, Wayne **Gann** of Smartville; a son, Joel **Gann** of Smartville; a daughter, Brenda Gann-**Campbell** of Smartville; four grandchildren and seven great grandchildren. The family suggests donations to The Sharing Place, 5105 F St., Sacramento, CA 95819. Services are scheduled for 11 a.m. Saturday at Lipp & Sullivan Chapel. The Rev. Joseph Rice will officiate. Visitation will be from noon to 8 p.m. Friday at the chapel. Burial will be in Smartville Cemetery.

2 Jun 1995 (Friday); Page C2 Col 2; Bertha F. **Collins**, 84, of Penn Valley, died May 30, 1995, in Grass Valley. Born in Spokane, Wash., she was a former resident of Marysville and worked for the J. J. Newberry Company on D Street. During World War II, she was instrumental in founding the Marysville United Service Organization. After moving to Smartville, she and her husband owned and operated the Hillcrest Inn Restaurant and Tavern for 18 years. She was a Yuba County 4-H leader for 20 years, founding member of the Mooney Flat Volunteer Fire Department, member of the Immaculate Conception Church Altar Society, supporter of the Penn Valley Fire Department and involved in many other civic and Catholic organizations. Survivors include a son, Daniel **Collins** of Penn Valley; and two grandsons. She was preceded in death by her husband, Eugene **Collins**. In

NEWSPAPER INFORMATION

lieu of flowers, the family suggests memorials to Sierra Services for the Blind. A service will be at 10 a.m. today at St. Patrick's Catholic Church in Grass Valley. Burial will be in the Immaculate Conception Catholic Cemetery in Smartville. Arrangements are under the direction of Hooper & Weaver Mortuary.

4 Jul 1995 (Tuesday); Page C2 Col 5; Roberta "Bobbey" Beatty **Yankey**, 62, of Merced, died June 24, 1995, at David Grand United States Air Force Medical Center, Travis Air Force Base. A native of Smartville, she was raised in Livermore and returned to the Yuba-Sutter area from 1968 to 1972 before moving to Merced. She was a 1950 graduate of Livermore High School and a 1954 graduate in home economics from the University of California, Berkeley. She taught in elementary and high schools with teaching credentials in general elementary, general secondary and special home economics, was a realtor associate and a volunteer of the Merced-Mariposa chapter of the American Cancer Society. She was a member of Saint Luke's Episcopal Church in Merced, chairman of the UC Alumni Association's Scholarship Committee for the Merced County district, vice president/president of the Officers' Wives' clubs at several Air Force bases, secretary of several bowling leagues and the Merced Roaming Elks camper group and the Merced Board of Realtors' delegate to the National Association of Realtors' convention. She was active in various Merced High School groups including the Parent-Teacher-Student Association and received the Boy Scouts' Wawona District Award of Merit. Survivors include her husband of 39 years, John S. **Yankey** III of Merced; a son, Paul D. **Yankey** of Merced; and two sisters, Frances **Hilke** of Livermore and Margaret **Bateman** of Burlingame. The family suggests donations to Saint Luke's Episcopal Church, 350 W. Yosemite Ave., Merced, Calif. 95348, the American Cancer Society or to the Susan G. Koman Breast Cancer Foundation, 5005 LBJ Freeway, Suite 370, Dallas, Texas, 75244. A memorial service is scheduled at 2 p.m. Friday at St. Luke's Episcopal Church in Merced. Private inurnment will be in the Masonic Cemetery, Smartville. Arrangements are under the direction of Bryan-Baker Funeral House, Fairfield.

1 Jan 1996 (Monday); Page C2 Col 2; Clarence Francis **Poole**, 90, of Smartsville, died Dec. 29, 1995, at The Fountains. Born in Marysville, he was a lifetime resident of the Yuba-Sutter area. He worked as a self-employed rancher for 30 years. Survivors include two sons, Dan **Poole** and William **Poole** both of Smartsville; one daughter Barbara **Tuttle** of Rodeo; eight grandchildren, 11 great grandchildren, and 2 great great-grandchildren. He was preceded in death by his wife, Hazel Mae **Poole**. The family suggests memorials to Smartville Cemetery District, P.O. Box 198, Smartville, Calif. 95977, or to Loaves and Fishes, P.O. Box 2161, Sacramento, Calif. 95814. Graveside service will be at 11 a.m. Thursday at Smartville Cemetery with Jerry Russell officiating. Visitation will be from noon until 7 p.m. Wednesday at the Ullrey Memorial Chapel.

12 Mar 1997 (Wednesday); Page C2 Col 2; Derrel Austin **Pate** Jr., 38, of Yuba City died March 10, 1997, in Madera. A native of El Monte, he was a two-year resident of the Yuba-Sutter area. He was a chain-link fence installer. Survivors include one daughter, Lenore **Pate** of Madera; his father and stepmother, **Derrel**

NEWSPAPER INFORMATION

and Kathy **Pate** of Upland; his mother, Karen **Audette** of San Gabriel; one brother, Steven **Pate** of Marysville; and one sister, Tina **Castillo** of Upland. He was preceded in death by one sister, Lisa **Pate**. A graveside service will be at 2 p.m. Saturday at Smartville Cemetery. The Rev. Jim Edward of Smartville Community Church will officiate. Visitation will be from 2 to 8 p.m. Friday at Lipp & Sullivan Chapel in Marysville.

6 Mar 1998 (Friday); Page C2 Col 1 & 2; Aldon Lee **Greever**, 79, of Marysville died March 4, 1998, at Rideout Memorial Hospital. Born in Hooker, Okla., he was a Yuba-Sutter resident for 36 years. He was a mechanic for 14 years at Daoust Chevrolet. Survivors include his wife, Maxine **Greever** of Marysville; four sons, Jack **Greever**, Joe **Greever** and Jimmy **Greever**, all of Marysville, and Jim **Griffith** of Peoria, Ill.; two daughters, Ruby Ann **Brown** of Marysville and Deloris **Rose** of South Gate; one brother, Kyle W. **Greever** of Nevada City; 17 grandchildren and 19 great-grandchildren. He was preceded in death by a daughter Billie Lee **Lessley**. The family suggests memorials to the American Cancer Society. A funeral service will be conducted at 2 p.m. Monday at Lipp & Sullivan Chapel with the Rev. Ray Spence of Hall Street Community Church of Marysville officiating. Visitation will be from 2 to 8 p.m. Sunday. Burial will be in the Smartville Cemetery.

21 Apr 1998 (Tuesday); Page C2 Col 2 & 3; Maxine M. **Greever**, 78, of Marysville died April 19, 1998, at Rideout Memorial Hospital. Born in Oklahoma, she was a Yuba Sutter resident for 36 years. She was a homemaker. Survivors include three sons, Jack **Greever**, Joe **Greever** and Jimmy **Greever**, all of Marysville; two daughters, Ruby Ann **Brown** of Marysville and Deloris **Rose** of South Gate; 17 grandchildren and 19 great grandchildren. She was preceded in death by her husband of 58 years, Aldon Lee **Greever**, and a daughter, Billie Lee **Lessley**. The family suggests donations to the Cancer Society. A service will be held at 10 a.m. Thursday at Lipp & Sullivan Chapel, with the Rev. Ray Spence of Hall Street Baptist Church of Marysville officiating. Visitation will be from 2 to 8 p.m. Wednesday at the chapel. Burial will be in the Smartville Cemetery, Smartville.

6 May 1998 (Wednesday); Page C2 Col 3; Picture shown in obituary; Dorothy Grace **Gilliam**, 58, of Smartville died May 2, 1998, at her residence. Born in San Francisco, she was a Yuba Sutter resident for 21 years. She worked as a senior clerk for Pacific Gas & Electric Co. for 27 years and was a member of Order of Eastern Star, Marysville. Survivors include her mother, Grace **Smith** of Alameda. The family suggests donations to Valley Hospice, 970 Plumas St., Yuba City, Calif., 95991. A memorial service will be held at 11 a.m. Saturday at the Penn Valley Community Church in Penn Valley, with the Rev. Ron Hartman officiating. Inurnment will be in Smartville Cemetery. Arrangements are under the direction of Lipp & Sullivan Chapel, Marysville.

25 Jul 1998 (Saturday); Page C2 Col 2; Jackie Warren **Greever**, 54, of Marysville died July 22, 1998, at Rideout Memorial Hospital. Born in Elk City, Okla., he was

NEWSPAPER INFORMATION

a Yuba Sutter resident for 18 years. He worked for many years in the service station business for Exxon Corp. Survivors include his wife, Martha **Greever** of Marysville; one son, Jeremy **Greever** of Marysville; two daughters, Jennifer **Greever** of Baltimore, Md., and Susan **Barnett** of Oyster Bay, N. Y.; two brothers, Joe **Greever** and Jimmy **Greever**, both of Marysville; two sisters, Ruby Ann **Brown** of Marysville and Delores **Rose** of St. Regis, Mont.; and a grandchild. He was preceded in death by his parents, **Aldon** and Maxine **Greever**, and by a sister, Billie Lee **Lessley**. The family suggests memorials to the Fremont Rideout Foundation, 970 Plumas St., Yuba City, Calif. 95991. A graveside memorial service will be conducted at 2 p.m. Monday at Smartville Cemetery with the Rev. Ray Spence of Hall Street Baptist Church officiating. Arrangements are under the direction of Lipp & Sullivan Chapel.

9 Nov 2000 (Thursday); Page C2 Col 3; Wayne Jack **Gann**, 83, of Smartville died Nov. 6, 2000, at Rideout Memorial Hospital. Born in Etowah, Tenn., he was a Yuba-Sutter resident since 1976. He was a chemist for Arco Petroleum for 35 years. Survivors include one daughter, B. C. **Gann**-**Campbell** of Astoria, Ore.; one son, Joel **Gann** of Linda; two sisters, Ruby **Wilson** of Etowah, Tenn.; and Anna Lou **Reed** of Estes Park, Colo.; five grandchildren and eight great-grandchildren. He was preceded in death by his wife, Gladys **Gann**. Memorial contributions may be made to the American Cancer Society. Visitation will be held from 2 to 8 p.m. today at Lipp & Sullivan Chapel. A graveside service will be conducted at 2 p.m. Friday at Smartville Cemetery.

27 Dec 2000 (Wednesday); Page C2 Col 4 & 5; Wilberta M. **Allinio**, 96, of Yuba City died Dec. 21, 2000, at Sun Bridge Care Center in Marysville. Born in Spencerville, she was a Yuba-Sutter resident for 25 years. She was the theater manager for the San Francisco Civic Opera for 20 years. Survivors include one daughter, Patricia **Olson** of Turlock; one sister, Edna **Bodle** of Yuba City; four grandchildren; 10 great-grandchildren; and 10 great great-grandchildren. She was preceded in death by her husband, Pierre **Allinio**, and one brother, Stewart **Strollias**. Private family services will be held at a later date. Burial will be in Smartville Cemetery. Arrangements are under the direction of Lipp & Sullivan Chapel.

Birth Notices

17 Aug 1853; **Black**, Robert born in Marysville, recorded in book 0 page B; Sex Male; Father William **Black**; Mother Agnes **Black**

14 Dec 1880; **Black**, Effie Agnes born in Marysville, recorded in delayed book V61OR page 43; Sex Female; Father Robert **Black**; Age 27 of Marysville; Residence Grass Valley, California; Mother Eveline Maria **Sanford**; Age 25 of W. Gore, Nova Scotia; Residence Smartsville, California; Signed by Uncle: A. B. **Sanford** of Smartsville Recorded on 8 Apr 1941.

1 Aug 1896; **Wheaton**, Mabel May born in Smartsville, recorded in book 1 page W; Sex Female; Father A. G. **Wheaton**; Mother May **Fraser**; Sister Martha E. **Wheaton** Spencer of Marysville signed the Supplemental Birth Certificate.

28 Feb 1903; **Gleason**, Harold James born in Marysville, recorded in delayed book 4 page 92; Sex Male; Father Elmer Percy **Gleason** of California; Residence Rough & Ready, California; Mother Lydia Lee **Vineyard** of California; Residence Rough & Ready, California.

19 Jul 1905; **Poole**, Charles Francis born in Marysville, recorded in delayed book 5 page 12; Sex Male; Father Frank David **Poole**; Native of Yuba Co; Mother Mary Ellen **Skehan**; Native of Nevada Co.; This certificate states there is a marriage certificate dated 11/28/1926 – 11/30/1926.

24 Oct 1905; **Magonigal**, Thomas George born in Smartsville, recorded in book 2 page 202; Sex Male; Father Thomas **Magonigal**; Age 42 of California; Residence Smartsville; Mother Hannah **Bushby**; Age 32 of California; Residence Smartsville.

28 Jan 1907; **Magonigal**, Frank Webster born in Smartsville, recorded in book 2 page 202; Sex Male; Father Thomas **Magonigal**; Age 41 of California; Residence Smartsville; Mother Hannah **Bushby**; Age 33 of California; Residence Smartsville.

27 Feb 1909; **Bushby**, Irma born in Smartsville, recorded in book 3 page 65; Sex Female; Father Wm. **Bushby**; Age 43 of California; Residence Smartsville; Mother Fanny **Owen**; Age 36 of Missouri; Residence Smartsville; No of child to this mother 3. No of child now living 3.

27 Dec 1911; **Murphy**, Gladys Lillian born in Smartsville, recorded in O.R. book 3 page 239; Sex Female; Father John L. **Murphy**; Age 24 of California; Residence Smartsville; Mother Ida **Wright**; Age 21 of California; Residence Smartsville; No of child to this mother 2; No of child now living 2.

1 May 1914; **Bach**, John Joseph born in Marysville, recorded in book 3 page 385A; Sex Male; Father John Edward **Bach**; Age 43 of California; Residence Smartsville; Mother Edith May **Widener**; Age 33 of California; Residence Smartsville; No of child to this mother 1st; No of child now living 1.

BIRTH NOTICES

5 Oct 1915; **Fippin**, John Allen born in Marysville, recorded in book 4 page 1; Sex Male; Father Asa D. **Fippin;** Age 30 of California; Residence Smartsville; Mother Emma **Wheaton**; Age 21 of California; Residence Smartsville; No of child to this mother 3, number of child now living 3.

11 Sep 1917; **Spencer**, Carl Allan born in Marysville, recorded in book 1M page 92; Sex Male; Father Arthur E. **Spencer**, Age 24 of Marysville; Mother Martha E. **Wheaton**, Age 18 of Marysville; Number of child to this mother 1st, number of child now living 1.

31 May 1919; **Allen**, Josephine A. born in Marysville, Yuba Co., California; recorded in book 1-M page 235; Father Charles **Allen**, Residence Marysville, Age 48 of Michigan, Occupation Service man; Mother Josephine Henrietta **Sivope**, Age 31 of California, Occupation Housewife, Number of child to this mother 3, number of child now living 3.

24 Jan 1925; **Barrett**, John Joseph born in Marysville, Yuba Co., California; recorded in book 2M page 249; Father John Joseph **Barrett**, Residence Marysville, Age 35 of California, Occupation Foreman P.G.&E. Co.; Mother Maude Rena **Gordon**, Age 33 of California, Occupation Housewife, Number of child to this mother 2, number of child now living 2.

16 Mar 1934; **Welch**, Clarence Roy born in Marysville, recorded in book 6 page 96A; Sex Male; Father Clarence Roy **Welch**; Age 38 of Yuba Co, California; Residence Yuba Co.; Mother Winifred Helen **Kennedy**; Age 20 of Plumas Co.; Residence Yuba Co.; No of child to this mother 2; No of child now living 2.

3 Sep 1934; **Fippin**, Sidney Wallis born in Smartsville, recorded in book 6 page 150A; Sex Male; Father Asa David **Fippin**; Age 49 of Rough & Ready, California; Residence Smartsville; Mother Emma Annabelle **Wheaton**; Age 41 of Mooney Flat; No of child to this mother 11; no of child now living 9.

Marriage Notices

5 Mar 1858 in Marysville, Yuba Co., California, recorded in book 1, page 82-A; Groom Isaac B. **Otis** of Marysville, California; Bride Anna **Keith** of Marysville, California; Witness Levi **Mann**

2 Apr 1863 in Marysville, Yuba Co., California, recorded in book 1, page 169; Groom Wright **Clarke**; Bride Mary **Beatty**

19 Feb 1865 in Yuba Co., California, recorded in book 1, page 335; Groom James **Cranshaw** of Smartville, California; Bride Catherine E. **Cross** of Smartville, California

5 Nov 1865 in Yuba Co., California, recorded in book 2, page 12; Groom John **Boyer** of Smartsville, California; Bride Elizabeth "Lizzie" **Linehan** of Smartsville, California

27 May 1867 in Yuba Co., California, recorded in book 2, page 137; Groom Jacob **Pritchett**; Bride Mrs. Julia **Walsh**

27 Feb 1870 in Yuba Co., California, recorded in book 2, page 282A; Groom Daniel **Cory** of Smartville, California: Bride Mary E. **Morrison** of Smartville, California

29 Oct 1873, in Marysville, Yuba Co., California, recorded in book 3, page 77; Groom James O. **Williams** of Browns Valley, California; Age 24 native of Missouri; Bride Lizzie **Lowe** of Browns Valley, California; Age 16 native of California; Witness Jas. **Stall** of Marysville, California; Comments The consent of the father and mother of said Lizzie Lowe having been filed in my office

5 Nov 1875, in Marysville, Yuba Co., California, recorded in book 3A, page 167A; Groom Thomas **Haworth** of Yuba Co., California; Age 65 native of England; Bride Elizabeth **Painten** of Yuba Co. California; Age 63 native of Tennessee; Witness Thomas **Melody** of Yuba Co., California

20 Dec 1875, in Smartsville, Yuba Co., California, recorded in book 3A, page 134; Groom: W. W. **Chamberlin** of Smartsville, California; Age 42 native of Canada; Bride Annie Keith Otis of Smartsville, California; Age 37 native of Scotland; Witness John McConnell of Smartsville, California; Witness Timothy Brier of Smartsville, California; Comments: Groom was previously married. Annie Otis is widowed of Mr. **Otis** (deceased).

24 May 1876, in Smartsville, Yuba Co., California, recorded in book 3B, page 191; Groom Allen **Presley** of Smartsville, California; Age 26 native of Ireland; Bride Rachael R. E. **Flint** of Smartsville, California; Age 19 native of California; Witness **Flint** of Smartsville, California

MARRIAGE NOTICES

29 Jan 1877, in Marysville, Yuba Co., California, recorded in book 3, page 218A; Groom James **Bowman** of Rose Bar, California; Age 25 native of New York; Bride Louisa **Jones** of Lincoln, California; Age 17 native of Illinois; Witness Miss G. J. **Barchette** of Marysville, California; Witness Miss Laura **Kelser** of Virginia Twp., California; Comments: The consent of the Father of said Louisa Jones to said marriage filed in my office

9 Jan 1881, in Marysville, Yuba Co., California, recorded in book 3, page 374A; Groom Thomas William **Peckham** of Marysville, California; Age 25 Native of California; Bride Sophrona E. **Wallis** of Marysville, California; Age 20 native of California; Witness G. C. **McDonald** of Yuba Co., California; Witness Mrs. A. W. **Todd** of Marysville, California

25 Jun 1887, in Marysville, Yuba Co., California, recorded in book 4, page 478; Groom William Henry **Colling** of Yuba Co., California; Age 28 native of California; Bride Bertha **Jeffery** of Yuba Co., California; Age 22 native of California; Witness Miss C.M. **Infraham** of Marysville, California; Witness Miss I.L. **Belcher** of Marysville, California

17 Nov 1887, in Marysville, Yuba Co., California, recorded in book 5, page 13; Groom Wallace J. **Sanford** of Yuba Co., California; Age 23 native of California; Bride Eva C. **Jones** of Yuba Co., California; Age 20 native of California; Witness Mrs. **Willis** of Marysville, California; Witness Mrs. **Scofield** of Marysville, California

20 Oct 1889, in Yuba Co., California, recorded in book 5, page 307; Groom Charles K. **Colling** of Butte Co., California; Age 36 native of California; Bride Edith A. **Douglas** of Butte Co., California; Age 21 native of California; Witness Edward B. **Warne** of Oroville, California; Witness W. T. **Nelson** of Butte Co., California

4 Nov 1889, in Smartsville, Yuba Co., California, recorded in book 5, page 144; Groom Robert **Beatty** of Yuba Co., California; Age 34 native of California; Bride Miss Frances **Tifft** of Yuba Co., California; Age 27 native of California; Witness Simon A. **Davis** of Smartsville, California; Witness Jennie E. **Harris** of Roseville, California; Comments: I did not ascertain the nativity of the parents of the parties married. Neither of the above parties was married before

29 Jan 1890, in Smartsville, Yuba Co., California, recorded in book 5, page 167; Groom William B. **Magonigal** of Smartsville, California; Age 32 native of California; Bride Ella **Bowman** of Smartsville, California; Age 23 native of California; Witness Thomas **Magonigal** of Yuba Co., California; Witness Lizzie **Bowman** of Yuba Co., California

1 Nov 1891, in Marysville, Yuba Co., California, recorded in book 5, page 315; Groom William **Bushby** of Yuba Co., California; Age 26 native of California; Bride Fannie **Owen** of Yuba Co., California; Age 18 native of Missouri; Witness

MARRIAGE NOTICES

Susan **Wimberley** of Linda Township, California; Witness Missouri **Owen** of Linda Township, California; Comments: The father of William **Bushby** was born in the United States & his mother in Ireland. The parents of Miss F. **Owen** were born in the United States.

8 Nov 1891, in Lone Tree, Yuba Co., California, recorded in book 5, page 320; Groom Benjamin **Adkins** of California; Age 22; Bride Clara E. **Nevels** of Yuba Co., California; Age 18; Witness George **Haling** of Pleasant Valley, California; Witness Frank **Murch** of Lone Tree, California

28 Sep 1894 in Marysville, Yuba Co., California, recorded in book 6, page 49; Groom James Mortimer **Hapgood** of Yuba Co., California; Age 22 native of California; Bride Fannie Elizabeth **Marple** of Yuba Co., California; Age 24 native of California; Witness Miss Julia **Murphy** of Yuba Co., California; Witness S. O. Gunning of Marysville, California; Comments: Bridegrooms father & mother Ed **Hapgood** and Elizabeth **Hapgood**, natives; the former of Illinois, the latter of England. Brides parents William & Elizabeth **Marple**, natives of Pennsylvania. Neither of the parties has been married before.

6 Feb 1895, in Yuba Co., California, recorded in book 6, page 77; Groom Joseph D. **Emery** of Smartsville, California; Age 28 native of New York; Bride Mary V. **Rose** of Smartsville, California; Age 27 native of California; Witness Lee H. **Newbert*** of Marysville, California; Witness Ethel **Rose** of Smartsville, California; Comments: *The 1st, 2nd, 3rd letters of the last name is not readable, this is my best guess.

10 Dec 1895, in Yuba Co., California, recorded in book 6, page 141; Groom Brady **Compton** of Smartsville, California; Age 23 native of California; Bride Mary J. **Burns** of Smartsville, California; Age 19 native of California; Witness Morris **Murphy** of Smartsville, California; Witness Ida **Compton** of Smartsville, California

30 Oct 1898, in Smartsville, Yuba Co., California, recorded in book 7, page 30; Groom Joseph B. **Welch** of Spenceville, California; Age 30 naïve of California; Bride Alice May **Reed** of Spenceville, California; Age 23 native of California; Witness Bert **Reed** of Spenceville, California; Witness Belle **Wright** of Spenceville, California

4 Dec 1898, in Marysville, Yuba Co., California, recorded in book 7, page 32A; Groom John W. **Monk** of Marysville, California; Age 30 native of Illinois; Bride Mamie E. **Ball** of Marysville, California; Age 29 native of California; Witness W. E. **Monk** of Meridian, California; Witness James W. **Monk** of Meridian, California

25 Oct 1899 in Marysville, Yuba Co., California, recorded in book 7, page 53A; Groom William B. **Magonigal** of Smartsville, California; Age 42 native of California; Bride Ida **Hughes** of Marysville, California; Age 21 native of

MARRIAGE NOTICES

California; Witness Lula V. **Baumgartner** of Marysville, California; Witness: A. C. **Baumgartner** of Marysville, California

24 Feb 1901, in Smartsville, Yuba Co., California, recorded in book 7A, page 111; Groom John **Toland** of San Francisco, California; Age 28 native of California; Bride Elizabeth **Bach** of Smartsville, California; Age 28 native of California; Witness John **Bach** of Smartsville, California; Witness Miss C. **Frigaskis** of Milwaukee, Wisconsin

11 May 1901, in Marysville, Yuba Co., California, recorded in book 7A, page 118A; Groom Alfred **Sanford** of Smartsville, California; Age 31 native of California; Bride Emily T. **Murdock** of Smartsville, California; Age 31 native of Nova Scotia; Witness Richard **Belcher** of Marysville, California; Witness Lizzie A. **Morrissey** of Marysville, California

7 Dec 1901 in Marysville, Yuba Co., California, recorded in book 7, page 143; Groom James W. **Monk** of Marysville, California; Age 24 native of California; Bride Virginia E. **DaShiell** of Marysville, California; Age 23 native of California; Witness Julia **DaShiell** of Marysville, California; Witness M. E. **Monk** of Marysville, California

19 Nov 1902 in Smartsville, Yuba Co., California, recorded in book 7, page 176A; Groom William B. **Jeffery** of Smartsville, California; Age 34 native of California; Bride Ellen H. **Perkins** of Erle, California; Age 25 native of California; Witness John **Cramsie** of Smartsville, California; Witness Bertha **Whitesides** of Erle, California

21 Dec 1902, in Waldo, Yuba Co., California, recorded in book 7A, page 186; Groom John Edward **Bach** of Smartsville, California; Age 32 native of California; Bride Alice **Wright** of Waldo, California; Age 22 native of California; Witness Mac **Wright** of Waldo, California; Witness Arthur **Mooney** of Smartsville, California

4 Jul 1904, in Lone Tree, Near Smartsville, Yuba Co., California, recorded in book 7B, page 246; Groom Joseph **Bushby** of Rose Bar Township, California; Age 21 native of California; Bride Nora **Murch** of Smartsville, California; Age 27 native of California; Witness Ellen **Caine** of Rose Bar Township, California; Witness M. E. **Vineyard** of Rose Bar Township, California

11 Dec 1904, in Smartsville, Yuba Co., California, recorded in book 7B, page 262; Groom R.S. **Gassaway** of Smartsville, California; Age 48 native of California; Bride Mrs. Bertha **Colling** of Smartsville, California; Age 38 native of California; Witness Ruth **Colling** of Smartsville, California; Witness Earle **Colling** of Smartsville, California

2 Jul 1905, in Waldo, Yuba Co., California, recorded in book 7B, page 296A; Groom William F. **Hacker** of Nevada City, California; Age 26 native of California;

MARRIAGE NOTICES

Bride Cora E. **Peckham** of Waldo, California; Age 24 native of California; Witness William L. **Jones** of Waldo, California; Witness Mabel F. **Peckham** of Waldo, California

8 Nov 1911, in Marysville, Yuba Co., California, recorded in book 8, page 301; Groom Asa David **Fippin** of Rough & Ready, California; Age 26 native of California; Bride Emma Annabelle **Wheaton** of Smartsville, California; Age 18 native of California; Witness Mr. Albert Ernest **McCrea** of Smartsville, California; Witness Miss Mabel May **Wheaton** of Smartsville, California; Comments: Color of both parties is white. Groom is a farmer.

28 Jan 1913 in Smartsville, Yuba Co., California, recorded in book 8, page 394; Groom Albert Ernest **McCrea** of Smartsville, California; Age 26 native of California; Bride Mabel M. **Wheaton** of Smartsville, California; Age 16 native of California; Witness A. D. **Fippin** of Smartsville, California; Witness Emma **Fippin** of Smartsville, California; Comments: Consent of Brides parents given in writing

5 Dec 1916, in Marysville, Yuba Co., California, recorded in book 9, page 347; Groom Frank **Foreman** of Smartsville, California; Age 19 native of California; Father Jacob Washington **Foreman** of California; Mother Ida **Gassaway** of Penn Valley, California; Bride Blanche Violet **Bushby** of Smartsville, California; Age 20 native of California; Father William **Bushby** of California; Mother Fannie **Woens** of Missouri; Witness Mrs. W. H. **Nelson** of Yuba City, California; Witness Mrs. Lydia **Buck** of Marysville, California; Comments: 1st marriage for both parties. Color of both parties is white.

19 Nov 1933, in Vineyard Residence, Yuba Co., California, recorded in book 13, page 354; Groom James H. **Gleason** of Nevada City, California; Age 30 native of California; Father Elmer P. **Gleason** of California; Mother Lydia L. **Vineyard** of California; Bride Mildred L. **Vineyard** of Wheatland, California; Age 22 native of California; Father George E. **Vineyard** of California; Mother Iola Leone **Pittman** of California; Witness D. E. **Gleason** of Oakland, California; Witness Alice **Vineyard** of Wheatland, California; Comments: 1st marriage of both parties. Color of both parties is white. Groom is a forester.

25 Aug 1961, in Marysville, Yuba Co., California, recorded in book 34, page 106; Groom James Radford **Welch** of Marysville, California; Born 23 Apr 1932 native of California; Father Ray **Welch** of California; Mother Winifred **Kennedy** of California; Bride Sharlene Joyce **McClain** of Marysville, California; Born 17 Aug 1943 native of California; Father Gordon **McClain** of Oklahoma; Mother Lillian **Duvall** of Canada; Witness Gordon K. **McClain** of Marysville, California; Witness Joe G. **Welch** of Browns Valley, California

11 Jan 1963, in Yuba Co., California recorded in book 36, page 12; Groom Jackie Warren **Greever** of Marysville, California; Born 22 Nov 1943 native of: Oklahoma; Father Aldon **Greever** of Oklahoma; Mother Minnie **Maples** of Oklahoma; Bride Dorothy Jean **Allen** of Olivehurst, California; Born 18 Jul 1948

MARRIAGE NOTICES

native of California; Father Dale **Allen** of Arkansas; Mother Ruby **Dewitt** of Arkansas; Witness Deloris **Rose** of Downey, California; Witness Frank **Martin** of Marysville, California; Comments: Neither party has been previously married.

26 Sep 1975, in Marysville, Yuba Co., California, recorded in book 48, page 265; Groom Carnel **Lessley** of Yuba City, California; Born 25 Jul 1928 native of Oklahoma; Father Jeff **Lessley** of Oklahoma; Mother Ruby **Amos** of Oklahoma; Bride Betty **Martin** of Marysville, California; Born 12 Dec 1942 native of Oklahoma; Father Aldon Lee **Greever** of Oklahoma; Mother Maxine **Maples** of Oklahoma; Witness Kyle **Greever** of Marysville, California; Witness Ruby **Lessley** of Yuba City, California; Comments: 2nd Marriage for both parties.

24 July 1983, in Smartsville, Yuba Co., California, recorded in book 56, page 185; Groom Homer Clyde **Mills** of Smartsville, California; Born 3 Jan 1924 native of Oklahoma; Father Johnnie Clifton **Mills** of Texas; Mother Ruby Bel **Crabtree** of Texas; Bride Maria **Betancourt** of Smartsville, California; Born 22 Jun 1959 native of Chihuahua, Chihuahua; Father Francisco **Betancourt** of Chihuahua, Chihuahua; Mother Eloisa Herrera **Betancourt** of Chihuahua, Chihuahua; Witness Naomie Ruth **Rice** of Yuba City, California; Witness Kathleen **Shain** of Smartsville, California

Death Notices

25 Oct 1858; **Holmes**, Alfred died in Yuba Co., California, recorded in book 0 page H; Sex Male; Age at death 3 yrs; Birthplace Nevada Co., California

8 Feb 1859; **Crankshaw**, Elizabeth died in Yuba Co., California, recorded in book 0 page C; Sex Female; Age at death 42 yrs; Birthplace England; Status Married.

15 Apr 1859; **Daly**, Michael died in Yuba Co., California, recorded in book 0 page D; Age at death 31 yrs; Birthplace Ireland; Status Single; Occupation Miner; Cause of death was caving in of bank.

30 Jan 1860; **Beatty**, Alice died in Yuba Co., California, recorded in book 0 page B; Sex Female; Age at death 4 yrs 10 mos 3 dys; Born Goodhue Creek, Sierra Co., California; Status: Single.

22 Feb 1860; **Campbell**, Daniel died in Yuba Co., California, recorded in book 0 page C; Age at death 35 yrs; Birthplace Deneglan County, Ireland; Status Single; Occupation Miner; Cause of death was caving in of bank.

20 Sep 1873; **Haworth**, Thomas died in Yuba Co., California, recorded in book 1 page H; Sex Male; Age at death 31 yrs; Birthplace England; Status Single; Residence Marysville.

10 Oct 1873; **Cory**, William died in Yuba Co., California, recorded in book 1 page C; Sex Male; Born in California.

3 Sep 1889; **Havey**, Amanda Veronica died in Yuba Co., California, recorded in book 1 page H; Sex Female; Age at death 5 yrs 6 mos 8 dys; Born in California; Informant R. W. **Tifft**, M. D.

27 Feb 1890; **Magonigal**, William died in Yuba Co., California, recorded in book 1 page M; Sex Male; Age at death 72 yrs; Birthplace Ireland; Status Married; Residence Smartsville.

26 Apr 1890; **Divver**, Mary A. died in Yuba Co., California, recorded in book 1 page D; Sex Female; Age at death 54 yrs 4 mos 1 dys; Born in Ireland; Status Widowed; Informant Ortis F. **Lee**, Physician

9 Nov 1890; **Colling**, Infant son of Mrs. **W. H.** died in Yuba Co., California, recorded in book 1 page C; Sex Male; Age at death 4 dys; Birthplace California; Status Single; Residence Yuba Co.; Informant R. W. **Tifft**.

16 Dec 1891; **Wallace**, Ellen died in Yuba Co., California, recorded in book 1 page W; Sex Female; Age at death 80 yrs; Birthplace Scotland; Status Widow.

DEATH NOTICES

6 Jan 1892; **O'Brien**, Phillip died in Yuba Co., California, recorded in book 1 page O; Age at death 80 yrs; Birthplace Ireland; Status Widower; Cause of death old age.

28 Jul 1892; **Pryor**, Michael died in Yuba Co., California, recorded in book 1 page P; Age at death 62 yrs 3 mos 4 dys; Birthplace Ireland; Status Married.

11 Nov 1892; **Magonigal**, Nancy died in Yuba Co., California, recorded in book 1 page M; Sex Female; Age at death 71 yrs 8 mos 26 dys; Birthplace Ireland; Status Widowed.

12 Jan 1893; **Gunning**, Infant son of Mrs. S. O. **Gunning** died in Yuba Co., California, recorded in book 1 page G; Born 12 Jan 1893 in California; Cause of death was premature birth.

25 Feb 1893; **Tifft**, R. W. died in Yuba Co., California, recorded in book 1 page T; Sex Male; Age at death 61 yrs 4 mos 2 dys; Birthplace Ohio.

30 Apr 1893; **Byrne**, Thos. P. died in Yuba Co., California, recorded in book 1 page B; Age at death 12 yrs 2 mos 10 dys; Birthplace Smartsville; Status Single; Gunshot wound.

22 Nov 1893; **Welch**, Mrs. J. F. died in Yuba Co., California, recorded in book 1 page W; Sex Female; Age at death 69 yrs 8 mos 26 dys; Birthplace Kentucky; Status Married.

13 Nov 1894; **Fisher,** Gracie E. died in Yuba Co., California, recorded in book 1 page F; Sex Female; Age at death 11 yrs 8 mos 4 dys; Birthplace California; Status Single.

13 Nov 1894; **O'Brien**, Mary died Yuba Co., California, recorded in book 1 page O; Age at death 62 yrs 6 mos 28 dys; Birthplace Ireland; Status Married.

15 Nov 1894; **Dewain (Dewan)**, John died in Yuba Co., California, recorded in book 1 page D; Age at death 16 yrs 4 mos 13 dys; Birthplace California; Status Single; Cause of death broken neck.

20 Feb 1895; **Flanagan**, M. (Michael) died in Yuba Co., California, recorded in book 1 page F; Age at death 57 yrs; Birthplace Ireland; Status Married.

9 Jun 1895; **Dewain (Dewan)**, John died in Yuba Co., California, recorded in book 1 page D; Age at death 56 yrs 9 mos; Birthplace Ireland; Status Married.

28 Nov 1895; **Pryor**, Mary died in Yuba Co., California, recorded in book 1 page P; Age at death 64 yrs; Birthplace Ireland; Status Widow; Cause of death pneumonia.

DEATH NOTICES

6 Jan 1896; **Quinkert**, Elizabeth died in Yuba Co., California, recorded in book 1 page Q; Age at death 67 yrs; Birthplace Germany; Status Married.

7 Jan 1896; **Magonigal**, Ellen J. died in Yuba Co., California, recorded in book 1 page M; Sex Female; Age at death 30 yr 9 dys; Birthplace California; Status Married.

27 Jan 1896; **Casey**, Margaret J. died in Yuba Co., California, recorded in book 1 page C; Age at death 24 yrs 10 mos; Birthplace California; Status Single.

27 Feb 1896; **Mooney**, Nellie died in Yuba Co., California, recorded in book 1 page M; Sex Female; Age at death 21 yrs 5 mos 3 dys; Birthplace California; Status Single.

2 Jun 1897; **Conboy**, Mrs. Margaret died in Yuba Co., California, recorded in book 1 page C; Age at death 46 yrs; Birthplace Ohio; Status Married; Chronic Brights Disease.

18 Sep 1897; **Horan**, Michael died in Yuba Co., California, recorded in book 1 page H; Residence Smartsville; Age 39 yrs in Ireland; Status Married; Died of chronic Brights disease; Informant Geo. S. **Holbrook**, physician.

9 Feb 1898; **Conlin**, Thomas died in Yuba Co., California, recorded in book 1 page C; Age at death 65 yrs 9 mos 9 dys; Birthplace Ireland; Status Married; Died of Exhaustion.

23 Apr 1898; **Connery**, Mary died in Yuba Co., California, recorded in book 1 page C.

24 May 1898; **Murphy**, Catherine died in Yuba Co., California, recorded in book 1 page M; Born in New Jersey; Status Married; Died of Exhaustion, F. F. **Hollbrook**, physician.

7 Aug 1898; **Smith**, Anthony died in Yuba Co., California, recorded in book 1 page S; Age at death Status Married.

5 Mar 1899; **McNutt**, J. F. died in Yuba Co., California, recorded in book 1 page M; Sex Male; Age at death 84 yrs 1 mos 25 dys; Birthplace Tennessee.

25 Apr 1899; **Bowman**, Jonie died in Yuba Co., California, recorded in book 1 page B; Sex Female; Age at death 73 yrs 3 mos 28 dys; Birthplace Ireland; Status Widowed.

20 Jun 1899; **McQuaid**, Mary died in Smartsville, Yuba Co., California, recorded in book 1 page M; Age at death 63 yrs; Birthplace Massachusetts; Status Married; Cause of death chronic Nephritis.

DEATH NOTICES

10 Jul 1899; **Henderson**, Bridget died in Yuba Co., California, recorded in book 1 page H; Age at death 57 yrs; Birthplace Ireland; Status Widow.

13 Sep 1899; **Colling**, Wm. H. died in Yuba Co., California, recorded in book 1 page C; Sex Male; Age at death 44 yrs; Birthplace California; Status Married; Residence Yuba Co.; Informant G. **Holbrook.**

15 Sep 1899; **Dempsey**, Annie died in Yuba Co., California, recorded in book 1 page D; Residence Smartsville; Age at death 60 yrs; Birthplace Ireland; Status Married; Rupture of gall bladder.

4 Nov 1899; **Pritchett**, J. H. died in Yuba Co., California, recorded in book 1 page P; Sex Male; Age at death 77 yrs; Birthplace New Jersey; Status Married.

5 Dec 1899; **Rooney**, Mary died in Yuba Co., California, recorded in book 1 page R; Age at death 43 yrs 4 mos 18 dys; Status Married.

20 May 1900; **McNalley**, Mary died in Yuba Co., California, recorded in book 1 page M; Age at death 55 yrs; Birthplace Ireland; Status Married; Died of heart disease.

30 Jan 1901; **Doyle**, James died in Yuba Co., California, recorded in book 1 page D; Sex Male; Age at death 28 yrs; Birthplace California; Status Single; Informant Geo. Holbrook.

15 Feb 1901; **Walsh**, Mary died in Yuba Co., California, recorded in book 1 page W; Age at death 72 yrs; Birthplace Ireland; Status Married.

18 Jan 1902; **Jeffrey**, M. L. died in Yuba Co., California, recorded in book 1 page J; Sex Female; Age at death 61 yrs; Birthplace Pennsylvania; Status Married.

9 Jun 1902; **Walsh**, J. (John) died in Yuba Co., California, recorded in book 1 page W; Age at death 74 yrs; Birthplace Ireland; Status Widower.

30 Sep 1902; **Byrne**, Mary Roseann died in Yuba Co., California, recorded in book 1 page B; Age at death 55 yrs; Birthplace New Jersey; Cancer.

29 Sep 1903; **Welch**, Thomas died in Yuba Co., California, recorded in book 1 page W; Sex Male; Age at death 56 yrs; Birthplace Missouri; Status Single.

8 Dec 1903; **Bowman**, Lizzie died in Yuba Co., California, recorded in book 1 page B; Sex Female; Age at death 40 yrs 11 mos 4 dys; Birthplace California; Status Single.

9 Jan 1904; **Walsh**, Edward died in Yuba Co., California, recorded in book 1 page W; Age at death 40 yrs; Birthplace California; Status Single; Cause of death was diabetes.

DEATH NOTICES

18 Jan 1904; **Linehan**, E. Thomas died in Yuba Co., California; recorded in book 1 page L; Age at death 33 yrs; Birthplace California; Status Single; Occupation Farmer; Tumor of Brain.

11 May 1904; **Flanagan**, Peter died in Yuba Co., California, recorded in book 1 page F; Age at death 59 yrs; Birthplace Ireland; Status Single; Occupation Merchant; Cause of death was paralysis.

2 Sep 1904; **Conrath**, Louis died in Yuba Co., California, recorded in book 1 page C; Age at death 49 yrs in Missouri; Status Married; Occupation Moulder; Cause of death was suicide cut throat.

29 Sep 1904; **Casey**, Peter died in Yuba Co., California, recorded in book 1 page C; Age at death 76 yrs; Birthplace Ireland; Status Married; Occupation Farmer; Died of Old age.

13 Nov 1904; **Dempsey**, Rose died in Yuba Co., California; recorded in book 1 page D; Age at death 1 year; Birthplace California; Status child; Cause of death was heart failure.

3 Nov 1905; **Smith**, Mrs. Bessie died in Yuba Co., California, recorded in book 1 page S; Sex Female; Age at death 31 yrs 23 dys; Birthplace Canada; Status Married.

21 Dec 1905; **Havey**, Anna Lucile died in Smartsville, Yuba Co., California, recorded in book 2 page 1; Sex Female; Born 13 Feb 1896 in Smartsville, California; Occupation School Girl; Father John C. **Havey** of New Jersey; Mother Mary Ann **Pryor** of New York; Informant Jno C. Havey of Smartsville, California.

29 Jan 1906; **Harriman**, Elsie died in Yuba Co., California, recorded in book 2 page 4; Sex Female; Born 30 Jun 1871 in New Jersey; Status Married; Residence Smartsville; Occupation Housewife; Father Walter **Platt** of New Jersey; Mother Emma **Pill** of New Jersey.

15 May 1906; **Leahy**, Mike died in Smartsville, Yuba Co., California, recorded in book 2 page 8; Age at death 66 yrs; Birthplace Ireland; Status Single; Occupation Laborer; Informant J. K. **Kelly** – Coroner.

23 Nov 1906; **Dougherty**, Ann died in Smartsville, Yuba Co., California, recorded in book 2 page 16; Age at death 71 yrs; Birthplace Co. Longford, Ireland; Status Married; Occupation Housewife; Father of Ireland; Mother of Ireland; Informant Danl **Dougherty** of Smartsville, California.

30 Jan 1907; **Marple**, Whitten died in Timbuctoo, Yuba Co., California, recorded in book 2 page 17; Sex Male; Age at death 84 yrs; Birthplace Pennsylvania; Status Widowed; Residence Timbuctoo; Occupation Miner; Father Witten **Marple**; Informant J. H. **Warne** of Smartsville, California.

DEATH NOTICES

31 Jul 1907; **Jeffery**, George Archibald died in Yuba Co., California, recorded in book 2 page 26; Sex Male; Born 21 Dec 1829 in Nova Scotia; Status Widowed; Occupation Blacksmith; Informant Ella **Jeffery** of Marysville, California.

21 Aug 1907; **Peardon**, John died in Yuba Co., California, recorded in book 2 page 27; Sex Male; Born 24 Aug 1843 of England; Status Widowed; Residence Smartsville; Occupation Carpenter; Father James **Peardon** of England.

9 Oct 1907; **Schmidt**, Veronica died in Smartsville, Yuba Co., California, recorded in book 2 page 30; Age at death 81 yrs 15 dys; Birthplace Germany; Status Widowed; Occupation Hotel Keeper; Father Joseph **Merz** of Germany; Informant Veronica **Manasco** of Smartsville, California.

10 Dec 1907; **Flanagan**, John Henry died in Yuba Co., California; recorded in book 2 page 31 A; Born 20 Jan 1868 of New Jersey; Status Divorced; Occupation Miner; Father Michael **Flanagan** of Ireland; Mother Ellen **Conell** of Ireland; Informant Thos. J. **Flanagan** of Smartsville, California.

3 Mar 1908; **Sweeney**, Annie died Yuba Co., California; recorded in book 1 page 151-A; Age at death 75 yrs; Birthplace Ireland; Status Widow; Father of Ireland; Mother of Ireland; Informant Mrs. Anna **Haggerty** of Marysville, California.

27 Jun 1908; **Young**, John J. died in Yuba Co., California, recorded in book 1 page 170-A; Age at death 67 year in New York; Status Widowed; Occupation Miner; Informant Mr. **Vanderpool** – Yuba Co. hospital records, Marysville, California.

4 Jul 1908; **Vineyard**, Mary Emily died in Smartsville, Yuba Co., California, recorded in book 2 page 36 A; Sex Female; Born 7 Feb 1842 in Platville, Wisconsin; Status Married; Residence Smartsville; Occupation Housewife; Father James **Hammonds** of Missouri; Mother Elizabeth **Meredith** of Missouri; Informant John T. **Vineyard** of Smartsville, California

4 Nov 1908; **Bowman**, Henry Gordon died in Earle, Yuba Co., California, recorded in book 2 page 44; Sex Male; Age at death 54 yrs; Birthplace New York; Status Single; Residence Earle, Yuba Co., California; Occupation Clerk; Father William **Bowman** of Ireland; Mother Jane **Hurst** of Ireland; Informant George **Akin** of Earle, California.

13 Nov 1908; **Conlin**, Margaret died in Smartsville, Yuba Co., California, recorded in book 2 page 45; Born 1 Aug 1835 in Ireland; Status Widow; Father Michael **Mulhare** of Ireland; Mother of Ireland; Informant Miss Josie **Conlin** of Smartsville, California.

14 Nov 1908; **Magonigal**, Frank Webster died in Yuba Co., California, recorded in book 2 page 42; Sex Male; Born 28 Jan 1907 in California; Residence Smartsville; Father T. G. **Magonigal** of Roses Bar, California; Mother Hannah **Bushby** of Roses Bar, California; Informant Mrs. W. B. **Magonigal** of Smartsville, California.

DEATH NOTICES

14 Feb 1909; **Mellarkey**, Edward died in Smartsville, Yuba Co., California, recorded in book 2 page 51A; Born 12 May 1814 in Ireland; Status Married; Occupation Miner.

21 Jul 1909; **Caine**, Daniel died in Marysville, Yuba Co., California, recorded in book 2 page 59; Born Apr (no dates given); Age at death 83 yrs; Birthplace Ireland; Status Married; Occupation Farmer; Father Richard **Caine** of Ireland; Mother of Ireland; Informant Mrs. Daniel **Caine** of Smartsville, California.

24 Jul 1909; **Rooney**, Eugene F. died in Marysville, Yuba Co., California, recorded in book 2 page 58; Age at death 28 yrs 1 mos 8 dys; Birthplace California; Status Single; Occupation Laborer; Father M. **Rooney** of Ireland; Mother Mary Sanders of New York; Informant Mr. John Sanders of Smartsville, California.

20 Aug 1909; **McCarty**, Andrew died in Smartsville, Yuba Co., California, recorded in book 2 page 60 A; Age at death 76 yrs; Birthplace Ireland; Status Married; Occupation Laborer; Informant Rose **McCarty** of Smartsville, California.

11 Jan 1910; **Divver**, William F. died in Marysville, Yuba Co., California, recorded in book CM 1 page 268; Sex Male; Born 3 Feb 1869 in California; Status Divorced; Occupation Harness Maker; Father Chas. **Divver** of Ireland; Mother Mary **Buller** of Alamba; Informant Mrs. Maggie **Hoskins** of Santa Rosa, California.

14 Dec 1910; **Gunning**, Samuel O. died in Marysville, Yuba Co., California, recorded in book 1 page CM 324: Residence Marysville; Age at death 75 yrs 11 mos 14 dys; Birthplace Ireland; Status Married; Occupation County Recorded; Father S. O. **Gunning** of Ireland; Mother of Ireland; Informant Mrs. S. O. **Gunning** of Marysville, California.

23 Dec 1910; **Ryan**, Patrick died in Marysville, Yuba Co., California, recorded in book 1 page 325; Age at death 79 yrs; Birthplace Ireland; Status Widower; Informant Mary J. **Ryan** of Marysville, California.

3 Sep 1911; **Divver**, George M. died in Yuba Co., California, recorded in book 2 page 28; Sex Male; Born 5 Aug 1871 in California; Status Single; Occupation Clerk; Father Charles **Divver** of Ireland; Mother is of Ireland; Informant B. B. **Divver** of Marysville, California.

1 Dec 1911; **Burns**, John died in Smartsville, Yuba Co., California, recorded in book 2 page 96; Age at death 70 yrs; Birthplace Ireland; Occupation Miner.

8 Jan 1912; **Linehan**, Ellen died in Yuba Co., California, recorded in book 2 page 97; Age at death 75 yrs; Birthplace Mayo, Ireland; Status Married; Occupation Housewife; Father Peter **Kelly** of Ireland; Mother **Madden** of Ireland; Informant Timothy **Linehan** of Smartsville, California.

DEATH NOTICES

18 May 1912; **Cramsie**, William died in Yuba Co., California, recorded in book 2 page 102; Born 18 Mar (no year given); Age at death 74 yrs 9 mos 3 dys; Birthplace Ireland; Father John **Cramsie** of Ireland; Mother Margaret Jane **Dougherty** of Ireland; Informant W. P. **Cramsie** of Marysville, California.

7 May 1912; **Jeffery**, Ella died in Yuba Co., California, recorded in book 2 page 85; Sex Female; Born 21 Mar 1863 in California; Status Single; Residence Yuba Co.; Father George **Jeffery** of Nova Scotia; Mother Mary **Davis** of Pennsylvania; Informant Mrs. T. J. **Tyrell** of Marysville, California.

1 Feb 1913; **Sanders**, Annie died in Yuba Co., California, recorded in book 2 page 118-A; Born 22 Nov 1853 in New Jersey; Status Single; Father Daniel **Sanders** of Ireland; Mother Beirne (should read Byrne) of Ireland; Informant John **Sanders** of Smartsville, California.

26 Apr 1913; **Dougherty**, Daniel John Ambrose died in Marysville, Yuba Co., California, recorded in book 2 page 120; Born 27 Mar (no year listed); Age at death 37 yrs 1 mos; Birthplace California; Status Single; Occupation Laborer; Father Daniel **Dougherty** of Ireland; Mother Anne **Brady** of Ireland; Informant Mrs. Grace **Dougherty** of Smartsville, California.

2 Jul 1913; **Murphy**, Gladys Lillian died in Marysville, Yuba Co., California, recorded in book 2 page 124; Sex Female; Born 27 Dec 1913 in California; Status Single; Residence Marysville; Father John J. **Murphy** of Nevada; Mother Ida **Wright** of California; Informant John J. **Murphy** of Waldo, California.

3 Sep 1913; **Driscoll**, Catherine died in Smartsville, Yuba Co., California, recorded in book 2 page 126; Age at death 75 yrs (1838); Birthplace County Cork, Ireland; Status Widowed; Father John **Driscoll** of County Cork, Ireland; Mother Mary **Sweney** of County Cork, Ireland; Informant John F. **Dempsey** of Smartsville, California.

5 Nov 1913; **McQuaid**, John died in Smartsville, Yuba Co., California, recorded in book 2 page 130; Born 10 Dec 1837 in Ireland; Occupation Retired; Informant Mrs. Wm. **Allread** of Smartsville, California.

6 Jul 1914; **Dougherty**, Daniel died in Marysville, Yuba Co., California, recorded in book 2 page 140; Age at death 89 yrs; Birthplace Ireland; Status Widowed; Occupation Retired; Father of Ireland; Mother of Ireland; Informant Kate **Dougherty** of San Francisco, California.

23 Oct 1914; **Mitchell**, Elvira Bell died in Yuba Co., California, recorded in book 2 page 244; Sex Female; Born 7 Oct 1879 in California; Status Married; Father Alden **Wright** of California; Mother Mary **Welch** of California; Informant Thomas **Mitchell** of Marysville, California.

DEATH NOTICES

23 Feb 1915; **Ford**, Patrick died in Yuba Co., California, recorded in book 2 page 268; Age at death 71 yrs; Birthplace Ireland; Status Single; Occupation Miner; Father Patrick **Ford** of Ireland; Mother of Ireland; Informant Chas. H. **Bell** of Marysville, California.

7 Jul 1915; **Sanders** (**Saunders**), Daniel died in Marysville, Yuba Co., California, recorded in book 2 page 291; Residence Smartsville; Age at death 90 yrs 2 mos 7 dys in Ireland; Status Married; Occupation Miner; Father of Ireland; Informant Yuba Co., Hospital records in Marysville, California.

6 May 1917; **Looney**, Catherine died in Yuba Co., California, recorded in book 3 page 35; Age at death 90 yrs; Birthplace Ireland; Status Widowed; Occupation At home; Father **Haggerty** of Ireland; Mother of Ireland; Informant Mrs. Chas **Sperbech** of Browns Valley, California.

21 Sep 1917; **Hickeson**, Katie Jane died in Yuba Co., California, recorded in book 3 page 42; Sex Female; Born 31 Dec 1869 in California; Status Married; Spouse William **Hickeson;** Occupation Housewife; Father Geo. **Bushby** of England; Mother **Dougherty** of Ireland; Informant Wm. **Bushby** of Hammonton, California.

5 Oct 1917; **Casey**, William Joseph died in Yuba Co., California; recorded in book CM 3 page 36; Born 6 Dec 1864 in California; Status Widowed; Occupation Miner; Father Peter **Casey** of Ireland; Mother Winifred **Howsaw** of Ireland; Informant Mrs. Jas. McWilliams of Marysville, California.

1 Feb 1918; **Foreman**, Violet Lillian died in Yuba Co., California, recorded in book 3 page 48; Sex Female; Born 2 Aug 1917 in California (Date on stone and here do not match); Father Frank **Foreman** of California; Mother Blanche **Bushby** of California; Informant Frank **Foreman** of Marysville, California.

5 Mar 1921; **Bowman**, William Henry died in Yuba Co., California, recorded in book 3 page 283; Sex Male; Born 24 Feb 1877 in California; Status Divorced; Residence Smartsville; Occupation Carpenter; Father James **Bowman** of New York; Mother Eliza **Jones** of Illinois; Informant F. W. **Bowman** of Marysville, California.

7 Sep 1921; **Sheehan**, Julia died in Smartsville, Yuba Co., California, recorded in book 3 page 116; Age at death 85 yrs; Birthplace Ireland; Status Widow; Father of Ireland; Mother of Ireland; Informant D. F. **McAuliffe** of Marysville, California.

11 May 1922; **Sheehan**, Daniel Francis died in Yuba Co., California, recorded in book 3 page 341; Born 22 Feb 1862 in California; Status Single; Occupation Mining Engineer; Father Patrick **Sheehan** of Ireland; Mother Julia **McAuliffe** of Ireland; Informant Josie C. **Sheehan** of Marysville, California.

8 Jun 1922; **Linehan**, John Henry died in Marysville, Yuba Co., California, recorded in book 3 page 128; Residence Smartsville; Born 27 Feb 1863 in

DEATH NOTICES

California; Status Married; Spouse Anna **Linehan**; Occupation Blacksmith; Father Timothy **Linehan** of Ireland; Mother Ellen **Kelly** of Ireland; Informant Mrs. Anna **Linehan** of Smartsville, California.

13 Jun 1922; **Poole**, Francis died in Yuba Co., California, recorded in book 3 page 129; Sex Male; Born 20 Feb 1832 in England; Status Married; Spouse Mary Ann Poole; Occupation Farmer; Father Wm. **Poole** of England; Mother of England; Informant Frank **Poole** of Marysville, California

23 Jan 1923; **Welch**, James died in Yuba Co., California, recorded in book 3 page 140; Sex Male; Born 28 Apr 1861 in California; Status Married; Spouse Lillie **Welch**; Occupation Farmer; Father R. **Welch** of Missouri; Mother Creen **Bast** of Missouri; Informant Lillie **Welch** of Smartsville, California.

1 Apr 1923; **Mellarkey**, James S. died in Marysville, Yuba Co., California, recorded in book 4 page 24; Born 4 May 1857 in Smartsville; Status Married; Occupation Miner, retired; Father Edward **Mellarkey** of Ireland; Mother **McLean** of Ireland; Informant Mrs. Kate **House** of Marysville, California.

31 Jul 1923; **Woodroffe**, Sherman E. died in Marysville, Yuba Co., California, recorded in book 4 page 362; Sex Male; Born 16 Apr 1880 in California; Status Married; Spouse Janie **Woodroffe**; Occupation Oilier Marysville Dredging Co.; Father John **Woodroffe** of Nova Scotia; Mother Sarah **Loomer** of Nova Scotia; Informant Mrs. Janie **Woodroffe** of Marysville, California.

19 Sep 1923; **Allread**, Margaret Ellen died in Marysville, Yuba Co., California; recorded in book 4 page 42; Born 12 Dec 1876 in Yuba Co., California; Status Married; Spouse William **Allread**; Occupation Housewife; Father John **McQuaid** of Ireland; Mother Mary **Mulvany** of Ireland; Informant Wm. **Allread** of Smartsville, California.

7 Oct 1923; **Mooney**, Lucy Jane died in Yuba Co., California, recorded in book 3 page 157; Sex Female; Born 13 May 1869 in Smartsville, Yuba Co., California; Status Single; Residence Smartsville, California; Occupation At home; Father Thomas **Mooney** of Ireland; Mother Mary Jane **Huling** of Indiana; Informant Thomas **Mooney** of Smartsville, California.

23 Mar 1924; **Barrett**, Katherine (Catherine) died in Marysville, Yuba Co., California, recorded in book 4 page 61; born 19 Jul 1864 in California; Status Widowed; Spouse John **Barrett**; Occupation at home; Father John **Driscoll** of Ireland; Mother Katherine **Driscoll** of Ireland; Informant John **Barrett** of Marysville, California.

10 Aug 1924; **Linehan**, Timothy died in Smartsville, Yuba Co., California, recorded in book 3 page 166; Residence Smartsville; Born Feb 1824 in Ireland; Status Widower; Occupation Retired; Father Patrick **Linehan** of Ireland; Mother Mary **Murphy** of Ireland; Informant Katherine **Linehan** of Smartsville, California.

DEATH NOTICES

11 Mar 1925; **Adkins**, Clara Elizabeth died in Yuba Co., California, recorded in book 3 page 176; Sex Female; Born 22 Sep 1873 in Placer Co., California; Status Married; Spouse Benjamin **Adkins;** Occupation At home; Father William H. H. **Nevels** of Arkansas; Mother Addie **Mans** of California; Informant Benjamin **Adkins** of Wheatland, California.

11 Mar 1925; **Adkins**, Frances M. died in Yuba Co., California, recorded in book 3 page 77; Sex Female; Age at death 89 yrs; Birthplace Kentucky; Status Widow; Occupation At home; Father John **Willis** of Virginia; Informant O. P. **Adkins** of Smartsville, California.

18 May 1925; **McQuaid**, Mary died in Smartsville, Yuba Co., California, recorded in book 3 page 179; Born 8 Apr 1866 in Smartsville; Status Single; Occupation School teacher; Father John **McQuaid** of Ireland; Mother Mary **Mulhare** of Ireland; Informant J. D. **McQuaid** of Smartsville, California.

1 Mar 1926; **Sanford**, Euphemia died in Yuba Co., California, recorded in book 3 page 201; Sex Female; Born 19 Mar 1830 in Nova Scotia; Status Married; Residence Smartville; Occupation At home; Father Thos. **Wallace** of Nova Scotia; Mother Ellen **Bruce** of Nova Scotia; Informant Benj. **Sanford** of Smartsville, California

22 Mar 1926; **Magonigal**, Lizzie died in Lone Tree, Yuba Co., California, recorded in book 3 page 204; Sex Female; Born 28 Nov 1854 in Sucker Flat, Yuba Co., California; Status Single; Occupation At home; Father William **Magonigal** of Ireland; Mother Nancy **Boyd** of Ireland; Informant Tom **Magonigal** of Smartsville, California.

8 Jun 1926; **McGovern**, Thomas James died in Hammonton, Yuba Co., California, Recorded in book 3 page 207; Born 20 Jan 1865 in California; Occupation Dredge man; Status Married; Father John **McGovern** of Ireland; Mother Alice **Flanagan** of Ireland; Informant Mrs. Mary **McGovern** of Hammonton.

8 Feb 1927; **Mooney**, Clara Louise died in Yuba Co., California, recorded in book 3 page 222; Sex Female; Born 4 Jul 1858 in Yuba Co., California; Status Single; Residence Smartsville; Occupation At home; Father Thomas **Mooney** of Ireland; Mother Mary J. **Huling** of Indiana; Informant Thomas **Mooney** of Smartsville, California.

30 Oct 1927; **McCarty**, Susan died in Timbuctoo, Yuba Co., California (the headstone shows a year date of 1928, this is incorrect, the correct date is 1927); recorded in book 3 page 237; Residence Timbuctoo; Born Jan 1839 in Ireland; Status Window; Father Matthew **Flanagan** of Ireland; Mother Mary **Smith** of Ireland; Informant A. J. **McCarty** of Hammonton, California.

1 Jul 1928; **Cooney**, William died in Smartsville, Yuba Co., California, recorded in book 3 page 250; Born 20 Jul 1865 in Illinois; Status Married; Spouse Mary

DEATH NOTICES

Cooney; Occupation Rancher; Informant Mrs. Mary **Cooney** of Smartsville, California.

5 May 1929; **Mooney**, Jessie Gardines died in Yuba Co., California, recorded in book 5 page 12; Sex Female; Born 10 Feb 1870 in California; Status Single; Occupation At home; Father Thomas **Mooney** of Ireland; Mother Mary J. **Huling** of Ireland; Informant Thomas **Mooney** of Smartsville, California.

3 Feb 1930; **Johnson**, William Allen died in Marysville, Yuba Co., California, recorded in book 5 page 56; Sex Male; Born 3 Feb 1930 in Marysville, California; Father William **Johnson** of California; Mother Mary **Murphy** of California; Informant Wm. A. **Johnson** of Smartsville, California.

13 Feb 1930; **Warne**, James Henry died in Yuba Co., California, recorded in book 3 page 282; Sex Male; Born 24 Mar 1851 in England; Status Married; Spouse Mary J. **Warne**; Residence Smartsville; Occupation Retired; Father James **Warne** of England; Mother of England; Informant Mary J. **Warne** of Smartsville, California.

6 Jan 1931; **Havey**, John C. died in Smartsville, Yuba Co., California, recorded in book 4 page 118; Sex Male; Born on 16 May 1849 in New Jersey; Status Married; Spouse Mary Ann **Havey**; Father Patrick **Havey** of Ireland; Mother Mary **Havey** of Ireland; Informant; Mrs. Mary Ann **Havey** of Smartsville, California.

16 Aug 1931; **Galligan**, Hugh Peter died in Marysville, Yuba Co., California, recorded in book 4 page 231; Born 26 Sep 1858 in Iowa; Status Married; Spouse Mary Elizabeth **Galligan**; Father Mathew **Galligan** of Ireland; Mother Susana **Ewing** of Ohio; Informant Clarence **Galligan** of Marysville, California.

17 Aug 1931; **Bach**, John Edward died in Yuba Co., California, recorded in book 4 page 223; Sex Male; Born 12 Aug 1869 in Sebastapool, California; Status Married; Spouse Ethel M. **Bach** Residence Smartsville; Occupation Land Foreman; Father John **Bach** of Germany; Mother Mary **Tregaskis** of Wisconsin; Informant Mrs. Ethel M. **Bach** of Smartsville, California.

26 Mar 1932; **McQuaid**, John Henry died in Yuba Co., California, recorded in book 4 page 339; Born 17 Apr 1870 in Smartsville; Status Married; Spouse Katherine R. **McQuaid**; Occupation Retired; Father John **McQuaid** of Co. Galway, Ireland; Mother Mary **Mullhar** of Ireland; Informant Mrs. Katherine **McQuaid** of Marysville, California.

8 Jan 1933; **Call**, Salmon M. died near Smartsville, Yuba Co., California, recorded in book 5 page 3 A; Sex Male; Born 16 Jul 1892 in Quebec, Canada; Status Married; Spouse Hattie **Call;** Residence Smartsville; Occupation Rancher; Father **Call** of Canada; Informant Mrs. Hattie **Call** of Smartsville, California

12 Feb 1933; **Marple**, Harry died in Timbuctoo, Yuba Co., California, recorded in book 5 page 9; Sex Male; Born 26 Apr 1876 in Timbuctoo; Status Single;

DEATH NOTICES

Residence Timbuctoo; Father Whitten **Marple** of Pennsylvania; Mother Elizabeth **Gilberthomp** of Pennsylvania; Informant Coroners Records of California

29 Jun 1933; **Bowman**, Emma died in Marysville, Yuba Co., California, recorded in book 5 page 50; Sex Female; Born 1860 in California; Status Single; Residence Marysville; Father William **Bowman** Mother Jane **Hurst;** Informant Mrs. **Swift** of Marysville, California

29 Apr 1934; **Walsh**, John E. died near Smartsville, Yuba Co., California, recorded in book 5 page 125 B; Sex Male; Born 20 Jul 1864 in California; Status Widowed; Residence Smartsville; Occupation Rancher; Father John **Walsh** of Ireland; Mother Mary **Dougherty** of Ireland; Informant Walter **Walsh** of Smartsville, California

15 Nov 1934; **Bowman**, Louisa died in Rose Bar, Yuba Co., California, recorded in book 5 page 176; Sex Female; Born 23 Oct 1859 in Illinois; Status Widowed; Spouse James **Bowman;** Residence Rose Bar; Occupation At home; Father John **Jones** of Brophy, Yuba Co., California; Mother Louisa **Jones** of Brophy, California; Informant Mrs. E. B. **Swift** of Marysville, California.

11 Apr 1935; **Peckham**, Thomas William died in Wheatland, Yuba Co., California, recorded in book 5 page 220; Sex Male; Born 16 Apr 1855 in San Francisco, California; Status Married; Spouse Saphrona Elizabeth **Peckham;** Residence Wheatland; Father **Peckham;** Mother **Marian Elizabeth** of London, English; Informant Mrs. Saphrona E. **Peckham** of Wheatland, California.

8 Jun 1935; **O'Brien**, Anna Josephine died in Smartsville, Yuba Co., California; recorded in book 5-B page 240; Residence Smartsville; Born 29 May 1864 in Smartsville, California; Status Single; Father James K. **O'Brien** of Ireland; Mother Mary **Kirby** of Ireland; Informant James K. **O'Brien** of Smartsville, California.

10 Sep 1935; **Spencer**, Carl A. died in Marysville, Yuba Co., California, recorded in book 6 page 17; Sex Male; Born 11 Sep 1937 in Yuba Co.; Status Single; Residence Hammonton; Occupation Student; Father Arthur **Spencer** of New York; Mother Martha **Wheaton** of California; Informant Mrs. Arthur **Spencer** of Hammonton, California.

9 Dec 1935; **Adkins**, Benjamin F. died in Yuba Co., California, recorded in book 6 page 37; Sex Male; Born 1868; Status Single; Residence Wheatland; Father **Adkins;** Mother **Willis** of Kentucky; Informant Coroner's Office of Yuba County, California.

6 Feb 1936; **Bach**, Ethel M. died in Marysville, Yuba Co., California, recorded in book 6 page 67A; Sex Female; Born 15 Nov 1878 in Smartsville; Status Widow; Spouse Mr. John E. **Bach**, deceased; Residence Smartsville; Occupation Housewife; Father Russell **Widener** of Tennessee; Mother Alpha **Atchley** of Tennessee; Informant Ethel K. **Bach** of Smartsville, California.

DEATH NOTICES

20 Jan 1937; **Welch**, Chester Andrew died in Cabbage Patch, Yuba Co., California, recorded in book 6 page 151A; Sex Male; Born 14 Apr 1936 in Wheatland; Status Single; Residence Wheatland; Occupation Infant; Father D. N. **Welch** of California; Mother Katherine **Walker** of California; Informant Mr. D. N. **Welch** of Wheatland, California.

8 Feb 1937; **Conrath**, Elizabeth died in Marysville, Yuba Co., California; recorded in book 6B page 174; Residence Marysville; Born 1 Jan 1860 in Detroit, Michigan; Status Widowed; Spouse Lois **Conrath**; Occupation Housewife; Father John **Turnkert**; Informant Mr. Lud **Conrath** of Marysville, California.

17 Feb 1937; **Cramsie**, John E. died in Marysville, Yuba Co., California; recorded in book 6 page 178A; Residence Smartsville; Born 7 Nov 1872 in Smartsville; Status Married; Spouse Anna **Cramsie**; Occupation Farmer; Father William **Cramsie** of Ireland; Mother Elizabeth **Harvey** of New Jersey; Informant Mrs. Anna **Cramsie** of Smartsville, California.

4 Mar 1937; **Quinkert**, Clara died in Marysville, Yuba Co., California; recorded in book 6-A page 190-A; Residence Smartsville; Born 5 Feb 1866 in Michigan; Status Married; Spouse John **Quinkert**; Occupation Housewife; Father Joseph **Campbell** of Michigan; Mother Delliah of Michigan; Informant Elva **Lowden**, R. N. of Yuba Co., hospital in Marysville, California.

21 Apr 1937; **Congdon**, George died near Smartsville, Yuba Co., California, recorded in book 6 page 198; Sex Male; Born 10 Oct 1854 in Connecticut; Status Single; Residence Smartsville; Occupation Farmer; Father Frank **Congdon** of Connecticut; Mother **Lois** of Connecticut; Informant Corners records of Yuba Co. of Marysville, California.

10 Aug 1937; **Hord**, Mary Wright died in Marysville, Yuba Co., California, recorded in book 6 page 243; Sex Female; Born 2 Aug 1862 in Central Gaither, Sutter Co., California; Status Widowed; Spouse Charles **Hord**; Residence Waldo District; Occupation Homemaker; Father Radford Ellis **Welch** of Missouri; Mother Serena Jane **Bast** of Missouri; Informant Mrs. Velma **Johnson** of Marysville, California.

11 May 1938; **Welch**, Lillie E. died in Yuba Co., California, recorded in book 7 page 126; Sex Female; Born 14 Jul 1866 in Waldo; Status Widowed; Spouse James **Welch**; Residence Waldo District; Occupation Housewife; Father David N. **Jones** of Vermont; Mother Elizabeth **Young** of Illinois; Informant Ira **Welch** of Lincoln, California.

16 Jun 1938; **Magonigal**, Clarence B. died in Yuba Co., California, recorded in book 7 page 158; Sex Male; Born 27 Sep 1909 in Smartsville, California; Status Single; Residence Smartsville; Father Thomas **Magonigal** of Smartsville, California; Mother Hanna **Rushily** of Smartsville, California; Informant Corner-Sheriff records of Yuba Co., California.

DEATH NOTICES

23 Oct 1939; **Gassaway**, Bertha died in Yuba Co., California, recorded in book 7 page 221; Sex Female; Born 25 Jan 1865 in Smartsville, California; Status Widowed; Residence Marysville; Occupation Housewife; Father George **Jeffery** of Nova Scotia; Mother Maria L. **Davies** of Pennsylvania; Informant Mrs. Ed **Davies** of Allegheny, California.

7 Feb 1940; **Poor**, Mark E. died in Marysville, Yuba Co., California, recorded in book 9 page 32; Sex Male; Born 11 Jan 1862 in Mormon Island, California; Status Widowed; Spouse Emma **Poor;** Residence Browns Valley; Occupation Farmer; Informant Robert **Poor** of Marysville, California.

6 Jan 1941; **Mooney**, Mary E. died in Yuba Co., California, recorded in book 10 page 13; Sex Female; Born 11 Feb 1860 in Yuba Co., California; Status Single; Residence Smartsville, California; Father Thomas **Mooney** of Ireland; Mother Mary J. **Huling** of Indiana; Informant Adelaide **Mooney** of Smartsville, California.

1 Sep 1941; **Quinkert**, John died in Marysville, Yuba Co., California, recorded in book 10 page 150; Sex Male; Born 12 Aug 1857 in Detroit, Michigan; Status Widowed; Spouse Clara **Quinkert**, deceased; Residence Smartsville; Informant L. J. **Conrath** of Marysville, California.

18 Nov 1941; **Sheehan**, Johanna C. died in Marysville, Yuba Co., California; recorded in book 10 page 199; Sex Female; Born 21 Nov 1855 in Boston, Massachusetts; Status Single; Father was born in Ireland; Mother Julia **McAuliffe** of Ireland; Informant Dennis **McAuliffe** of Marysville, California.

4 Jul 1942; **Murdock**, Mary died in Smartsville, Yuba Co., California, recorded in book 11 page 114; Sex Female; Born 25 Oct 1847 in Nova Scotia; Status Widowed; Spouse James **Murdock;** Residence Smartsville; Occupation Housewife; Father Alexander **Lavers** of Nova Scotia; Mother Melinda **Sanford** of Nova Scotia; Informant Mrs. Emily **Sanford** of Smartsville, California.

31 Jan 1943; **Newbert**, William L. died in Marysville, Yuba Co., California, recorded in book 11 page 26; Sex Male; Born 10 Nov 1873 in Yuba Co., California; Status Married; Spouse Nellie Wright **Newbert**; Residence Marysville; Occupation Carpenter; Father Thomas A. **Newbert** of Maine; Mother Millicent **Jackson** of New York; Informant Mrs. Nellie **Newbert** of Marysville, California.

25 May 1943; **Jeffery**, Ellen H. died in Smartsville, Yuba Co., California, recorded in book 12 page 76; Sex Female; Born 1 Apr 1877 of Yuba Co., California; Status Married; Spouse William B. **Jeffery,** who is 74 yrs old; Residence Smartsville; Father Joseph **Perkins** of New Hampshire; Mother Jennie **Magonigal** of Pennsylvania; Informant William B. **Jeffery** of Smartsville, California.

15 Nov 1944; **Mead**, Leo Garrett died in Marysville, Yuba Co., California, recorded in book 13 page 213; Residence Marysville; Born 27 Aug 1883 in

DEATH NOTICES

Smartsville; Status Single; SS#566-10-8554; Father Garrett **Mead** of Ireland; Mother Annie **Leahy** of Ireland; Informant Miss Mary **Mead** of Marysville, California.

28 Mar 1945; **Cann**, Thomas James died in Yuba Co., California, recorded in book 14 page 31; Sex Male; Born 24 Jan 1886 in England; Status Married; Spouse Maude **Cann** aged 60 yrs; Residence Smartsville for 30 yrs; Occupation Machinist; Father James **Cann** of England; Mother Annie **Harris** of England; Informant Mrs. W. **Orr** of Yuba City, California.

19 Aug 1945; **Cooney**, Mary died in Marysville, Yuba Co., California, recorded in book 14 page 134; Sex Female; Born 1 May 1868 in Smartsville, California; Status Widowed; Spouse William **Cooney**; Father John **Walsh** of Ireland; Mother Mary **Dougherty** of Ireland; Informant Walter L. **Walsh** of Marysville, California.

19 Dec 1945; **Rigby**, Albert died in Smartsville, Yuba Co., California, recorded in book 14 page 208; Sex Male; Born 27 Sep 1866 of Brooklyn, New York; Status Married; Spouse Ida **Rigby** who is 73 yrs; Occupation Goldmine; Father George **Rigby** of Ireland; Mother **Campbell** of Ireland; Informant George **Rigby** of Smartsville, California.

25 Apr 1946; **Hapgood**, Fannie died in Timbuctoo, Yuba Co., California, recorded in book 15 page 62; Sex Female; Born 28 Mar 1862 in California; Status Married; Spouse James M. **Hapgood**; Residence Timbuctoo; Occupation Housewife; Father Witten **Marple** of Pennsylvania; Mother Elizabeth **Silverthorn** of Pennsylvania; Informant Mrs. Lavinia Ray of Grass Valley, California.

2 Jan 1947; **French**, Joseph died in Marysville, Yuba Co., California; recorded in book 16 page 12; Sex Male; Born 6 Feb 1854 in Michigan; Informant Patients records in Yuba County Hospital.

21 Jan 1947; **Byrne**, Mary Josephine Ryan died in Marysville, Yuba Co., California, recorded in book 16 page 19; Residence Marysville; born 8 Dec 1882 in Smartsville; Status Married; Spouse Robert E. **Byrne**; Occupation Housewife; Father Patrick **Ryan** of Ireland; Mother Hannah **Myers** of Ireland; Informant Robert E. **Byrne** of Marysville, California.

5 Jul 1950; **Parkison**, Chloe Mae died in Yuba Co., California, recorded in book 19 page 136; Sex Female; Born 7 Feb 1888 in Iowa; Status Married; Spouse John **Parkison;** Residence Smartsville; Occupation Housewife; Father Edward **Parrish** of Ohio; Mother Melinda **DePue** of Iowa; Informant Yuba Co. Hospital Records in Marysville, California

16 Oct 1950; **Marple**, George Washington died in Marysville, Yuba Co., California, recorded in book 19, page 209; Sex Male; Born 22 Feb 1871 in California; Status Never Married; Residence Marysville; Occupation Blacksmith

DEATH NOTICES

for the Rail Road; Father William Whitten **Marple** of Pennsylvania; Mother Elizabeth **Silverthorn** of Pennsylvania; Informant Mrs. Lavinia E. Ray.

25 Nov 1950; **Whitmore**, Alice Edna died in Yuba Co., California, recorded in book 19, page 236; Sex Female; Born 29 Mar 1909 in Iowa; Status Married; Spouse Jess Franklin **Whitmore;** Residence Smartsville; Occupation Housewife; Father Horace **Montgomery** of Iowa; Mother Cleo **Freend** of Iowa; Informant Yuba Co. Hospital Records in Marysville, California.

24 Jan 1951; **McClure**, James Alexander died in Smartsville, California, recorded in book 20 page 28; He was "just passing through"; Born 27 Jul 1878 in Ireland; Status Never Married; Occupation Miner of Gold mines; Father Alexander **McClure** of Ireland; Mother Jane **McKinley** of Ireland; Informant Joseph Patrick **McClure** of Smartsville, California.

9 Feb 1951; **McCarty**, Matthew Henry died in Timbuctoo, Yuba Co., California, recorded in book 20 page 36; Sex male; Born 1 Jan 1869 in California; Status Never Married; Residence Smartsville; Occupation Miner of Gold; Father Andrew **McCarty** of Ireland; Mother Susan **Flanagan** of Ireland; Informant Clarence **Galligan**.

8 Apr 1951; **Rigby**, Ida May died in Smartsville, Yuba Co., California, recorded in book 20 page 75; Residence Smartsville; Sex Female; Born 16 May 1871 in California; Status Widowed; Occupation Housewife; Father Daniel **Dougherty** of Ireland; Mother Ann **Brady** of Ireland; Informant George **Rigby**.

4 Dec 1951; **McClure**, Joseph died in Marysville, Yuba Co., California, recorded in book 20 page 233; Born 7 Apr 1886 in Massachusetts; Residence Marysville; Status Single; Occupation Laborer; Father Alexander **McClure** of Ireland; Mother Jane **McKinley** of Ireland; Informant Yuba Co. Hospital Records.

21 Jan 1953; **Sanford**, Emily died in Yuba Co., California, recorded in book 22, page 19; Sex Female; Born 27 Jan 1870 in Nova Scotia; Status Married; Spouse A. B. **Sanford;** Occupation Housewife; Father James **Murdock** of Nova Scotia; Mother Mary **Lavers** of Nova Scotia; Informant A. B. **Sanford** of Smartsville, California.

5 Feb 1953; **McGanney**, Georgie Elizabeth died in Marysville, Yuba Co., California, Recorded in book 22 page 28; Sex Male; Born 10 Mar 1879 in California; Status Widowed; Occupation Housewife; Father George **Gray** of California; Mother Elizabeth **Burke** of California; SS#567-24-3531; Informant Noreen **Facer** – Daughter.

13 Jul 1953; **Mooney**, Adelaide died in Yuba Co., California, recorded in book 22, page 148; Sex Female; Born 19 Nov 1865 in California; Status Never Married; Residence Smartsville, California; Father Thomas **Mooney** of Ireland; Mother **Huling;** Informant Yuba County Hospital Records in Marysville, California.

DEATH NOTICES

14 Jul 1953; **Jeffery**, William Baras died in Yuba Co., California, recorded in book 22, page 154; Sex Male; Born 27 Aug 1868 in California; Status Widowed; Residence Smartsville, California for 84 yrs; Occupation Blacksmith; Father George **Jeffery** of Nova Scotia; Mother Maria **Davies** of Pennsylvania; Informant Elmer **Collings** of Smartsville, California..

31 Jul 1953; **McCarty**, Robert Emmet died in Marysville, Yuba Co., California; recorded in book 22 page 171; Sex Male; Born 30 Sep 1880 in California; Status Married; Spouse Ella **McCarty**; Occupation Dredger man; Father Andrew **McCarty** of Ireland; Mother Susan **Flanagan** of Ireland; Informant Ella **McCarty** of Smartsville, California.

25 Aug 1953; **Linehan**, Peter Francis died in Smartsville, Yuba Co., California; recorded in book 22 page 195; Residence Smartsville; Born 17 Oct 1865 in California; Status Widower; Occupation Cook; Father Timothy **Linehan** of Ireland; Mother Ellen **Kelly** of Ireland; Informant Mrs. Mary **McGovern** of Smartsville, California.

31 Jul 1955; **Compton**, Clarence Albert died in Marysville, Yuba Co., California; recorded in book 24 page 159; Residence Smartsville; Born 14 Sep 1898 in California; Status Divorced; Occupation Ground man for P G & E.; Father Brady F. **Compton** of California; Mother Mary Jane **Byrne** of California; Informant William **Compton**.

19 Apr 1955; **Hughes**, Alois died in Marysville, Yuba Co., California, recorded in book 24 page 84; Residence Smartsville; Born 17 Dec 1880 in Switzerland; Citizen of Switzerland; Status Never Married; Occupation Laborer; Father Alois **Hug** of Switzerland; Mother Anna **Truttmann** of Switzerland; Information Public Administration of Marysville.

13 May 1957; **Galligan**, Mary E. died in Marysville, Yuba Co., California, recorded in book 26 page 100; Residence Marysville; Born 14 May 1864 in California; Status Widowed; Occupation Housewife; Father Andrew **McCarty** of Ireland; Susan **Flanagan** of Ireland; Informant George **Galligan**.

15 Dec 1957; **Niles**, Theresa Jane died in Smartsville, Yuba Co., California, recorded in Book 26 page 241; Residence Smartsville; Born 21 Jun 1883 in Ireland; Status Married; Spouse George W. **Niles**; Occupation Housewife; Father **Kerwin** of Ireland; Mother **Jane** of Ireland; Informant George W. **Niles** of Smartsville, California.

2 Oct 1958; **Whitmore**, Jesse Franklin died in Yuba Co., California, recorded in book 27, page 194; Sex Male; Born 9 Jun 1891 in Texas; Status Widowed; Occupation Laborer; Mother Anne **Davis**; Informant Virginia **Whitmore** of Marysville, California.

DEATH NOTICES

17 Jan 1959; **Pisani**, Bertha E. died in Yuba Co., California, recorded in book 28, page 18; Sex Female; Born 10 Nov 1903 in California; Status Married; Spouse Lesley **Pisani;** Father Thomas G. **Magonigal** of California; Mother Hannah **Bushby** of California; Informant George **Magonigal** of Smartsville, California.

12 Jun 1959; **Pisani**, Leslie Pomp died in Yuba Co., California, recorded in book 28, page 136; Sex Male; Born 28 Feb 1904 in California; Status Widowed; Father Joseph **Pisani** of Italy; Mother Amelia **Ronkey** of California; Informant Lorraine **Pisani** of Loma Rica, California.

8 Apr 1961; **Martein**, Elizabeth Jane died in Marysville, Yuba Co., California, recorded in book 30 page 80; Residence Smartsville; Born 2 Apr 1871 in California; Status Widowed; Occupation Housewife; Father Daniel **Dougherty** of Ireland; Mother Ann **Brady** of Ireland.

2 Jul 1961; **Johnson**, William Franklin died in Marysville, Yuba Co., California, recorded in book 30 page 132; Residence Marysville; Born 14 Sep 1864 in New York; Status Widowed; Occupation Patrolman P. G. & E.; Father Charles **Johnson**; Mother was born in England; Informant Mrs. Evelyn **Miles** of Marysville, California.

15 Nov 1961; **Magonigal**, Emma E. died in Sutter Co., California, recorded in book 1961 page 5100-214; Sex Female; Born 3 Apr 1870 in Nova Scotia, Canada; Status Widowed; Residence Yuba City, California; Occupation Housewife; Father Daniel **Fraser** of Nova Scotia, Canada; Mother Elizabeth **Loomer** of Nova Scotia, Canada; Informant Mrs. Olive **Sweetland** of Yuba City, California.

22 Mar 1963; **Magonigal**, William Daniel died in Yuba Co., California, recorded in book 32, page 74; Sex Male; Born 13 Sep 1898 in California; Status Married; Spouse Eleanor G. **Magonigal;** Father John **Magonigal** of California; Mother Emma **Fraser** of Nova Scotia; Informant Eleanor G. **Mogonigal** of Smartsville, California.

14 May 1963; **Murphy**, John Joseph died in Yuba Co., California, recorded in book 32, page 124; Sex Male; Born 14 Mar 1886 in Nevada; Status Married; Spouse Ida E. **Murphy**; Residence Marysville; Occupation Constable; Father Michael **Murphy**; Informant Ruth M. **McCowell** of Marysville, California.

30 May 1963; **Foreman**, Frank Gilbert died in Yuba Co., California, recorded in book 32, page 137; Sex Male; Born 18 Jun 1895 in California; Status Married; Spouse Frieda **Foreman;** Residence Marysville; Occupation Trucker; Father Jake **Foreman.**

2 Dec 1963; **Newbert**, Nellie Wright died in Yuba Co., California, recorded in book 32, page 288; Sex Female; Born 26 Oct 1887 in California; Status Widowed; Residence Yuba Co.; Father Aden **Wright** of Kentucky; Mother Mary B. **Wright** of California; Informant Ida E. **Murphy** of Marysville, California.

DEATH NOTICES

3 Jan 1964; **Rose**, John Oscar died in Yuba Co., California, recorded in book 33, page 11; Sex Male; Born 2 Mar 1872 in California; Residence Browns Valley; Occupation Laborer; Father John **Rose** of Scotland; Mother Anne V. **Dougherty**; Informant Raymond **Burris** of Browns Valley, California.

6 Apr 1964; **Herboth**, George died in Marysville, Yuba Co., California, recorded in book 33 page 72; Residence Marysville; Born 11 Nov 1889 in California; Status Married; Spouse Elizabeth E. **Herboth**; Occupation Owner Herboth Machine Shop; Father John **Herboth** of Illinois; Mother Elizabeth **Kulker**; Informant Elizabeth E. **Herboth** of Marysville, California.

25 Jun 1964; **Welch**, Winifred Helen died in Yuba Co., California, recorded in book 33, page 142; Sex Female; Born 9 Apr 1914 in California; Status Married; Spouse Clarence R. **Welch**, Sr.; Residence Loma Rica; Father John **Kennedy** of California; Mother Sarah **Groves** of California; Informant Clarence R. **Welch**, Sr. of Loma Rica, California.

13 Feb 1965; **Hinkle**, Noble Francis died in Yuba Co., California, recorded in book 34, page 43; Sex Male; Born 1 Jun 1911 in Washington; Status Divorced; Residence Smartsville; Occupation Electrician; Father Archie **Hinkle** of Illinois; Mother Maude **Franks** of Oregon; Informant Mrs. Thelma M. **Quilty** of Smartsville, California.

4 Jul 1967; **Fitzhugh**, James Monroe died in Yuba Co., California, recorded in book 36, page 189; Sex Male; Born 5 Jan 1880 in California; Status Married; Spouse Grace **Fitzhugh;** Residence Smartsville; Occupation Cattleman; Father Henry **Fitzhugh** of Kentucky; Mother Martha **Mayben** of Missouri; Informant Gary J. **Miller** – Sheriff-Corner of Marysville, California.

17 Dec 1967; **LeBoeuf**, Cecilia Ethelyn died in Marysville, Yuba Co., California, recorded in book 36 page 328; Residence Smartsville; Born 22 Nov 1907 in Oregon; Status Never Married; Father William **LeBoeuf** of Rhode Island; Mother Ethelyn **Burroughs** of California; Informant Bertha **Collins** of Smartsville, California.

25 Nov 1968; **Davis**, Bert Benjamin died in Yuba Co., California, recorded in book 37, page 281; Sex Male; Born 19 May 1890 in Montana; Status Married; Spouse Adeline **Harris;** Residence Smartsville; Occupation Forest Ranger; Father Ervin **Davis;** Mother Selena R. **Mollett** of Wisconsin; Informant Adeline **Davis** of Smartsville, California.

11 Dec 1968; **Fippin**, Asa David died in Yuba Co., California, recorded in book 37, page 294; Sex Male; Born 18 May 1885 in California; Status Married; Spouse Emma **Wheaton;** Occupation Gold Miner; Father John **Fippin** of Ohio; Mother Julia **Single;** Informant Julia **Tremewan** of Grass Valley, California.

DEATH NOTICES

29 Oct 1969; **Cann**, Maud died in Yuba Co., California, recorded in book 38, page 286; Sex Female; Born 3 Apr 1884 in California; Status Widowed; Residence Marysville; Occupation Homemaker; Father Abraham **Tisher** of Pennsylvania; Mother Margaret **Rouse** of Pennsylvania; Informant George **Stuckey** of Yuba City, California.

2 Jun 1970; **Welch**, Clarence Ray died in Yuba Co., California, recorded in book 39, page 151; Sex Male; Born 13 Jan 1896 in California; Status Widowed; Residence Loma Rica; Occupation Laborer; Father James **Welch** of California; Mother Lillian **Jones** of California; Informant: Joseph G. **Welch** of Browns Valley, California.

6 Nov 1970; **Naglee**, Joseph C. died in Marysville, Yuba Co., California, recorded in book 39 page 289; Residence Marysville; Born 17 May 1917 in New York; Status Married; Spouse Elizabeth **Bennett**; SS#122-03-6900; Occupation Insurance Life Underwriter; Father Samuel **Naglee** of New York; Mother Rilla **Reed** of Pennsylvania; Informant Elizabeth **Naglee** of Marysville, California

19 Dec 1971; **Carlson**, Kim Loring died in Marysville, Yuba Co., California, recorded in book 40, page 363; Sex Male; Born 28 Feb 1951 in Washington; Status Never married; Residence Mooney Flat Road; Occupation Student; Father James **Carlson** of Pennsylvania; Mother Reva **Burns** of California; Informant Reva **Rouse** of Marysville, California.

23 Dec 1971; **Brett**, Walter Nelson died in Yuba Co., California, recorded in book 40, page 367; Born 12 Dec 1891 in California; Status Married; Spouse Rose **Cardo;** Mother Louise **Nelson** of Denmark; Informant Rose **Brett** of Smartsville, California.

30 May 1972; **LeBourveau**, Ernest died in Yuba Co., California, recorded in book 41, page 154; Sex Male; Born 15 Aug 1877 in Canada; Status Widowed; Residence Marysville; Occupation Carpenter; Father John F. **LeBourveau** of Canada; Mother Theresa M. **Page** of Canada; Informant Public Conservator in Yuba Co. Courthouse, Marysville, California.

13 Jul 1972; **Herboth**, Elizabeth E. died in Marysville, Yuba Co., California; recorded in book 41 page 202; Residence Marysville; Born 11 Jul 1890 in California; Status Widowed; SS#571-44-6991D; Occupation Housewife; Father Louis **Conrath** of Michigan; Mother Elizabeth **Quinkert** of Michigan; Informant Louis **Conrath** of Marysville, California.

29 Sep 1974; **Welch**, James Radford died in Yuba Co., California, recorded in book 43, page 286; Sex Male; Born 23 Apr 1932 in California; Residence Marysville; Occupation Rancher; Father Clarence **Welch** of California; Mother Winifred **Kennedy** of California; Informant Mr. Joe **Welch** of Browns Valley, California.

DEATH NOTICES

11 Mar 1976; **Williams**, James Paul died in Marysville, Yuba Co., California, recorded in book 45 page 78; Residence Smartsville; Born 7 Nov 1971 in California; Status Single; Father Joe Mac **Williams** of Texas; Mother Christine **Vinsonhaler** of California; Informant Mr. And Mrs. Joe Mac Williams of Smartsville, California; Cause of death was a victim of house fire.

11 Mar 1976; **Williams**, Joseph Matthew died in Smartsville, Yuba Co., California, recorded in book 45 page 77; Residence Smartsville; Born 13 Jun 1968 in California; Status Single; Father Joe **Mac Williams** of Texas; Mother Christine **Vinsonhaler** of California; Informant Mr. & Mrs. Joe **Mac Williams** of Smartsville, California; Cause of death was a victim of house fire.

11 May 1976; **Williams**, Juliana Rebecca died in Smartsville, Yuba Co., California, recorded in book 45 page 76; Residence Smartsville; Born 28 Jan 1975 in California; Status Single; Father Joe Mac **Williams** of Texas; Mother Christine **Vinsonhaler** of California; Informant Mr. And Mrs. Joe Mac **Williams** of Smartsville, California; Cause of death was a victim of house fire.

15 Apr 1976; **Bartlett**, Nellie Gladys died in Yuba Co., California, recorded in book 45, page 117; Sex Female; Born 6 Jan 1903 in Oregon; Status Widowed; Residence Smartsville; Occupation At home; Father **Fisher;** Informant Melba **Parks** of Smartsville, California.

26 Aug 1976; **Carlson**, Ric died in Marysville, Yuba Co., California, recorded in book 45, page 227; Sex Male; Born 6 Feb 1946 in California; Status Married; Spouse Angela **Elder**; Residence El Dorado Co.; Occupation Pastor; Father Jim **Carlson**; Mother Reva **Burns**; Informant Angela **Carlson** of South Lake Tahoe, California.

29 Sep 1979; **Hudson**, LaVerne J. died in Smartville, Yuba Co., California, recorded in book 48, page 250; Sex Female; Born 15 Jan 1909 in Utah; Status Married; Spouse Ralph R. **Hudson;** Father John Peter **Jensen** of Utah; Mother Clara **Petersen** of Utah; Informant Ralph R. **Hudson** (Husband) of Smartsville, California.

26 Sep 1981; **Greever**, Kyle Aldon died in Yuba Co., California, recorded in book 50, page 275; Sex Male; Born 21 Aug 1893 in Kansas; Residence Marysville; Occupation Baker; Father Charles T. **Greever** of Tennessee; Mother Drussilla **Fuller** of Virginia; Informant Aldon **Greever** (Son) of Marysville, California.

13 Nov 1981; **Smith**, Frank Earl died in Yuba Co., California, recorded in book 50, page 325; Sex Male; Born 8 Dec 1912 in California; Status Married; Spouse Catherine **Smith;** Residence Smartsville; Father John **Smith** of England; Mother Florence **Thomas** of Nevada; Informant Catherine **Smith** (Wife) of Smartsville, California.

DEATH NOTICES

19 Feb 1984; **Grace**, John Thomas died in Marysville, Yuba Co., California, recorded in book 53 page 54; Residence Smartsville; Born 25 Oct 1894 in Pennsylvania; Status Single; SS#560-07-4777; Occupation Mechanic – Self Employed; Father John Thomas **Grace** of England; Mother Mary **Alice** ? of England; Informant Rose Ellen **Grace** (Sister) of Smartsville, California.

9 Apr 1983; **Lessley**, Betty Lee died in Yuba Co., California, recorded in book 52, page 144; Sex Female; Born 12 Dec 1942 in Oklahoma; Residence Marysville; Father Aldon Lee **Breever** of Oklahoma; Mother Maxine **Maples** of Oklahoma; Informant Frank A. **Martin** (Son) of Marysville, California.

15 Jan 1985; **Manford**, James Bunch died in Yuba Co., California, recorded in book 54, page 18; Sex Male; Born 2 Feb 1904 in California; Status Widower; Residence Marysville; Occupation Plumber; Father William B. **Manford** of Missouri; Mother Loretta **Bunch** of Missouri; Informant Lloyd **Hughes** (Nephew) of Marysville, California.

28 Aug 1985; **Welch**, Georgina Louise died in Yuba Co., California, recorded in book 54, page 313; Sex Female; Born 28 Dec 1924 in California; Status Widowed; Residence Marysville; Occupation Clerk; Father Frederick **Gnadig** of California; Mother Frances **Nathan** of California; Informant Carolyn **Hassler** (Daughter) of Dobbins, California.

30 Sep 1985; **Mills**, Homer Clyde died in Yuba Co., California, recorded in book 54, page 370; Sex Male; Born 3 Jan 1924 in Oklahoma; Status Married; Spouse Maria Isabel **Betancourt;** Residence Smartsville; Father Johnnie C. **Mills** of Texas; Mother Ruby B. **Crabtree** of Texas; Informant Maria Isabel **Mills** (Wife) of Smartsville, California.

7 Oct 1985; **Murphy**, Ida Elsie died in Yuba Co., California, recorded in book 55, page 381; Sex Female; Born 3 Apr 1890 in Waldo District, California; Status Widowed; Residence Yuba Co.; Occupation Rancher; Father Aden **Wright** of Kentucky; Mother Mary **Welch** of Sutter Co., California; Informant Ruth M. **McDowell** (Daughter) of Marysville, California.

3 Oct 1986; **Likens**, Alma Lou died in Yuba Co., California, recorded in book 55, page 387; Sex Female; Born 20 Apr 1913 in Texas; Status Married; Spouse Hance B. **Likens;** Residence Smartsville; Occupation Homemaker; Father Eddie E. **Palmer** of Missouri; Mother Laura Alice **Conway** of Texas; Informant Hance B. **Likens** (Husband) of Smartsville, California.

10 Nov 1986; **Likens**, Hance Benjamin died in Sutter Co., California, recorded by Registrars # 38651000306.

22 May 1989; **Gilliam**, William Glenn died in Yuba Co., California, recorded in book 58, page 235; Sex Male; Born 24 Sep 1913 in Kentucky; Status Married; Spouse Dorothy G. **Vega;** Residence Marysville; Occupation Heavy Equip; Father

DEATH NOTICES

Virgil **Gilliam** of Kentucky; Mother Beulah **Price** of Kentucky; Informant Dorothy G. **Gilliam** (Wife) of Smartsville, California.

24 Apr 1991; **Gruber**, William Henry died in Marysville, Yuba Co., California; recorded in book 60 page 221; Residence Smartsville; Born 7 Aug 1900 in Illinois; Status Divorced; SS#554-16-1169; Occupation Prospector – Mining; Father William Patrick **Gruber** of Germany; Mother Antonetta **Cramer** of Germany; Informant Public Guardian of Yuba Co., Marysville, California.

26 Jan 1992; **Rigby**, George Albert died in Marysville, Yuba Co., California; recorded in book 61 page 37; Sex Male; Born 18 Jul 1909 in California; Status Married; Spouse Frances **Collins**; Residence Smartsville; Occupation Trouble shooter for pacific Telephone; Father Albert **Rigby** of New York; Mother Ida **Dougherty** of California; Informant Frances **Rigby** (Wife) of Smartsville, California.

20 Sep 1993; **Vinsonhaler**, Howard F. died in Smartsville, Yuba Co., California; recorded in book 62 page 393; Residence Smartsville; Born 8 Sep 1914 in Kansas; Status Married; Spouse Mildred L. **Manie**; SS#520-14-9237; Occupation Truck driver – heavy equipment; Military service from 1932 – 1949; Father Arthur **Vinsonhaler** of Kansas; Mother Maybelle **Hutchinson** of Kansas; Informant Mildred L. **Vinsonhaler** (Wife) of Smartsville, California.

4 Oct 1993; **Hudson**, Ralph Rule died in Yuba Co., California, recorded in book 62, page 419; Sex Male; Born 8 Dec 1905 in Wyoming; Status Widowed; Residence Smartsville; Father Edwin **Hudson** of Utah; Mother Leona **Obert** of Iowa; Informant Richard **Hudson** (Son) of Smartsville, California.

30 May 1994; **Rigby**, Frances Mary died in Smartsville, Yuba Co., California; recorded in book 63 page 223; Sex Female; Born 2 Oct 1917 in Illinois; Status Widowed; Residence Smartsville; Occupation Waitress; Father Patrick Francis **Collins** of Ireland; Mother Nora **Kellcher** of Illinois; Informant Frances Nora **Compton** – Daughter of Marysville, California.

23 Aug 1994; **Gann**, Gladys Dupes died in Yuba Co., California, recorded in book 63, page 352; Sex Female; Born 19 Apr 1917 in Tennessee; Status Married; Spouse Wayne Jack **Gann;** Residence Smartsville; Occupation Nurse; Father Earl Charles **Dupes** of Tennessee; Mother Dovie Anne **Hahn** of Tennessee; Informant Wayne J. **Gann** of Smartsville, California.

26 Feb 1998; **Lane**, Esther Theodora died in Yuba Co., California, recorded in book 67, page 125; Sex Female; Born 17 Apr 1927 in California; Status Married; Spouse Lawrence L. Lane; Residence Smartsville; Father Raymundo **Arguello** of Colorado; Mother Emelia **Vihel** of Colorado; Informant Lawrence L. **Lane** (Husband) of Smartsville, California.

DEATH NOTICES

4 Mar 1998; **Greever**, Aldon Lee died in Yuba Co., California, recorded in book 67, page 131; Sex Male; Born 15 Feb 1919 in Oklahoma; Status Married; Spouse Minnie Maxine **Maples**; Residence Marysville; Occupation Mechanic; Father Kyle Aldon **Greever** of Oklahoma; Mother Hazel Claire **Kramer** of Kansas; Informant Minnie Maxine **Greever** (Wife) of Marysville, California.

19 Apr 1998; **Greever**, Maxine Minnie died in Yuba Co., California, recorded in book 67, page 209; Sex Female; Born 11 Jan 1920 in Oklahoma; Status Widowed; Residence Marysville; Occupation Homemaker; Father Joe **Maples** of Oklahoma; Mother Eveline **Whited** of Oklahoma; Informant Ruby A. **Brown** (Daughter) of Marysville, California.

22 Jul 1998; **Greever**, Jackie Warren died in Yuba Co., California, recorded in book 67, page 365; Sex Male; Born 22 Nov 1943 in Oklahoma; Status Married; Spouse Martha Belle **Cooper**; Residence Marysville; Occupation Service attendant; Father Aldon Lee **Greever** of Oklahoma; Mother Maxine Minnie **Maples** of Oklahoma; Informant Martha B. **Greever** (Wife) of Marysville, California.

6 Nov 2000; **Gann**, Wayne Jack died in Yuba Co., California, recorded in book 69, page 575; Sex Male; Born 18 Dec 1916 in Tennessee; Status Widowed; Residence Smartsville; Occupation Chemist; Father James Joseph **Gann** of Ireland; Mother Minnie Mary **Shoffiett** of Switzerland; Informant B. C. **Gann-Campbell** (Daughter) of Smartsville, California.

21 Dec 2000; **Allinio**, Wilberta Maude died in Yuba Co., California, recorded in book 69, page 684; Sex Female; Born 3 Dec 1904 in California; Status Widowed; Residence Yuba Co.; Occupation Theater Manager; Father William C. **Adkins** of California; Mother Emma **Stewart** of Ohio; Informant Nancy **Kauk** (POA) of Browns Valley, California.

Appendix

For Birth, Marriage, & Death Records of Butte County   **Record 1850 to present**
    Butte County Recorder
    25 County Center Drive   Oroville, CA 95965   (530) 538-7691

For Birth, Marriage, & Death Records of Nevada County   **Records 1860 to present**
    Nevada County Recorder
    950 Maidu Avenue   Nevada City, CA 95959   (530) 265-7264

For Birth, Marriage, & Death Records of Sutter County   **Records 1850 to present**
    Sutter County Recorder
    Yuba City, CA 95991   (530) 822-7134

For Birth, Marriage, & Death records of Yuba County   **Records 1858 to present**
    Yuba County Clerk Office
    935 14th Street   Marysville, CA 95901   (530) 741-6341

For Probate & Will records:
    Yuba County Superior Court
    County Clerk's Office
    215 5th Street   Marysville, CA 95901   (530) 749-7600

For Newspaper records: **Records 1851 to present**
    Yuba County Library
    303 Second Street   Marysville, CA 95901   (530) 749-7380

For the Appeal-Democrat Newspaper:
    1530 Ellis Lake Drive
    Marysville, CA 95901   (530) 741-2345

For Pictures of Headstones:
    Renie Riccobuano   e-mail: jjr@inreach.com
    P. O. Box   Yuba City, CA 95993

Index

A

Addington
  Adelaide, 152, 154
Adkins, 197
  B., 107, 116
  Benjamin, 8, 123, 126, 181, 195, 197
  C., 107
  Clara, 8, 123, 195
  F., 107
  Frances, 8, 116, 195
  J., 107, 116
  O., 116, 195
  Oliver, 8
  Owen, 8, 107
  R., 107
  Robert, 116, 126
  W., 86
  W., Mrs., 123
  William, 7, 209
Ah
  Day, 27
Akin
  George, 190
Alexander
  Donna, 169
Allen
  Charles, 178
  Dale, 184
  Dorothy, 183
  Edward, 14
  George, 15, 39
  Josephine, 15, 144, 178
  Laura, 15
Allinio
  Pierre, 175
  Wilberta, 7, 175, 209
Allread
  Helen, 122, 124
  Margaret, 16, 122, 194
  Mary, 122, 124
  Mrs., 97
  William, 122, 123, 124, 194
  William, Mrs., 104, 192
Altschul
  Harold, 17
Amos
  Ruby, 184
Anderson
  August, 1, 130, 131
  Emma, 1, 116, 117
  Mable, 143, 148
Angelisch
  Dolores, 169
Ardizone
  Grace, 161
Arguello
  Raymundo, 208
Armstrong
  Charles, Mrs., 42
Atchley
  Alpha, 197
  Harrison, 118
  McCord, 118
Audette
  Karen, 174
Austin
  Bert, 1, 131
  Ella, 1, 116, 117, 131
  Fred, 131
  John, 131
  Mrs., 131

B

Bach
  Alice, 4, 78, 116
  Elizabeth, 68, 182
  Ethel, 4, 118, 126, 196, 197
  J., 85
  John, 4, 54, 55, 68, 78, 116, 118, 126, 177, 182, 196, 197
  John, Mrs., 118

INDEX

Katie, 4
M., Mrs., 68
Mary, 4, 116
Vernon, 126
Bacus
  Veca, 139
Bahling
  Naida, 160
Baker
  Berna, 163
Baldwin
  J., Mrs., 67, 90, 113
Ball
  Mamie, 181
Barchette
  G., 180
Barnes
  Glenn, 6
  Ida, 131
Barnett
  Susan, 175
Barrett
  Catherine, 18, 194
  Dan, 122
  Dick, 122
  Jack, 122
  James, 169
  Jim, 122
  John, 18, 66, 122, 169, 178, 194
  Joseph, 18
  Kate, 103
  Katherine, 122
  Mamie, 122
  Marjorie, 122
  Mary, 169
  Maud, 169
  Richard, 127
Barrie
  J., 68
  Wil, 131
  William, Mrs., 102, 117
Barry
  Louisa, 134, 158
Bartlett
  Ed, 126
  John, 159
  Nellie, 7, 159, 206

Bast
  Creen, 194
  Serena, 198
Bateman
  Margaret, 173
Bates
  L., Mrs., 95
Baumgartner
  A., 182
  Lula, 182
Beasley, 18
  Elizabeth, 18, 63
  Richard, 18, 63, 119
  William, 119
  William, Mrs., 101
Beatty, 146
  Alice, 16, 31, 185
  Ann, 16, 58, 59
  Bobbie, 140
  F., 4
  Frances, 4, 51
  Grace, 4, 140
  Infant Son, 4
  John, 58, 79, 81
  Margaret, 140
  Mary, 179
  R., 4, 115, 126
  R., Mrs., 120
  Richard, 4, 58, 59, 78, 79, 81, 140
  Robert, 4, 40, 51, 55, 58, 60, 66,
    79, 81, 110, 146, 180
  Robert, Mrs., 44
Bebout
  Charles, 5, 165
  Hilary, 165
  James, 165
  Robert, 165
Becker
  Kate, 106
Beghtel
  Jessie, 136
Beirne
  Bartholomew, 20
  Charles, 20
  James, 128
  Margaret, 19, 20, 75
  Patrick, 19, 20, 75

INDEX

Terence, 19
Belcher
  I., 180
  Richard, 182
Bell
  Charles, 193
  Miss, 68
Belveal
  Verna, 153
Bennett
  Elizabeth, 205
  Grace, 14, 127
  Katherine, 115
  R., Mrs., 91
Bennion
  Margaret, 4, 159
  Peggy, 165
  Richard, 4
  Robert, 4
Berkeley
  Doris, 129, 135
Bertha, 145
Betancourt
  Eloisa, 184
  Francisco, 184
  Maria, 207
  Marin, 184
Bevan
  T., 96
Binet
  La Tosca, 4
Birmingham
  Abbie, 26
  G., 26
  Ida, 26
Bitner
  C., 116
  Captain, 117
  Cyrus, 1
  John, 1
  Mary, 1, 117
Black
  Agnes, 177
  Alfred, 148
  Effie, 6, 60, 149
  Eva, 6
  Grace, 141

  Mae, 142
  Robert, 6, 177
  William, 149, 177
Blake
  Christine, 125
  Dollie, 125
Boardman
  C., 68
  William, 50
Boatwright
  Alice, 6
Bodle
  Edna, 175
Bone
  Thomas, 14
Bowen
  Charles, Mrs., 97, 104
Bowman
  Alexander, 5
  Ella, 180
  Emma, 11, 102, 197
  F., 102, 117, 193
  George, 11
  Gordon, 12, 47, 58, 59, 78, 102
  Harry, 11
  Henry, 190
  J., 102, 117
  James, 12, 58, 101, 180, 193, 197
  James, Mrs., 112
  Jane, 58, 59
  Jonie, 187
  Lizzie, 11, 78, 180, 188
  Louisa, 12, 197
  Mother, 11
  William, 11, 29, 102, 117, 190, 193, 197
Boyd
  Nancy, 195
Boyer
  Eliza, 25
  John, 25, 179
  Lizzy, 25
Brady
  Ann, 192, 201, 203
Breen
  John, 70
  William, Mrs., 159

INDEX

Breenan
  M., 48
Breever
  Aldon, 207
Bregnan
  Michael, 75
Breir
  Timothy, 179
Brett
  Nelson, 13
  Rose, 13, 156, 163, 205
  Walter, 156, 205
Brickell
  Catherine, 29, 31
  Jennie, 29, 31
  John, 29, 31, 89
Bridge
  Betty, 9, 35
  John, 35
  Thomas, 9, 35
Brislin
  Mary, 157
Bristow
  Elmer, 19, 105
  George, 19
  Virginia, 19
Broken
  Fred, 101
Broom
  Bertha, 14
Brott
  Virginia, 136
Brown
  Charlene, 156
  Elizabeth, 1
  Jean, 153, 163
  Kate, 130
  Laura, 39
  Ruby, 164, 174, 175, 209
Bruce
  Ellen, 195
Brunckhorst
  Elmer, 1
  Fred, 1
  Oscar, 1
Bryant
  B., 127

Buck
  Elijah, 25
  Lydia, 183
Buckley
  Ann, 23
  Michael, 23
  Patrick, 23
Buller
  Mary, 191
Bunch
  Loretta, 207
Burke
  Ann, 24, 40
  Elizabeth, 201
  Frank, 23
  Genevieve, 23
  Patrick, 41
  Teresa, 125
  William, 23, 125, 129
Burne
  John, 70
Burnes
  John, 75
  Kate, 69
  Mary, 181
Burns
  Charles, Mrs., 100
  James, 23, 31, 49
  John, 23, 54, 98, 191
  Kate, 70
  Margaret, 23
  Reva, 205, 206
Burris
  Bessie, 110
  Byron, 110
  C., Mrs., 68
  Charles, Mrs., 95
  Ethel, 150
  Raymond, 204
Burroughs
  Ethelyn, 204
Bushby
  Blanche, 183, 193
  Earl, 137, 147
  Edward, 7, 50, 62, 71
  Fannie, 7, 141, 147
  Frank, 7, 71

INDEX

George, 7, 71, 79, 110, 114, 124, 154, 193
George, Mrs., 67
Hannah, 177, 190, 203
Irma, 7
J., 7
Joe, 71
Joseph, 79, 110, 114, 182
Katherine, 7, 110
Nora, 7, 137
Sarah, 7, 110, 114, 141, 142
William, 7, 42, 71, 110, 114, 124, 141, 177, 180, 183, 193
Butler
  C., Mrs., 121
  Chris, Mrs., 138
Byer
  Henry, 131
Byrne, 192
  A., Mrs., 130
  Adie, 52
  Henry, 76, 90, 113, 140, 142
  J., 63
  J., Mrs., 94, 114
  James, 18, 75, 76, 90, 113
  James, Mrs., 52
  John, 18, 76, 90, 113
  Margaret, 18
  Mary, 17, 18, 128, 135, 188, 200, 202
  Robert, 18, 63, 76, 90, 113, 127, 135, 140, 200
  Rose, 18, 76, 90
  Sarah, 18, 76, 90
  Selma, 18, 160
  Thomas, 18, 186

C

Cady
  W., Mrs., 97
Cain
  Bridget, 20
  Daniel, 20, 70, 91, 122
  Jane, 39
  Richard, 50
Caine
  Daniel, 79, 191
  Ellen, 182
  Richard, 191
Calaghan
  Michael, 24
  Patrick, 24
Call, 196
  George, 136
  Harriett, 6, 136
  Hattie, 196
  S., 7
  Salmon, 136, 196
Callaghan
  Patrick, 144
Callahan
  P., 66
  Patrick, 76
Calligan
  H., Mrs., 137
Calvin
  John, 25
  Margaret, 25
Campbell, 200
  Alan, 1, 170
  B., 175, 209
  Brenda, 170, 172
  Brent, 170
  Christopher, 170
  Daniel, 22, 35, 185
  Donald, 170
  George, 170
  Infant, 22
  James, 22, 116
  Joseph, 198
  Michael, 170
  Mildred, 170
  Mrs., 22
  Patrick, 146
Cann
  Bert, 12, 133
  Charles, 133, 151, 153
  Edna, 12
  James, 133, 151, 200
  Maud, 12, 133, 151, 153, 200, 205
  Phillip, 12, 133, 151, 153
  Thomas, 12, 133, 200

INDEX

Winona, 12
Canning
  David, Mrs., 116, 122
Cardo
  Rose, 205
Carew
  Malachy, 45
Carey
  Leo, 124
  W., Mrs., 104
  William, Mrs., 59
Carlson
  Andrew, 160
  Angela, 160
  James, 156, 205
  Jim, 160, 206
  Judy, 156, 160
  Kim, 25, 156, 205
  Rebekah, 160
  Ric, 25, 156, 160, 206
  Serene, 160
Carney
  Cora, 2, 162, 169
  James, 156
  Richard, 2, 155
  W., 155
  William, 2, 155, 161, 169
Carpenter
  William, 25
Carroll
  E., 124
Carson
  D., 70
Carter
  Ida, 7, 122
  W., 122
Casey, 85
  Abigail, 112
  F., 85
  John, 19, 85, 110, 112
  Margaret, 19, 50, 187
  Mary, 19, 36
  P., 85
  Peter, 19, 36, 50, 80, 85, 189, 193
  William, 19, 85, 193
  Winifred, 19, 36, 113
Castillo
  Tina, 174
Ceresa
  Janet, 162
  Jeffrey, 162
Chadwick
  Jacob, 27
Chamberlain
  William, 146
Chamberlin
  W., 2, 82, 179
  William, 82
Chapman
  Wilma, 162
Charles, 92
Clark
  Abigale, 9
  Henry, 9
  J., 9
  Jane, 9
  L., 9, 37
  Levi, 8
  M., 12
  Mary, 9, 12
  Sherman, 9
  W., 12
  Wright, 12
Clarke
  Annie, 12
  Wright, 179
Clayton
  Charles, 8, 170
  Esther, 8, 170
Clifford
  Keith, 14
Coburn
  F., Mrs., 107
  Mary, 116, 126
Cocking
  William, Mrs., 101
Coffield
  Crystal, 158
Cole
  Dorothy, 149
Collbran
  Arthur, 25
  Florence, 25
  James, 25

INDEX

John, 25
Mabel, 25
Virginia, 25
Colletti
   Matilda, 149
Collier
   Bruce, 161, 162
   Gene, 162
   John, 12, 161, 162
   Leo, 162
   Marvin, 162
   Pat, 155
   Patricia, 12, 161
   Virginia, 162
Colling
   Aurelia, 146
   Bertha, 72, 81, 182
   Charles, 61, 180
   Earle, 182
   Elmer, 128, 139, 202
   Infant son, 185
   Lola, 147
   Ruth, 182
   W., Mrs., 185
   William, 10, 60, 61, 180, 188
Collins
   Bertha, 15, 152, 172, 204
   Daniel, 172
   Emma, 169
   Eugene, 172
   Frances, 208
   Patrick, 172, 208
Compton
   B., 103
   B., Mrs., 76, 90, 140
   Brady, 7, 128, 142, 181, 202
   Brady, Mrs., 113
   Charles, 75, 128, 141, 142
   Clarence, 15, 128, 141, 142, 202
   Farrell, 141
   Frances, 208
   Fred, 128
   Gale, 128
   Ida, 181
   Mary, 7, 128, 141, 142, 143
   Nellie, 128, 146
   Nora, 172
   William, 141, 202
Conboy
   Margaret, 24, 53, 187
Conell
   Ellen, 190
Congdon
   Amelia, 26
   F., 26, 43
   Frances, 6
   Frank, 198
   Fredrick, 6
   George, 6, 26, 198
   Infant Daughter, 26
   L., 26
   Lucretia, 6, 26, 43
   Sarah, 26
Conlin
   John, 17, 91, 100
   Josephine, 91
   Josie, 190
   Margaret, 17, 91, 190
   Sarah, 17
   Thomas, 17, 38, 39, 47, 91, 146, 187
   Thomas, Mrs., 59
   William, 54, 77, 91, 100
Connery
   Mary, 187
Conrath, 49
   Byron, 150, 157
   Clara, 80
   Elizabeth, 16, 100, 198
   Frank, 80, 129
   Gustav, 16, 80
   L., 129, 199
   L., Mrs., 65
   Lizzie, 80
   Lois, 198
   Louis, 16, 79, 80, 100, 150, 156, 157, 189, 205
   Lud, 198
   Ludwig, 80
Conroy
   Thomas, 55
Conry
   Mary, 22
   Thomas, 22, 54

INDEX

Conway
  Laura, 207
Cooley
  Janie, 6
Cooney
  Mary, 17, 134, 196, 200
  William, 17, 134, 195, 200
Cooper
  Martha, 209
Coopmann
  Fredrick, 22
Corcoran
  Patrick, 45
Corey
  W., Mrs., 122
Cory
  Daniel, 29, 33, 55, 179
  Elizabeth, 29, 33
  William, 29, 33, 185
Cottrell
  Beulah, 136, 161
Coughlan
  William, 20, 79
Councilman
  Eloise, 155
Countryman
  Frank, 137
Crabtree
  Ruby, 207
Cramer
  Antonetta, 208
  Earl, 157
Cramp
  Cora, 119
  Wilfred, 82
  William, 6, 82
Cramsie
  Anna, 198
  Annie, 127
  F., 99
  Francis, 98
  Frank, 126
  J., 63, 126
  J., Mrs., 87, 124
  Joe, 52
  John, 24, 52, 54, 84, 98, 99, 110, 126, 182, 192, 198
  Joseph, 24, 98
  Sarah, 98, 99
  W., 84, 126, 192
  Will, 52
  William, 24, 48, 52, 54, 84, 85, 98, 99, 110, 146, 192, 198
  William, Mrs., 52, 76
Crandall
  Raymond, 160
Cranshaw
  Elizabeth, 185
  James, 26, 35, 179
Crary
  L., 146
Craun
  E., Mrs., 125
  Viola, 163
Crawford
  Charles, 27
Creden
  Daniel, 23
Creedon
  Daniel, 34
Creps
  H., Mrs., 130
  Henry, Mrs., 76, 93, 94, 114
  Ida, 132
Crocker
  Fredona, 14
  Freedom, 114
Crockert
  H., 54
Crosett
  J., 14
Cross
  Catherine, 179
Crouch
  H., 14
Culvert
  Maude, 112
Cunningham
  Frank, 84
Curl
  Maud, 151
Curran
  Daniel, 51
  Dennis, 40

INDEX

John, 21, 45
Mark, 21, 33, 41, 51, 52
Sarah, 21, 40
Susan, 21, 37
Curry
   Ada, 133
   Artie, 14
   Bertram, 14
   Flora, 14
   Wilbur, 14
Curtis
   Thomas, 14, 41

D

Daley
   Winifred, 153, 163
Dalsheim
   Florence, 106
Dalton
   Harry, 45
Daly
   Ann, 19
   Cecelia, 19
   Infant, 19, 31
   James, 19, 31
   Jane, 18
   John, 19, 54, 83
   Kate, 19, 83
   Lawrence, 19, 41
   M, 19
   Marcus, 99
   Michael, 19, 185
   Patrick, 53
   T., 19
   Will, 83
Daniel
   Elizabeth, 2, 32
DaShiell
   Julia, 182
   Virginia, 182
Daugherty
   Daniel, 137
Davey
   Carrie, 14
   Ruth, 128, 139
   S., Mrs., 93

   William, 14
Davies
   Ed, Mrs., 199
   Elizabeth, 9
   Maria, 199, 202
   Samuel, 9, 32
Davis
   Adeline, 3, 153, 204
   Anne, 202
   Bert, 3, 153, 204
   Edward, 6, 142
   Elvon, 149
   Ervin, 204
   Mae, 6, 148
   Maria, 31
   Mary, 192
   S., 3
   S., Mrs., 44
   Simon, 40, 180
Day
   Elsie, 167
Dehes
   Nellie, 152
Dempsey
   Ann, 17, 61
   Annie, 188
   Catherine, 17, 103
   Daniel, 66, 92, 103
   J., Mrs., 122
   John, 17, 61, 70, 92, 103, 127, 192
   John, Mrs., 103
   Joseph, 17
   Kenneth, 103, 127, 132
   Michael, 92
   Rose, 17, 189
   W., 66, 127
   William, 54, 92, 103
Denehey
   Cornelius, 22
   Mary, 22
   Neil, 77
Denney
   Albert, 151
   Dean, 151
   Dorothy, 12, 151
Dennis

INDEX

Vincent, Mrs., 91
DePue
  Malinda, 200
Detrick
  Albertine, 149
  Eddington, 115, 149
  Elsie, 115
DeVaudrevil
  Charlotte, 165
Dewan
  Charles, 24
  Ellen, 24, 101
  James, 24, 69, 70, 101
  Jessie, 69
  John, 24, 48, 70, 186
  Nellie, 69
  William, 24, 63, 66, 69
Dewitt
  Ruby, 184
Dillon
  Peter, 22
Dittmer
  James, 29, 32
  John, 29, 32
  Katie, 29, 32
  Mary, 29, 32
Divver, 39
  Alvin, 93
  B., 191
  Bernard, 65
  Byron, 74, 93, 97
  Charles, 23, 64, 65, 191
  George, 23, 65, 74, 93, 97, 191
  Mary, 23, 41, 185
  William, 23, 65, 93, 98, 191
Dixon
  Lucille, 167
Donahue
  D., 43
  Daniel, 47
  Daniel, Mrs., 47
Dooley
  James, 2, 95
  James, Mrs., 90
  Louisa, 2, 95
Doubt
  Elizabeth, 18, 48

J., 18
John, 49
Dougherty, 193
  Ann, 22, 85, 189, 204
  Catherine, 22
  Daniel, 22, 39, 74, 85, 102, 105, 189, 192, 201, 203
  Ezekiel, 14, 37, 94
  Grace, 22, 103, 105, 192
  Ida, 208
  John, 22, 39
  Kate, 22, 103, 105, 192
  Lizzie, 103
  Margaret, 192
  Mary, 197, 200
  Michael, 74
  Sadie, 23, 74
  Thomas, 22, 50
  Virginia, 95
Douglas
  Edith, 180
Doyle
  Alexander, 22, 34, 42
  E., 22, 63, 90
  Edward, 67, 113
  Elizabeth, 22, 38
  J., 63, 90
  James, 20, 66, 67, 90, 113, 188
  John, 66, 67, 113
  Mary, 20, 90, 113
  Mrs., 67
  Olive, 22
Driggs
  Nellie, 153, 163
Driscoll
  Catherine, 18, 103, 192
  D., 122
  J., 122
  James, 54, 103
  John, 18, 54, 88, 103, 192, 194
  Katherine, 194
  Mary, 18, 42
  Sarah, 18
  Theodore, 18
  Timothy, 102
Dugan
  Bridget, 63

INDEX

Patrick, 63
Duggins
  Carrie, 153
Dunham
  D., 8
  Lucinda, 8
Dunning
  Iola, 111
Dupes
  Earl, 208
DuShane
  L., 13
Duvall
  Lillian, 183
Dykes
  Grace, 6, 149
  Hugh, 6, 141

E

Earley
  Frank, 97
  John, 97
  Timothy, 97
Early
  Frank, 89
  M., 23
  Margaret, 23
  Mary, 23
  T., 23
  Timothy, 23, 89
Eastman
  Clerk, 96
Edward
  John, 67
Ehmann
  George, 4
  Paul, 4
  Rosa, 4
Eiden
  Mable, 10
Eisen
  F., Mrs., 125
Elder
  Angela, 206
Ellsworth
  Dianne, 171

Richard, 171
Emery
  Joe, Mrs., 57
  Joseph, 68, 181
  Mary, 7
  V., Mrs., 68
Escobedo
  Mary, 152
Escovedo
  Mary, 154
Estmon
  Alden, 27
Eugenia
  Mary, 56
Ewing
  Susana, 196

F

Fabela
  Elizabeth, 161
Facer
  Noreen, 139, 201
Farish
  Adam, 25
  Etta, 25
  Helen, 25
  John, 25
  Mary, 25
Fay
  Frederick, 77
Fenton
  Lois, 5
  Louis, Mrs., 130
Ferguson
  Katherine, 172
Ferrier
  Louis, 130
Fevier
  Mrs., 63
Fiene
  Annie, 2
  Henry, 2, 90, 100
  Henry, Mrs., 99
  Margaret, 95
  W., 95
  William, 90

INDEX

Filcher
  Eugenia, 11
  W., Mrs., 104
  William, 62
Fippin
  A., 126
  Asa, 5, 120, 153, 163, 178, 183, 204
  Asa, Mrs., 120
  Berniece, 120
  Betty, 171
  Eleanor, 120
  Emma, 5, 125, 153, 163, 183
  Jess, 153, 163
  John, 5, 120, 204
  Julia, 120
  Nellie, 120
  Robert, 120, 153, 163
  Sidney, 3, 153, 163, 171
Firebaugh
  C., 127
Fisher, 206
  Grace, 1, 47, 186
  Rose, 2
  Samuel, 2, 76, 110
  William, 159
Fisk
  Amelia, 145
Fitts
  Amos, 131
Fitzhugh
  Grace, 2, 152, 154, 204
  Henry, 204
  James, 2, 152, 204
  Olive, 2
Fitzpatrick
  Mrs., 104
Flanagan
  Alice, 195
  James, 88
  John, 15, 88, 190
  Mary, 15
  Matthew, 195
  Michael, 16, 186, 190
  Peter, 15, 189
  Susan, 201, 202
  Thomas, 15, 88, 125, 190

Flatt
  Jessie, 166
Flint, 179
  George, 26
  Rachael, 34, 179
Flowers
  Hazel, 13
  Marcle, 138
  Murle, 148
Flynn
  Morris, 21
  P., 21
  Patrick, 21
  S., 21
Fogarty
  John, Mrs., 97
Forbes
  F., Mrs., 123
  Gerald, 123
  Rose, 139
Ford
  Patrick, 24, 193
Foreman
  Blanche, 141, 147
  Ella, 62
  Frank, 8, 64, 149, 183, 193, 203
  Frieda, 149, 203
  Ida, 8
  J., 64
  Jacob, 8, 34, 67, 99, 183
  Jake, 203
  Jermie, 8
  Pearl, 64
  Violet, 7, 193
  William, 64
Forges
  G., Mrs., 137
Fortune
  Martin, 18, 42
  Mary, 18
Foust
  Elizabeth, 65
  Louis, 6, 65
  Nelson, 6
Fox
  Donna, 168
Franks

INDEX

Maude, 204
Fraser
  Daniel, 9, 113, 148, 203
  Edward, 134
  Edwin, 113
  Elizabeth, 9
  Emma, 203
  G., 113
  James, 9, 66, 113, 134
  May, 177
Frazier
  Benjamin, 8, 75
  Carrie, 8
Freeman
  Billy, 55
Freend
  Cleo, 201
French
  Joseph, 24, 135, 200
Frigaskis
  C., 182
Frouke
  Colleen, 162
Fullcher
  William, 127
Fuller
  Drussilla, 206

G

Gaffney
  Ellen, 24, 144
Galligan
  Andrew, 137, 143
  Clarence, 137, 143, 196, 201
  George, 137, 143, 202
  H., Mrs., 139
  Hugh, 17, 143, 196
  Mary, 17, 143, 196, 202
  Mathew, 196
  Pete, 73
  Peter, Mrs., 92
Gann
  B., 209
  Gladys, 1, 172, 175, 208
  James, 209
  Joel, 172, 175

Wayne, 1, 172, 175, 208, 209
Garley
  Mrs., 93
Garson
  Ruth, 151
Gassaway
  Bertha, 10, 128, 199
  Ida, 34, 183
  Mary, 88
  R., 182
  Robert, 10, 81
  Robert, Mrs., 87
  Sadie, 120
Gavin
  Harold, Mrs., 122
Geraghty
  Alvina, 20, 138, 159
  John, 20, 105
Germann
  Charles, 134
Gerth
  Arelea, 154
Gianella
  T., 127
Gilbert
  Absalom, 4, 159, 164
  Carl, 159, 164
  Pauline, 4, 134, 157, 158, 159
Gilberthomp
  Elizabeth, 197
Gilham
  Mary, 126
Gilliam
  Dorothy, 4, 169, 174, 208
  Leslie, 169
  V., 170
  Virgil, 208
  William, 4, 169, 207
Gleason
  D., 183
  Elmer, 177, 183
  Harold, 7
  James, 183
Glyce
  Carrie, 139
Gnadig
  Ed, 166

INDEX

Frederick, 207
Gordon
  Maude, 178
Govne
  Ann, 26
  John, 26
  Maria, 26
  Thomas, 26
  William, 27
Grace
  Jack, 123
  John, 15, 164, 165, 207
  June, 123
  Robert, 123
  Rose, 15, 165, 207
  W., Mrs., 123
  Will, 123
Gratiot
  Eliza, 14, 41
Gray
  George, 201
  Georgia, 52
Greely
  Fred, 96
Green
  Elizabeth, 113
  Mrs., 81
Greever
  Aldon, 2, 157, 160, 162, 164, 174, 175, 183, 184, 206, 209
  Charles, 206
  Hazelle, 2, 157
  Jack, 2, 164, 174
  Jackie, 174, 183, 209
  Jennifer, 175
  Jeremy, 175
  Jim, 164
  Jimmy, 174, 175
  Joe, 164, 174, 175
  Kyle, 2, 157, 160, 162, 174, 184, 206, 209
  Martha, 175, 209
  Maxine, 2, 164, 174, 175, 209
  Minnie, 209
Griffith
  Butch, 164
  Jim, 174

Grimes
  Carrie, 165
Grose
  Harry, 14
Grove
  Lillian, 123
Groves
  Sarah, 204
Gruber
  Eveline, 170
  Gene, 170
  William, 15, 170, 208
Gunderson
  Patricia, 159, 165
Gunning
  Alvina, 96, 100, 102, 105, 138
  Baby, 20
  Ella, 96
  Ellen, 20
  Fremon, 20, 38
  Infant son, 186
  Jennie, 96
  Mary, 20
  Robert, 96, 138, 159, 163
  S., 54, 96, 102, 138, 159, 191
  S., Mrs., 186
  Samuel, 20, 37, 45, 95, 96, 138, 159, 163, 191
  Thomas, 96, 138

H

Hacker
  Cora, 6, 138, 144, 148
  Erma, 11
  Francis, 148
  Harvey, 138, 148
  William, 6, 138, 182
Hagerty
  William, Mrs., 89
Haggerty, 193
  Anna, 190
  William, Mrs., 97
Hahn
  Dovie, 208
Haling
  George, 181

INDEX

Hall
    Ed, Mrs., 131
    J., 7
    John, 70
    Lilas, 122
Hammerschmidt
    Donna, 166
Hammon
    Thomas, 64
Hammonds
    James, 190
Hampton
    Clara, 166
Handy
    Ethel, 126
    J., Mrs., 128
Hanley
    James, Mrs., 49
    Peter, 63, 67
Hansen
    Ralph, 161
Hanson
    Raymond, 149
Hapgood
    Annie, 11, 36
    Carol, 129, 135
    E., Mrs., 82
    Ed, 181
    Elizabeth, 11, 36, 82, 181
    Eugene, 11, 36, 104
    Eugenia, 43, 62
    Fannie, 27, 129, 134, 145, 200
    J., 136
    J., Mrs., 86
    James, 27, 104, 129, 135, 136, 145, 181, 200
    Lester, 136
Harriman
    Edward, 83
    Elsie, 6, 83, 189
    Stephen, 80
    Susan, 6, 80
Harris
    Adeline, 204
    Angel, 13
    Annie, 200
    J., Mrs., 115
    Jennie, 180
    Lee, 13
    Mrs., 44
Harryman
    Irene, 150, 151, 152, 154, 157
Hartley
    Mary, 10
    W., 10
    William, 37
Hartman
    Donna, 152
Harvey
    Elizabeth, 198
    John, 76
    Margaret, 76
Hassler
    Carolyn, 166, 207
Havey, 22
    Agnes, 105
    Amanda, 40, 185
    Anna, 189
    Chester, 119
    Elizabeth, 99
    Gertrude, 22
    J., Mrs., 135
    John, 22, 40, 52, 105, 119, 189, 196
    John, Mrs., 49, 119, 122
    Katie, 34
    Linn, 116
    Louis, 22, 119
    Lucille, 22
    Maggie, 52
    Margaret, 22, 52
    Mary, 2, 22, 52, 196
    Pat, 18
    Patrick, 37, 52, 196
    Veronica, 22
Hawkins
    Eldon, 13
    Gloria, 13
Haworth
    Abraham, 9
    Ann, 9
    Thomas, 9, 33, 179, 185
Hay
    Leroy, 140

INDEX

Hays
  Isaac, 11, 93
Heggerty
  Mary, 15
  Maurice, 15
  Thomas, 49
  Thomas, Mrs., 49
  William, 15
Henderson
  Bridgett, 24, 59, 188
  T., Mrs., 61, 92
Hendricks
  M., 73
Henke
  Frances, 140
Henry
  Joanne, 166
Herboth
  Alice, 137, 143
  David, 150
  Elizabeth, 16, 150, 156, 204, 205
  George, 16, 100, 150, 156, 204
  George, Mrs., 129
  John, 204
  Joseph, 150
  Louise, 150
Hermann
  Carl, 4
  Caroline, 4, 134
  Charles, 4, 157
  Frank, 134, 158, 159
Heyne
  Bernard, 4
  Ethel, 4
Hickeson
  Katie, 8, 114, 193
  Minnie, 114
  W., Mrs., 110
  William, 114, 193
Hickey
  Serena, 160
Hicks
  Joseph, 9
Higenbotham
  Viola, 167
Hilke
  Frances, 173

Hill
  J., Mrs., 44
  John, Mrs., 115
Hinkle
  Archie, 3, 204
  Gene, 151, 168
  Noble, 3, 151, 204
  Pat, 151
Hiscox
  Irwin, Mrs., 113
Hite
  America, 116
  D., Mrs., 107
  David, 9
  H., Mrs., 106, 120
  Herbert, 9
  Ida, 9, 140, 160
Hofer
  Marguerite, 137
Hogarth
  Harry, 76
  M., 10
  Rose, 10
Holbrook
  E., 48
  Elise, 16, 149
  George, 48, 187
  James, 16, 115, 149, 158
  Katherine, 115
  Kathleen, 16, 141
  Katie, 109
Holcomb
  David, 14
Holmes
  Alfred, 185
  Karen, 155
  Thatcher, 26, 32, 33
  Winslow, 26
Holt
  Mrs., 85
Hoogland
  Eileen, 167
Hopkins
  Margaret, 64, 74, 93, 97
  Suzanne, 160
Horan
  Michael, 18, 54, 187

INDEX

Mike, 49
Hord
   Charles, 198
   Mary, 11, 198
Hoskins
   Maggie, 191
House
   Anthony, 14
   Kate, 194
   W., Mrs., 91
Housekeeper
   Unknown Name, 27
Howsaw
   Winifred, 193
Hudson
   Edwin, 208
   Laverne, 12, 161, 206
   Ralph, 12, 161, 171, 206, 208
   Richard, 161, 171, 208
   Robert, 161, 171
Huffman
   Mary, 36
Hug
   Alois, 141
Hughes
   Alois, 21, 202
   Ida, 181
   Lloyd, 207
   Michael, 21
   Winifred, 21
Huling, 201
   Mary, 194, 195, 196, 199
   W., 76, 126
Hunt
   Jasper, Mrs., 117
Hurling
   Charles, 14
   George, 14
   Lucille, 14
Hurst
   Jane, 190, 197
Hutchinson
   David, 14
   Maybelle, 208
Hutchison
   Louise, 149
Hyatt
   Jacob, 10
Hymens
   George, Mrs., 67

I

Infraham
   C., 180
Ingram
   Mary, 14

J

Jackson
   M., 5
   Millicent, 199
   Minnie, 76
   William, 5
James
   Bud, 158
   Chester, 158
   Donald, 158
   Larry, 158
   William, 14, 158
Jeffery
   Bertha, 180
   Della, 72
   Ella, 10, 87, 99, 132, 190, 192
   Ellen, 11, 199
   George, 10, 31, 42, 72, 87, 128, 139, 146, 190, 192, 199, 202
   George, Mrs., 42
   Harry, 10, 42
   Justin, 139
   Lizzie, 72
   M, 188
   Maria, 10, 72, 139
   Robert, 41, 72, 87
   W., 103, 128
   William, 11, 54, 72, 87, 132, 139, 182, 199, 202
Jeffries
   Alice, 101
Jennings
   Martin, Mrs., 52
Jensen
   John, 206

INDEX

Jewell
  E., 8, 32
  Elizabeth, 8, 32
  Henry, 8, 32
Johnson
  Carrie, 16, 134, 148
  Charles, 203
  Clarence, 123, 134
  Cora, 123
  Donald, 148, 167
  F., Mrs., 67, 90, 113
  Francis, 16, 119, 123
  J. R., Mrs., 123
  Ruth, 161
  Velma, 198
  W., 123
  William, 16, 134, 147, 196, 203
Johnston
  Nora, 155
Jones
  Alma, 170
  D., 70, 122
  D., Mrs., 91
  David, 7, 198
  Eliza, 193
  Elizabeth, 7
  Eva, 39, 180
  J., 7
  Jason, 117
  John, 197
  Lillian, 205
  Lillie, 36
  Louisa, 132, 180, 197
  Lurline, 165
  Simon, 27
  W., 122
  William, 183
Juardo
  Margaret, 144

K

Kauk
  Nancy, 209
Kay
  Lavania, 129
  Samuel, 12, 59

Keating
  John, 169
Keegan
  Ann, 20
  Annie, 20
  Charles, 20
  Henry, 39
  Infant, 20, 39
  J., Mrs., 103
  Jack, 147
  James, 20
  John, 39
  John, Mrs., 105
  Mary, 20, 39
Keith
  Anna, 179
  Granville, 6, 83
Kellcher
  Nora, 208
Kelly
  Ellen, 194, 202
  George, Mrs., 138
  Harry, 100, 101
  J., 85, 189
  James, 24, 66, 67, 92
  John, 100, 101
  Kelly, 39
  Lizzie, 93
  Mary, 130
  Peter, 100, 101, 191
  Thomas, 39
Kelser
  Laura, 180
Kennedy
  John, 204
  Winifred, 178, 183, 205
Kerrigan, 20
  Ambrose, 20
  Annie, 60
  Eugenia, 60
  Katherine, 60
  Mary, 20, 56, 60, 120
  P., 56, 60
  Patrick, 20, 120
  Peter, 20, 33, 60, 120
  Thomas, 20, 56, 60
Kershaw

INDEX

Diane, 25
Henrietta, 25
Theodore, 25, 26
Kerwin, 202
Kibbe
  June, 151
Kibbee
  June, 152, 157
Kilbourn
  Minnie, 123
Kildahl
  Katie, 91
Kirby
  John, 17
  Mary, 46, 109, 197
Kneebone
  Bud, 131
  Mary, 145
Kramer
  Hazel, 209
Kraus
  Mrs., 93
Kulker
  Elizabeth, 204
Kuster
  D., Mrs., 121
  Martin, Mrs., 122

L

Lambert
  Sheldon, 29
Lane
  Esther, 13, 208
  Larry, 13
  Lawrence, 208
  Linda, 13
  Louisa, 13, 34
  Riley, 34
Larson
  L., 131
Laughlin
  Helen, 136, 161
Lavelle
  Mary, 24
Lavers
  Alexander, 199

Mary, 201
Le Bourveau
  Ernest, 156
Leahy
  Annie, 200
  Michael, 24
  Mike, 189
LeBoeuf
  Cecilia, 15, 152, 204
  Merrill, 152
  William, 204
LeBourveau
  Ernest, 4, 205
  John, 205
Ledwich
  James, 70
Lee
  Agnes, 22, 78
  James, 23, 36
  Mary, 36
  Thomas, 22, 36, 78, 84, 85
Lehman
  John, 52
Leigh
  Kenna, 23
Leslie
  George, Mrs., 90
Lessley
  Betty, 2, 164, 207
  Billie, 174, 175
  Carnel, 184
  Jeff, 184
  Ruby, 184
Lienke
  Freddy, 140
Likens
  Alma, 13, 166, 207
  Ben, 166, 167
  Donald, 166, 167
  Hance, 13, 166, 167, 207
  R., 166, 167
  Reginald, 13
Linehan, 15
  Anna, 119, 194
  Daniel, 15
  E., 189
  Edward, 15, 78

INDEX

Elizabeth, 179
Ellen, 15, 98, 191
J., 54
John, 15, 98, 119, 193
Katherine, 15, 98, 119, 140, 194
Mary, 15, 98
P., 119
Patrick, 194
Peter, 15, 98, 140, 202
T., 119
Timothy, 15, 54, 75, 98, 119, 191, 194, 202

Link
  Sarah, 128
Linscott
  James, 27
Livingston
  Nellie, 149
Lois, 198
Loomer
  Elizabeth, 203
  Henry, 5
  J., 5
  Maggie, 5
  N., 5
  Sarah, 5, 194
Looney
  Catherine, 22, 114, 193
Lotsen
  Alvina, 37, 96
Lowden
  Elva, 198
Lowe
  Lizzie, 179
Lucey
  P., 54
Lyons
  Jeremiah, 17, 36

M

Maas
  Blanche, 112
MacWilliams
  Joe, 23, 206
Madden, 191
  Joan, 165

Magganotta
  Peter, 147
Magonigal
  Alice, 13, 166
  Clarence, 12, 198
  Clayton, 166
  Eleanor, 169, 203
  Eleanore, 16
  Ellen, 11, 187
  Emma, 12, 134, 148, 203
  Frank, 177, 190
  Frankie, 12
  George, 137, 145, 154, 203
  Gerald, 13
  Hannah, 12, 133, 141, 142, 145, 154
  Henry, 13, 148, 166
  Jennie, 199
  Jerry, 166
  John, 12, 148, 164, 203
  John, Mrs., 113
  Lizzie, 11, 195
  Nancy, 11, 43, 186
  S., 11
  Samuel, 54
  Sarah, 114
  T., 12, 190
  T., Mrs., 110, 124
  Thomas, 133, 177, 180, 195, 198, 203
  Vera, 12, 164
  W., 11
  W., Mrs., 190
  William, 11, 12, 42, 148, 180, 181, 185, 195, 203
Malm
  Dorcas, 165
Manasco
  Amelia, 106
  C., 140
  C., Mrs., 88
  Calton, 106
  Clayton, 9
  George, 106, 120
  Grace, 106
  Veronica, 9, 120, 190
Manford

INDEX

Dorothy, 2, 164
James, 2, 164, 165, 207
William, 207
Manie
   Mildred, 208
Mann
   Levi, 179
Mans
   Addie, 195
Maples
   Joe, 209
   Maxine, 184, 207, 209
   Minnie, 183, 209
Marian, 197
Marks
   Charles, Mrs., 90, 100
Marlin
   J., 48
Marple
   Barney, 86
   Brant, 27
   Elizabeth, 27, 181
   Fannie, 181
   Fred, 27, 86
   George, 27, 86, 129, 135, 136, 200
   Harry, 27, 196
   Infant, 27
   John, 27
   Sam, 86, 129, 135
   Samuel, 27, 135
   Whelton, 27, 86
   Whitten, 189, 197, 200
   William, 181, 201
Martein
   Charles, 15, 137
   Elizabeth, 15, 137, 147, 203
   Lizzie, 137
Martin
   Betty, 184
   Danny, 164
   Frank, 164, 184, 207
   Jenny, 164
   Mary, 164
   Suzy, 164
   Tony, 164
Martinez
   Delores, 161
Mayben
   Martha, 204
McAllis
   F., 26
   Fannie, 26, 32
   J., 26
   John, 32
McAuliffe
   D., 193
   Dennis, 130, 199
   Julia, 193, 199
McCarthy
   Andrew, 72
   James, 73
McCarty
   A., 63, 92, 195
   Andrew, 17, 49, 92, 143, 191, 201, 202
   Ella, 139, 202
   James, 17, 39, 72, 73
   John, 17, 31
   M., 52, 63
   Matthew, 17, 92, 137, 201
   R., 103
   Robert, 17, 92, 137, 139, 140, 202
   Rose, 92, 191
   Susan, 17, 143, 195
   William, 41
McCay
   Margaret, 138
McClain
   Gordon, 183
   Sharlene J., 183
McClaskey
   John, 27
   Margaret, 36
McClure
   Alexander, 24, 71, 201
   James, 24, 136, 137, 201
   Joseph, 24, 137, 138, 201
McConnell
   James, 10
   John, 10, 179
   M., 146
   Mary, 10

INDEX

W., 103
William, 52
McCormick
   Pearl, 144, 148
McCoughlan
   William, 79
McCoughlin
   Margaret, 24
McCowell
   Ruth, 203
McCrea
   Albert, 10, 183
   Arthur, 10
McCune
   Lizzie, 35
McDaniel
   E., 96
McDivitt
   Tom, 137
McDonald
   Daniel, 53
   G., 180
   Margaret, 113
   Mary, 119, 122
   Mrs., 81
McDowell
   Ruth, 149, 167, 207
McFate
   T., Mrs., 91
McGanney
   Anna, 24, 84, 126
   Annie, 24, 47, 72, 129
   Baby, 24
   Charles, 61
   D., 61
   Daniel, 47, 48, 72, 84, 87, 124, 125, 127, 129, 146
   Edward, 24, 47, 72, 87, 124, 127
   F., 52
   Father, 24
   Frank, 47, 72, 87, 125, 127, 139
   Georgie, 24, 139, 201
   James, 24, 47, 50, 72, 87, 127, 129
   Mary, 72
   May, 47
   Mother, 24
   Ned, 24
McGinnis
   Tessie, 145
   Walter, Mrs., 104
McGloughlan
   Peter, 20
McGovern
   Helen, 140, 142
   Jack, 140, 142
   John, 15, 42, 195
   John, Mrs., 15, 44
   Mary, 15, 140, 142, 195, 202
   P., 41
   T., Mrs., 119
   Thomas, 15, 142, 195
McGrath
   J., 175
   John, 39
McKaig
   James, 81
   Robert, 81
McKeague, 8
   Robert, 113
McKeel, 6
   John, 6, 52
   Thomas, 6
McKenna
   Elizabeth, 24
   James, 47
   John, 24, 36
   Margaret, 47
   Thomas, 47
McKinley
   Jane, 201
McKinnon
   Hollister, Mrs., 121
McKinsey
   Edith, 150
McLaughlin
   W., Mrs., 123
McLean, 14, 194
McLeod
   Mary, 124
McLeon
   Mary, 127
McNally
   Edwin, 19

INDEX

F., 70
James, 63
Kate, 63
Mary, 19, 63, 130, 188
Thomas, 19, 63, 75
McNamara
  Kate, 40
McNutt
  Alice, 9, 51
  J., 56, 187
  John, 9
McPhillips
  Barney, 23
  Owen, 23, 31
McQuaid, 146
  Donald, 124
  Frank, 16, 40, 59, 97, 124
  J., 54, 70, 85, 103, 115
  Jack, 122, 124
  Joe, 38, 59
  John, 16, 48, 54, 59, 66, 68, 77, 91, 97, 100, 104, 110, 122, 123, 124, 146, 192, 194, 195, 196
  Katherine, 16, 196
  Margaret, 59
  Mary, 16, 59, 97, 104, 122, 123, 124, 169, 187, 195
  Thomas, 59, 97, 104, 122, 124
McWilliams
  Bridget, 22, 63
  Charles, 50
  James, 85
  James, Mrs., 193
Meade, 18
  Agnes, 118
  Annie, 103, 132
  Garrett, 118, 132, 200
  Helen, 118, 132
  John, 18, 103, 118, 132
  Leo, 18, 66, 103, 110, 118, 127, 132, 199
  Mary, 18, 103, 118, 132, 200
  W., 103
  William, 18, 103, 118, 132
Meager
  Dennis, 52

Mellarkey, 38
  Edward, 21, 91, 191, 194
  James, 21, 36, 91, 194
  John, 21, 39, 63
  Mary, 21, 106
Melody
  Bridget, 20
  Thomas, 20, 179
Meredith
  Elizabeth, 190
  George, 67
  W., 14
  William, 35
Merriam
  H., Mrs., 67, 90
  Henry, Mrs., 113
Merz
  Joseph, 190
Mette
  Loretta, 132
Meyers
  George, Mrs., 125
Michaels
  Walter, 25
Michel
  Ruth, 137, 143
Miles
  E., 16
  Evelyn, 134, 148, 203
  Frank, Mrs., 123
  Lois, 166
Miller
  Alfred, 168
  Anthony, 168
  Ethel, 89
  Gary, 204
  George, 13, 21, 168
  J., 48
  Jane, 21
  John, 21, 27, 31
  Joseph, 21
  Leo, 168
  Mark, 168
  Mary, 21, 159
  May, 21
  Mildred, 168
Mills

INDEX

Billy, 167
Brenda, 167
Claudia, 167
Harry, 167
Homer, 13, 167, 184, 207
J., 167
Jack, 167
Johanna, 167
Johnnie, 184, 207
Lorena, 167
Louis, 167
Maria, 167, 207
Steve, 167
Mitchell
    Bell, 105
    Catherine, 11, 70
    Elvira, 11, 192
    John, 11, 70, 75
    Thomas, 192
Moffatt
    J., 127
Mohn
    Joseph, 26
    Mary, 26
Mollett
    Selena, 204
Monasco
    George, 9
Moncur
    R., Mrs., 100
    Robert, Mrs., 100
Monk
    Betsy, 11
    Elizabeth, 12
    James, 11, 12, 181, 182
    John, 181
    M., 182
    W., 181
Montgomery
    Horace, 136, 201
    Horace, Mrs., 136
    Robert, 136
Moody
    Ellen, 29, 31
    Thomas, 31
Mooney
    A., 41
    Adelaide, 8, 129, 139, 199, 201
    Arthur, 8, 66, 182
    Clara, 8, 195
    Jessie, 8, 196
    Lucy, 8, 194
    Mary, 8, 129, 199
    Nellie, 8, 50, 187
    Ruby, 8
    Thomas, 8, 37, 41, 50, 129, 139, 194, 195, 196, 199, 201
Moran
    William, Mrs., 97
Morgan
    J., 127
Morrell
    Ellen, 114
Morrill
    F., Mrs., 110
Morris
    Marie, 128
Morrison
    Mary, 179
Morrissey
    Lizzie, 182
Morton
    Norman, 25
Mourill
    F., Mrs., 124
Mulhare
    Mary, 195
    Michael, 190
Mullhare
    Mary, 196
Mulligan
    James, 23
Mullin
    J., 20
    Julia, 20
    Mary, 20
Mullins
    Mary, 35
Mulvany
    Mary, 194
Munday
    T., 84
Murch
    Charles, 76

INDEX

Charles, Mrs., 107
Frances, 8, 116
Frank, 181
Nora, 79, 182
Murchie
  Kate, 153
Murdock
  Emily, 69, 142, 182
  Everett, 130, 138
  Henry, 10
  James, 199, 201
  Mary, 5, 130, 199
Murphey
  Irene, 119
Murphy
  Catherine, 18, 187
  Charles, 18, 39
  Charlie, 39
  Elizabeth, 36
  Gladys, 11, 177, 192
  Henry, 106
  Ida, 11, 149, 150, 167, 203, 207
  John, 11, 39, 42, 43, 106, 127, 131, 149, 167, 177, 192, 203
  John, Mrs., 49
  Julia, 181
  Mary, 194, 196
  Michael, 203
  Morris, 17, 34, 39, 106, 181
  Morris, Mrs., 52
  Rosana, 11, 43
  Rose, 106
  Stephen, 11, 42, 106
Myers
  Hannah, 200

N

Nagel
  Annie, 120
Naglee
  Arthur, 155
  Barbara, 155
  Brian, 155
  Bruce, 155
  Elizabeth, 155, 205
  Joseph, 15, 155, 205
  Mary, 155
  Peter, 155
  Samuel, 205
Nathan
  Frances, 207
Nelson
  G., 85
  Louise, 205
  W., 180, 183
Nevels
  Clara, 181
  William, 195
Newbert
  Ada, 76
  Chester, 76
  Edna, 76
  Ella, 11, 64
  H., 64, 76, 94, 130
  Hood, Mrs., 130
  Horace, 19, 62, 64, 76, 114, 139
  James, 94
  Lee, 57, 76, 181
  Louis, 139
  Mary, 19
  Millicent, 76
  Minnie, 10
  N., 63
  Nellie, 11, 130, 149, 199, 203
  P., 130
  Rose, 76
  S., 55
  T., 48, 63, 76, 94, 114
  T., Mrs., 76, 114
  Thomas, 10, 76, 84, 93, 94, 114, 139, 199
  Thomas, Mrs., 76
  W., 126, 149
  William, 11, 76, 83, 94, 114, 130, 199
Niles
  Anthony, 15
  George, 15, 144, 202
  Theresa, 15, 144, 202
  William, 144
Nugent
  P., Mrs., 45
Nutley

INDEX

William, 68

O

O'Banion
  Ida, 125
O'Brien, 144
  Agnes, 16, 48, 132, 133, 140
  Annie, 128, 144
  Bridget, 21
  Catherine, 48
  Helen, 16, 132, 133
  Isabel, 140
  James, 16, 17, 45, 48, 89, 107, 115, 132, 133, 146, 197
  Josephine, 16, 197
  Mary, 16, 46, 47, 132, 140, 186
  Philip, 17, 21
  Phillip, 186
  W., 66
  William, 132, 133, 158
O'Connell
  Maurice, 52
O'Connor
  Ann, 132
  John, 52
  Loretta, 132
  Thomas, 132
  Thomas, Mrs., 118
  Tommy, 132
  William, 54
O'Donnell
  Kate, 83
O'Sullivan
  E., 23
  Margaret, 23, 82
  Michael, 23
  P., 23
  Simon, 23
Obert
  Leona, 208
O'Brien
  Agnes, 109
  Gertrude, 132
  Helen, 109, 140
  Isabelle, 109
  James, 109, 132, 140
  Josephine, 140
  Josie, 109
  Mary, 133
  William, 110, 140
Olson
  Bertha, 143, 148
  Patricia, 175
Orr
  W., Mrs., 133, 200
Otis
  Anna, 31
  Annie, 82, 179
  Elizabeth, 2
  Eugene, 2
  Isaac, 2, 179
  James, 82
  L., 70
  M., 2
  T., 31
  T., Mrs., 2, 31
  Walter, 2
Owen
  Clarence, Mrs., 76, 113
  Fannie, 42, 177, 180
  H., Mrs., 90
  Missouri, 181

P

Page
  Olive, 166, 172
  Theresa, 205
Painten
  Elizabeth, 179
Palmer
  Eddie, 207
Parkison
  Bernard, 136, 161
  Bert, 136, 161
  Bob, 161
  Charlie, 161
  Chloe, 3, 136, 200
  John, 3, 136, 161, 200
  Robert, 136
Parks
  Melba, 206
Parrish

INDEX

Edward, 200
Herman, 136
Passni
  Les, 137
Pate
  Derrel, 14, 173
  Kathy, 174
  Lenore, 173
  Lisa, 174
  Steven, 174
Pearce
  Mary, 133
Peardon
  Ada, 10, 114
  Bertha, 83, 84, 88, 94
  Carmelita, 94
  Elizabeth, 10
  F., 10
  Francis, 88
  Frank, 10, 83, 84, 94, 125, 157
  Fred, 83, 84, 88, 94
  Gladys, 10
  Helen, 114
  J., 88
  James, 83, 84, 88, 94, 157, 190
  John, 10, 55, 83, 84, 87, 93, 146, 190
  John, Mrs., 84
  W., 125
  W., Mrs., 94
  Walter, 94
  William, 83, 84, 88, 94, 110, 114, 157
Peckham, 197
  Cora, 183
  Mabel, 183
  Marian, 64
  Mirriam, 11
  Sophronia, 11, 143, 197
  T., 143
  Thomas, 11, 148, 180, 197
  W., 144
  William, 148
Pendlebury
  Mrs., 14
Perez-Triana
  Margaret, 25

Perkins
  Ellen, 182
  Joseph, 132, 199
  M., Mrs., 8, 35
Pesano
  Les, 131
Peske
  Edna, 138
Petersen
  Clara, 206
Pettit
  Catherine, 22
  John, 22
  Katherine, 114
  Nicholas, 22
Pierce
  J., 146
  Mamie, 132
  Mary, 109
  R., 48
  Richard, Mrs., 141
  William, 41
Pill
  Emma, 189
Pine
  George, 96
Pisani
  Bertha, 12, 145, 203
  George, 145
  Joseph, 145, 203
  Leslie, 12, 145, 203
  Leslie, Mrs., 133
  Lorraine, 145, 203
Pitmar
  Marvin, 137
Pittman
  Ernest, Mrs., 143
  Hattie, 132
  Ida, 183
  Myrtle, 148
Plamond
  Mary, 65
Plamondon
  Wilhelmina, 129
Platt
  Walter, 189
Poirier

INDEX

R., Mrs., 91
Poncherelli, 13
Poole
   Bill, 172
   Charles, 177
   Clarence, 12, 135, 172, 173
   Clarence, Mrs., 148
   Daisie, 7, 135
   Dan, 172, 173
   Francis, 7, 119, 194
   Frank, 7, 120, 122, 135, 177, 194
   Harry, 135
   Hazel, 12, 166, 172, 173
   James, 120, 122
   John, Mrs., 130
   Mary, 7, 119, 122, 194
   William, 119, 122, 173, 194
Poor
   Ed, 126, 128
   Emma, 2, 126, 199
   Mark, 2, 126, 128, 199
   Mary, 119
   Robert, 126, 128, 199
Pott
   Darrel, 13
Powers
   Amos, 14
   Infant, 14
   Mrs., 14
Pratt
   Lucy, 121
Presley
   Allen, 9, 34, 49, 179
   Gertie, 9
   Infant, 9
   Mary, 9, 41
   Rachel, 9
Price
   Beulah, 208
Pritchett
   J., 188
   Jacob, 26, 61, 62, 179
   Susanna, 26
Proyer
   Essie, 166
Pryor
   Mary, 24, 49, 186, 189
   Michael, 24, 43, 49, 186
Puff
   Edmond, 1, 154, 158
   Edmund, 158
   Ethel, 1, 154, 158
   Leo, 154
Puttman
   Ruth, 140

Q

Quilty
   Patrick, 23, 150
   Thelma, 23, 150, 151, 168, 204
   Walter, 150
Quinkert
   Clara, 16, 129, 198, 199
   Elizabeth, 16, 49, 187, 205
   John, 16, 65, 79, 129, 198, 199

R

Rach
   Clara, 162
Rada
   Bernice, 159, 164
Radworth
   A., 25
   Charles, 25
   Marcus, 25
Ragby
   R., Mrs., 103
Ramos
   Tirso, 161
Ranberg
   Marjorie, 169
Randal
   Harry, Mrs., 128
   Ida, 128
Ray
   Charles, 4, 144
   Charles, Mrs., 136, 145
   Lavania, 135
   Lavinia, 4, 136, 145, 200, 201
Rearden
   Katherine, 77
   Louisa, 77

INDEX

N., Mrs., 77
Reddy
  Patrick, 99
Reed
  Alice, 181
  Anna, 175
  Bert, 181
  Rilla, 205
Reese
  Kevin, 171
  Kyle, 2, 171
  Mikala, 171
  Ramona, 171
  Richard, 171
Reid
  Mary, 10, 110
  Samuel, 10, 36
Reigh
  Alexander, 14
Reilly
  William, 49
  William, Mrs., 89
Revell
  Joseph, 10
  Martha, 10
Reynolds
  Nellie, 162
Rice
  Naomie, 184
Rickey
  John, 14, 77
  Peter, 14
Rigby
  A., 134
  Albert, 16, 137, 200, 208
  Albert, Mrs., 105
  Frances, 16, 172, 208
  George, 16, 134, 137, 147, 171, 172, 200, 201, 208
  Ida, 16, 137, 200, 201
  Joseph, 16, 137, 147
  Robert, 172
Riley
  William, 48
  William, Mrs., 97
Robbins
  Nathaniel, 14
Roberts
  C., 5
  Charles, 34
  Martha, 136
Robinson
  A., 32
  Bessie, 136, 161
  Sarah, 25
Ronkey
  Amelia, 203
Rooney
  Eugene, 19, 92
  Eugene F., 191
  Mary, 19, 62, 188
Rose
  Ann, 7
  Belle, 57
  Delores, 164, 174, 175, 184
  Ethel, 57, 181
  Frank, 57, 68, 95
  J., Mrs., 68
  James, 7, 57, 68, 76
  John, 7, 56, 95, 150, 204
  Mary, 181
  Oscar, 7, 57, 68, 95, 150
  Virginia, 94
Rossi
  L., Mrs., 117
Rouch
  Mary, 144
Rouse
  Margaret, 205
  Reva, 156, 160, 205
Royal
  Lloyd, 118
Royer, 21
  A., 21
  W., 21
Rushily
  Hanna, 198
Russell
  Alpha, 166
Russi
  Lewis, Mrs., 102
Ruth
  Margaret, 61
Ryan. *See* Patrick

INDEX

Amanda, 36
Ellen, 17, 37
Hannah, 17, 87
Johanna, 37
Mary, 96, 191
Minnie, 36
Nellie, 17
Patrick, 17, 37, 66, 68, 96, 200
Thomas, 17, 66

S

Sanders
  Annie, 19, 102, 192
  Catherine, 19, 127
  Charles, 102, 107
  Dale, 48
  Daniel, 19, 49, 68, 102, 107, 127, 192, 193
  John, 19, 102, 107, 127, 192
  O., 75
Sanford
  A., 126, 131, 177, 201
  A., Mrs., 130
  Alfred, 5, 69, 138, 142, 182
  Benjamin, 6, 69, 131, 142, 146, 195
  Emily, 5, 138, 199, 201
  Euphemia, 6, 131, 142, 195
  Eva, 6, 107
  Evalene, 177
  Malinda, 199
  N., 5
  Nathan, 131
  Sadie, 5
  Sarah, 6
  Thomas, 6
  W., 131
  Wallace, 5, 6, 39, 180
Saunders
  Amelia, 6, 81
  Dan, 92
  John, 6, 81, 92, 96
  Karen, 4
Scafire
  Mary, 169
Schellenger
  Charles, Mrs., 77, 116, 122
Schmidt
  George, 19, 64, 88
  Henry, 120
  Hiney, 88
  Veronica, 19, 88, 190
  William, 19, 88, 120
Schoeder
  Henry, 96
Schulze
  Dorothy, 133
Scofield
  Mrs., 180
Scogland
  Grace, 12
  Thomas, 12
Shain
  Kahleen, 184
Shea
  M., 19
  Simon, 19
Sheehan
  Daniel, 20, 130, 193
  Johanna, 20, 129, 199
  Josie, 193
  Julia, 20, 193
  Patrick, 193
Sheppard
  George, Mrs., 117
  Mary, 102
Sheward
  Howard, 164
Shields
  Donovan, 155
  Joe, 155
  Lloyd, 12
  Mary, 12
  Myrna, 155
  Paul, 155
  Ted, 155, 161
  Vernon, 155
Shilling
  Charles, 14
Shoffiett
  Minnie, 209
Silva
  Erna, 23

INDEX

Louis, 23
Silverthorn
  Elizabeth, 200, 201
Simpson
  A., Mrs., 104
  John, 29
  Mary, 29
  Thomas, 29
Sims
  Jane, 26
  Louise, 26
  Richard, 26
  Robert, 26
Single
  Julia, 204
Singleton
  Billie, 2, 157, 160
  Pete, 2, 160
Sink
  Sally, 141
  Sarah, 142
Sivope
  Josephine, 178
Skehan
  Mary, 177
Slattery
  C., 54, 146
Smart
  William, 145
Smethurst
  Elizabeth, 16
  G., 16
Smith, 52
  Agnes, 21
  Anthony, 21, 49, 187
  Bertha, 116
  Bessie, 10, 83, 189
  Bill, 162
  C., Mrs., 102
  Catherine, 10, 162, 206
  Eugenia, 62
  Floyd, 162
  Frank, 10, 162, 206
  Gloria, 20
  Grace, 174
  John, 21, 52, 53, 162, 206
  Margaret, 21

  Mary, 195
  Philip, 83
  Susan, 21
  T., 52
  Thomas, 21, 63
  W., 11, 43, 85
  William, 21
Snelt
  Mrs., 14
Spargo
  T., 14
  Thomas, 34
Spencer
  Arthur, 178, 197
  Arthur, Mrs., 197
  Carl, 3, 197
  Martha, 163
Sperbeck
  Charles, 193
  Charles, Mrs., 114
Sproule
  Allen, 115
Squires
  Evelyn, 154
Stafford
  Lola, 122
Stall
  James, 179
Stanton
  William, 21
Stering
  Grace, 136
  Gracie, 161
Stewart
  Emma, 209
Stone, 95
  Charlie, 26
  P., 26
  W., 26
  Willie, 26
Strollias
  Stewart, 175
Stubbe
  Ann, 169
  Dick, 169
Stukey
  George, 205

INDEX

George, Mrs., 153
Winona, 151
Suffin
   J., Mrs., 121
Sullivan
   Mary, 89
Sutherland
   Ruth, 171
Sutliff
   G., 37
Sweeney
   Ann, 22
   Anne, 89
   Annie, 22, 190
   John, 22, 45
   Mary, 192
   Michael, 22, 89
   Richard, 45
   Robert, 45
Sweetland
   Olive, 148, 203
Swift
   E., Mrs., 197
   Mrs., 197
   O., Mrs., 117

T

Tandy, 115
Taylor
   Audrey, 112
   Charles, Mrs., 110
   Dorothy, 128, 141, 142
   Edward, 1
   Fern, 164
   Frank, 11, 112
   Frank, Mrs., 102
   Jack, 112
   Joanne, 152
   Joseph, 112
   Roy, 152
   Sarah, 112
   Sterling, 112
Teske
   Edna, 14, 148
Thomas
   Florence, 206
   James, 45
Thompson
   H., 127
Thrush
   G., Mrs., 37
   George, 14
   Sarah, 14
Tifft
   Alice, 3, 39
   Frances, 40, 180
   George, 3
   Julia, 3, 40, 115
   R., 51, 115, 185, 186
   Ray, 3, 44
Tisher
   Abraham, 205
Todd
   A., Mrs., 180
Toland, 8
   Elizabeth, 3
   Hugh, 81, 113, 115
   John, 3, 50, 68, 81, 113, 115, 182
   John, Mrs., 116
   Joseph, 4
   Martha, 4, 113
   Mary, 68, 81, 113, 115
   Robert, 3, 4, 68, 81, 82, 113
   W., 113
   William, 4, 50, 81, 114
Tomlinson
   Elizabeth, 167
Totten
   Merna, 166
Trauger
   D., 127
Travena
   Ann, 86
   Mrs., 5
   N., 34
   Nicholas, 5, 83
Tregaskis
   Mary, 196
Trembly
   Kay, 12
   Keith, 12
Tremewan
   Julia, 153, 163, 204

INDEX

Truttmann
  Anna, 202
Tufts
  A., Mrs., 103, 122
Turner
  Kristi, 171
  Thomas, 171
Turnkert
  John, 198
Tuttle
  Barbara, 172, 173
  Sue, 115
Twomey
  Andrew, 19, 73, 74, 147
Tyrell
  T., Mrs., 87, 128, 192

V

Van Tiger
  Roy, Mrs., 126
Van Winkle
  Grace, 136
Vanderpool, 190
Vargas
  Albert, 3
  M., 3
  Marian, 161, 162
Vega
  Dorothy, 207
Verdier
  Ernest, 6
  Lucille, 6, 149
Vetkos
  Ore, 158
  Thelma, 158
  Winifred, 158
Vetour
  Marian, 142
Victoria
  Bernice, 168
Victour
  Marian, 128
Vierria
  Carolyn, 152
Vihel
  Emelia, 208

Vineyard
  Alice, 7, 42, 183
  Allen, 7
  Almora, 7
  George, 183
  J., 90
  J., Mrs., 122
  John, 7, 104, 116, 190
  Lydia, 177, 183
  M., 182
  Mary, 7, 90, 190
  Mildred, 183
  Miles, 116
  Sarah, 5
  William, 42, 116
Vinsonhaler
  Arthur, 208
  Christine, 206
  Howard, 23, 208
  Mildred, 208
Vitor
  Marian, 141
Vitt
  Adeline, 123
Voss
  George, 96

W

Wah
  Ying, 27
Walden
  Ruby, 167
Waldron
  Henry, 29, 32
Walker
  E., 55
  Hazel, 131
  John, 14, 55
  Katherine, 198
Wallace
  Ellen, 6, 42, 185
  James, 118
  Michael, 5, 118
  Thomas, 195
  Walter, 119
Wallis

INDEX

J., 143
Saphrona, 180
Walsh
  Anna, 17, 100, 121
  Annie, 5
  Charles, 17, 121
  Edward, 17, 78, 188
  Francis, 159
  Jim, 55
  Joe, 165
  John, 5, 17, 36, 62, 68, 72, 75, 97, 100, 121, 188, 197, 200
  John, Mrs., 67
  Julia, 179
  Margaret, 97, 104
  Mary, 17, 67, 68, 188
  Maud, 97, 104
  Michael, 100
  Phillip, 17, 62
  Richard, 17, 22, 35
  Sadie, 5, 159, 165
  Thomas, 17, 40
  W., Mrs., 91
  Walter, 5, 121, 159, 165, 197, 200
  William, 67
Ward
  Amanda, 37
  Anna, 98
  Caroline, 21, 37
  Infant son, 21
  Mary, 21, 34
  W., 21, 34
  William, 35, 36, 37, 98
Warne
  Edward, 180
  J., 6, 67, 93, 189
  James, 196
  Mary, 6, 93, 196
Warner
  C., 36
Washburn
  E., 41
Watkins
  Irene, 139
Weaver
  Marie, 3

Warren, 3
Webb
  Alice, 3
  Sandy, 3
Welch
  Arlene, 148
  Benjamin, 13
  Bill, 154
  Chester, 1, 198
  Clarence, 1, 121, 151, 152, 154, 204, 205
  D., 198
  David, 121
  Donald, 151, 152, 154, 157
  Gale, 78
  Georgia, 152
  Georgina, 1, 166, 207
  I., 138
  Ira, 121, 198
  J., 119
  J., Mrs., 186
  James, 1, 36, 78, 119, 121, 138, 151, 152, 154, 157, 183, 194, 198, 205
  Joe, 151, 157, 183, 205
  Joseph, 13, 45, 70, 78, 121, 152, 154, 181, 205
  Katherine, 1
  Lillie, 1, 121, 122, 194, 198
  Mary, 151, 192, 207
  N., 1
  Neal, 151
  Neil, 152, 154, 157
  Nevens, 139
  Nevin, 154
  R., 194
  R., Mrs., 45
  Radford, 13, 198
  Ray, 139, 183
  Sally, 157
  San, 157
  Serena, 13
  Susie, 157
  Thomas, 13, 78, 151, 154, 157, 188
  Willard, 121, 139
  William, 13, 119

INDEX

Winifred, 1, 150, 151, 204
Weldon
    Clarence, 123, 134
    Dan, 123
Welsh
    Arlene, 138
Westerfield
    W., 8
Whalen
    Jeremiah, 24
    Jeremiah, Mrs., 40
Wheaton
    A., 55, 66, 67, 76, 125, 177
    A., Mrs., 113
    Al, 70
    Allen, 3, 125, 163
    Emma, 178, 183, 204
    Evelyn, 3
    Florence, 3
    James, 125
    Lucille, 3
    Mabel, 177, 183
    Martha, 177, 178, 197
    Mary, 163
    Robert, 10, 125
Whited
    Eveline, 209
Whiteside
    A., Mrs., 138
    Ruth, 154
Whitesides
    A., Mrs., 121
    Bertha, 182
    Nancy, 132
Whitmore
    Alice, 3, 136, 201
    Donald, 136, 144
    Elvin, 136, 144
    Jesse, 3, 136, 144, 201, 202
    Virginia, 136, 144, 202
Widener
    Alpha, 7, 89, 118
    Edith, 177
    Hazel, 118
    Joe, 126
    Joseph, 89, 118
    Russell, 7, 89, 118, 197
    Ruth, 118
    William, 7, 89, 118
Wilbur
    J., Mrs., 113
    W., Mrs., 90
Wilkinson
    Jeanne, 170
Williams
    James, 9, 23, 36, 179, 206
    Joe, 206
    Joseph, 23, 206
    Juliana, 23, 206
    L., Mrs., 125
    T., Mrs., 83, 84, 88, 94
    Vivian, 168
Williamson
    Donald, 151
    Linda, 14
    Robert, 14, 151
Willis, 197
    John, 195
    Mrs., 180
Wilson, 70
    Linda, 6
    Mrs., 88
    Roy, 6
    Ruby, 175
    Theodore, 27
Wimberly
    Edward, 147
    Susan, 181
Winegar
    Madge, 3
    Marvin, 3
    Patricia, 3
    Petra, 3
Wirth
    Fredrick, 13, 152
    Mable, 13, 152, 161
Wiseman
    Sara, 13
Woehler
    Otto, 5
Woens
    Fannie, 183
Woodroffe
    Annie, 77

INDEX

Bruce, 5, 122
Earl, 5, 116, 122, 148
James, 122
Janie, 122, 194
John, 5, 77, 194
Pearl, 5
Robert, 55
Sara, 104
Sherman, 6, 121, 194
Woods
  Vera, 147
  W., 14
Woodworth
  Anita, 148
  Dean, Mrs., 144
Worl
  John, Mrs., 105
Wotherspoon
  Al, 158
  Kenneth, 158
Wright
  Aden, 11, 117, 150, 167, 203, 207
  Aden, Mrs., 78
  Alden, 192
  Alice, 182
  Belle, 181
  Ida, 177, 192
  M., Mrs., 119
  Mabel, 25
  Mac, 182
  Mary, 121, 167, 203
  Minnie, 84
  Nellie, 83

Y

Yankey
  John, 4, 173
  Paul, 4, 173
  Roberta, 4, 173
Yarboraugh
  Frances, 144
Yore
  Bernice, 153
Young
  Elizabeth, 198
  Ella, 138
  Gladys, 162
  Jack, Mrs., 122
  James, 14, 91
  John, 24, 190
  Mildred, 167
  Robert, Mrs., 159
  William, 169
Your
  Bernice, 163

Z

Zeisloft
  Christopher, 15, 171
  Dianna, 171
  Martha, 171
  Marvin, 171
  Shannon, 171
  Stephanie, 171
  Tiffany, 171

Smartville Cemeteries
Combined Information Analyses

Name on Headstone	P	Name in Records	Father's Name	Mother's Nm	Death Died	County	Born	Birth Place	Marriage	Relationships
Adkins, Benjamin F.	P	Benjamin Franklin		Willis	9 Dec 1935	Yuba	14 Nov 1868	California	8 Nov 1891	H-Clara E Nevels Adkins
Adkins, Clara Elizabeth	P		Nevels	Mans	11 Mar 1925	Yuba	22 Sep 1873	California	8 Nov 1891	W-Benjamin Adkins
Adkins, Frances M.	P	Frances Maria	Willis		1 Mar 1919	Yuba	14 Sep 1829	Kentucky		W-Oliver Adkins, "Mother"
Adkins, Oliver	P				1 Dec 1883		28 May 1826	Kentucky		H-Frances Adkins, "Father", Co. A. Inf. Inf. Mex War
Adkins, Oliver P.	P				4 Jan 1934		16 Oct 1866	California		S-Oliver & Frances M. Adkins
Adkins, Owen "Jeff"	P				12 Jun 1915	Butte	27 Jan 1860	California		S-F.M. Adkins
Adkins, Wm. B.	P	William B			20 Jun 1907	Yuba	31 May 1858		Married	
Ah, Day Toon					12 May 1908		1863			
Allen, Edward					24 Mar 1876		19 Feb 1854			
Allen, Geo. B.									23 Nov 1887	H-Laura Allen
Allen, Josephine A.	P		Allen	Sivope	1997		31 May 1919	California	23 Nov 1887	D-Charles & Josephine Allen
Allen, Laura D.									Widowed	W-Geo. B. Allen
Allinio, Wilberta Maude		Wilberta Maude	Adkins	Stewart	21 Dec 2000	Yuba	3 Dec 1904	California	Widowed	W-Pierre Allimio
Allread, Margaret E.		Margaret Ellen	McQuaid	Mulvany	19 Sep 1923	Yuba	12 Dec 1876	California	Married	W-William Allread, D-John & Mary McQuaid
Alighel, Harold Chas.	P				9 Jun 1910		1910	California	Child	Native of San Francisco
Anderson, August	P			Johnson	31 Jul 1942	Yuba	28 Aug 1863	Sweden	Widowed	H-Emma Anderson
Anderson, Emma	P		Bitner		2 Aug 1960		1865			W-August Anderson, D-Captain & Mary Bitner
Austin, Bert C.	P				1931		1881			
Austin, Ella	P	Ella M.	Bitner		4 Mar 1943	Nevada	21 Jul 1858	Pennsylvania	1879	W-John H. Austin, D-Captain & Mary Bitner
Bach, Alice Baby	P				8 Dec 1903		1903		Child	
Bach, Ethel M.	P	Ethel May	Widener	Atchley	6 Feb 1936	Yuba	15 Nov 1879	California	Widow	W-John E. Bach.
Bach, John					28 Jan 1898	Yuba	19 Jun 1830	Germany		
Bach, John E.	P	John Edward			17 Aug 1931	Yuba	13 Aug 1869	California	21 Dec 1902	H-Alice Wright Bach, S-Mary Ann Bach
Bach, John Joseph	P			Widner	16 Nov 1991		1 May 1914	California		S-John Edward & Edith May Bach
Bach, Katie	P	Katherine Louise			28 Aug 1916		7 Jan 1863			
Bach, Mary A.	P	Mary Alice			19 Feb 1929		1855			
Barnes, Glenn F.	P		Tregaskis		24 Dec 1918	Yuba	18 Jan 1837	California	Child	
Barrett, Catherine	P	Katherine	Driscoll	Driscoll	10 Jan 1897		27 Jun 1892		Widowed	W-John Barrett, D-John, & Katherine Driscoll
Barrett, John					20 Dec 1900	Yuba	19 Jul 1864	California		Accidental Death at Pennsylvania mine in Browns Valley, CA
Bartlett, Joseph Jr.	P	John Joseph Jr.	Gordon		24 May 1989	Santa Clara	1852		Single	CPL U. S. Army WWII, S-John Joseph & Maude Rene Barrett
Bartlett, Nellie G.	P	Nellie Gladys	Fisher	Parks	15 Apr 1976	Yuba	24 Jan 1925	California	Widowed	
Beasley	P	Elizabeth E.	Dugan		17 Apr 1900	San Francisco	6 Jan 1903	Oregon		W-Richard Beasley, D-Patrick & Bridget Dugan
Beasley	P	Richard			20 Apr 1922	Alameda	1861			H-Elizabeth E. Dugan Beasley
Beatty, Alice					30 Jan 1860	Yuba	1855	California	Child	
Beatty, Ann	P				15 Apr 1899	Yuba	1817	Ireland		W-Richard Beatty
Beatty, Frances E.	P		Tifft		25 Oct 1896	Yuba	1861	California	4 Nov 1889	W-Robert Beatty, D-R. W. Tifft
Beatty, Grace V.	P	Grace Veronica	Manasco		16 Sep 1954	Alameda	25 Apr 1893	California	Married	W-Richard Ray Beatty, D-C. Manasco
Beatty, Infant Son	P								Child	S-R. & F. E. Beatty
Beatty, Richard	P				16 Mar 1904	Yuba	1822	Ireland		Naturalized 27 Oct 1860, F-Robert & John Beatty, H-Ann
Beatty, Richard Ray	P			Tifft	28 Feb 1963	Alameda	19 Jun 1891	California	14 Dec 1917	H-Grace Manasco Beatty

Smartville Cemeteries
Combined Information Analyses

Name on Headstone	p	Name in Records	Father's Name	Mother's Name	Death	Died County	Born	Birth	Birth Place	Marriage	Relationships
Beatty, Robert	p				17 Mar 1926		22 Mar	1855	California	4 Nov 1889	H-Frances E Beatty, S-Richard Beatty
Bebout, Charles W.	P	Carles William		Welch	7 Apr 1984	Nevada	13 May	1921	Kentucky	Married	H-Hilary Bebout, S-Carrie Grimes
Beine, Bartholomew	P				24 Feb 1867			1867		Child	S-Patrick & Margaret Beine
Beine, Charles	P				13 Oct 1866			1833	Ireland		Native of Co. Leitrim
Beine, Margaret	P				26 Dec 1878			1845	Ireland		W-Patrick Beine, Native of Co. Kerry
Beine, Patrick	P				8 Oct 1884			1835	Ireland		H-Margaret Beine, Native of Co. Leitrim
Beine, Patrick J.	P				18 May 1902	Yuba		1873	California		Native of San Francisco, S-Patrick & Margaret Beine
Beine, Terence	P				26 Jan 1867		26 Jul	1864		Child	D-Patrick & Margaret Beine
Bennett, Grace					18 Jul 1937	Yuba	6 May	1886			
Bennion, Margaret Ann Walsh	P	Margaret Ann	Walsh	Kane	9 Aug 1997	Nevada	10 Aug	1917	Canada		
Bennion, Richard B.	P			Bringhurst	2 Jun 1996	Nevada	5 Jan	1913	Utah		
Bennion, Robert W.	P				27 Sep 1998		25 Mar	1943			
Binet, LaTosca	P		Ehmenn		10 Feb 1989			1901			D-Paul & Rosa Ehmenn
Birmingham, Abbie C. L.	P		Birmingham		3 Jun 1885		15 Apr	1854		Child	D-G. S. Birmingham
Birmingham, Ida May	P		Birmingham		22 Jul 1864		1 Jan	1856		Child	D-G. S. Birmingham
Bitner, Cyrus C.	P	Cyrus Currington			7 Aug 1917	Yuba	28 Oct	1836		Married	H-Mary Bitner, F-Ella Austin, F-Emma Anderson
Bitner, John	P				2 Jul 1871		2 Sep	1831			
Bitner, Mary	P				18 Oct 1920	Yuba		1839	Pennsylvania		W-Captain Cyrus C. Bitner
Black, Effie A.	P	Effie Agnes	Black	Sanford	30 May 1978	Alameda	14 Dec	1880	California	Single	S-Robert Black
Black, Eva	P	Eva M.			Jun 1890			1855			
Black, Robert	P		Black	Bicknell	1935		17 Aug	1853	Yuba		S-William & Agnes Black
Boatwright, Alice Gertrude	P		Criss		5 Oct 1988	Nevada	22 Feb	1917	Indiana		M-Linda Wilson
Bone, Thomas					30 Jan 1878						
Bowman, Alexander	P				19 Feb 1883		28 Jun	1813	Scotland		Native of Aberdeen Shaire, Scotland
Bowman, Emma	P		Bowman	Hurst	29 Jun 1933	Yuba	21 Nov	1860	California	Single	
Bowman, George	P				1925			1851			
Bowman, Gordon	P	Henry Gordon		Hurst	4 Nov 1908	Yuba		1853	New York	Single	
Bowman, Harry	P	Henry L.			8 Apr 1903			1822			
Bowman, James	P				17 Jan 1913	Nevada		1851	New York	29 Jan 1877	H-Louisa Bowman
Bowman, Lizzie	P			Hurst	8 Dec 1903	Yuba	4 Jan	1863	California	Single	Sister to Gordon Bowman
Bowman, Louisa	P		Jones		15 Nov 1934		23 Oct	1859	Illinois	29 Jan 1877	W-James Bowman
Bowman, Mother	P	Jane or Jonie	Hurst		23 Apr 1899	Yuba	28 Jan	1826	Ireland	Widowed	
Bowman, William	P				29 Aug 1867			1818			
Bowman, William H.	P				5 Mar 1921	Yuba	44 yrs	1880	California		Wife, Frank L Taylor
Boyer, Eliza Lenihan	P	Elizabeth	Lenehan		25 May 1872		3 May	1835	Ireland	5 Nov 1865	W-John Boyer
Boyer, John	P				24 Dec 1866			1866		Child	S-John & Eliza Boyer
Boyer, Lizzie	P		Lenincham		28 Jul 1869		17 Oct	1868		Child	D-John & Eliza Boyer
Brett, Rose Mary			Cardo	Mix	24 Dec 1981	Sutter	26 Sep	1900	Wisconsin	Widow	W-Walter N Brett, In Wirth Fam Plot
Brett, Walter Nelson				Nelson	23 Dec 1971	Yuba	12 Dec	1891	California	Married	H-Rose M Cardo Brett
Brickell, Catherine Estelle	P		Brickell		21 Dec 1866	Yuba	21 Jun	1866		Child	D-John & Jennie Brickell
Bridge, Betty Oldham	P				28 Feb 1879	Yuba	19 Jan	1814	England		W-Thoamas Bridge, M-John W. Bridge

Smartville Cemeteries
Combined Information Analyses

Name on Headstone	P	Name in Records	Father's Name	Mother's Name	Death Died	Died County	Born Birth	Birth Place	Marriage	Relationships
Bristow, Elmer					10 Oct 1871		1832			H-Agnes Virginia Havey Bristow
Bristow, Geo W.	P									
Bristow, Virginia										
Broom, Bertha										
Brown, Elizabeth A.	P				10 Sep 1908		10 Jun 1821			
Brunckhorst, Elmer W.	P	Elmer William			28 Feb 1979	Alameda	2 Nov 1912	Missouri		
Brunckhorst, Fred James	P				24 Nov 1972	Alameda	2 Oct 1879			
Brunckhorst, Oscar		Oscar Edward		Shay	26 May 1956	Alameda	31 Oct 1914	Montana		
Buck, Elijah					5 Feb 1878		1828			
Buckley, Ann	P				Oct 1884					
Buckley, Michael Jr.	P				27 Apr 1912					
Buckley, Michael Sr.	P				5 Jan 1927					
Buckley, Patrick H.	P				14 Mar 1875					S-M. & A. Buckley
Burke, Ann					25 Dec 1888	Yuba	4 Nov 1866		Married	
Burke, Genevieve	P						1812			
Burke, Patrick					1 Aug 1890	Yuba		1806 Ireland		
Burke, William B.	P				7 Aug 1928	Yuba		Pennsylvania	Married	H-Teresa Burke
Burns, James					15 Jan 1861	Yuba				
Burns, John	P				1 Dec 1911	Yuba	1841	Ireland	Single	
Burns, Margaret	P									
Busdby, Edward					9 Jan 1902	Yuba	1868	California		S-George Busdby
Busdby, Fannie	P	Fnannie	Wimberly	Owen	27 Oct 1960	Butte	19 Jan 1873	Missouri	1 Nov 1891	W-William Busdby
Busdby, Frank					2 Aug 1901	Yuba	29 Mar 1876	California		S-George Busdby, Suicide
Busdby, George Sr.					18 Jul 1925	Yuba	20 Sep 1830	England		H-Katherine Busdby
Busdby, Irma Jewel					19 Jul 1911		27 Feb 1909	California	Child	D-Fanny Owen
Busdby, J. M.	P	Joseph March "Joe"	Wimberly	Owen	30 Jul 1924		21 Jan 1883	California	4 Jul 1904	H-Non Murch Busdby, S-George Busdby
Busdby, Katherine			Dougherty		2 Dec 1915	Yuba	10 Sep 1838	Ireland		W-George Busdby
Busdby, Nora	P		Murch		30 Mar 1951	Placer	6 Apr 1872	California	4 Jul 1904	W-Joseph Busdby
Busdby, Sarah				Dougherty	8 Dec 1955	Yuba	4 Sep 1874	California	Single	D-William & Fannie Busdby
Busdby, Wm.	P	William			12 May 1955	Placer	29 Aug 1865	California	1 Nov 1891	H-Fannie Busdby, S-George Busdby
Byrne, James					24 Dec 1916	Sacramento	1843			H-Rose Anna Byrne
Byrne, James D.	P	James Daniel		Havey	8 Dec 1943	San Mateo	16 Dec 1873			U. S. Marines, Vet WWI & Spanish AM.
Byrne, John					1911		1841			
Byrne, John B.	P				1924		1875			
Byrne, Margaret										
Byrne, Mary R.	P	Mary Josephine	Ryan	Myers	21 Jan 1947	Yuba	8 Dec 1882	California	Married	W-Robert E. Byrne, D-Patick & Hannah Ryan
Byrne, Mary Roseann		Rose Anna	Ryan	Havey	30 Sep 1902	Yuba	1847	New Jersey	(1872)	W-James Byrne
Byrne, Robert E.	P	Robert Emmett		Havey	8 Apr 1954	Butte	24 Feb 1879	California	Widowed	H-Mary Ryan Byrne
Byrne, Rose B.	P	Rose Bell	Newbert	Jackson	23 Nov 1943	Sacramento	24 Nov 1878	California		
Byrne, Sarah M.		Sarah Frances	Byrne		13 Aug 1908	San Francisco	6 Jul 1884	California	Single	D-James & rose Anna Byrne
Byrne, Selna F. E.	P	Selna Frances Esthelda			8 Apr 1977	Butte	24 Mar 1885	California	Single	

Smartville Cemeteries
Combined Information Analyses

Name on Headstone	P	Name in Records	Father's Name	Mother's Name	Death	Died	County	Born	Birth	Birth Place	Marriage	Relationships
Byrne, Thomas P.					30 Apr	1893	Yuba	20 Feb	1881	California	Single	
Cain, Bridget	P				11 Feb	1869			1836			
Cain, Daniel	P	Daniel Caine			21 Jul	1909	Yuba	Apr	1826	Ireland	Married	S-Richard Caine
Call, Harriett A.	P	Harriet Amelia	Congdon		16 Jan	1950	Sutter	8 Feb	1864	Canada	Widow	W-S. M. Call, M-Salmon M. Call
Call, S. M.	P	Salmon M.				1932			1861	Canada	Married	H-Harriett A Call, F-Salmon M Call
Calvin, John			Calvin			1870						
Calvin, Mar?												
Campbell, Alan McKenzie	P				24 Apr	1991	Kitsap, WA	16 Jun	1947	Washington, DC	Married	W-Brenda Campbell, S-George & Mildred Campbell
Campbell, Daniel	P				22 Feb	1860	Yuba		1825	Ireland	Single	
Campbell, Daniel	P				31 Dec	1878	Yuba		1833	Ireland		Native of Co. Down, Ireland
Campbell, Daniel T.					6 Nov	1875		2 Nov	1874		Child	
Campbell, Infant					15 Aug	1878					Child	
Campbell, James					6 Nov	1918	Yuba		1843			
Campbell, Mrs.												
Cann, Bert	P			Tishes	19 Nov	1949	Lake	17 Jan	1914	California		H-Winona F Cann, S-Thomas J. Cann
Cann, Edna M.	P					1984			1906			W-Philip Cann
Cann, Maud	P	Maud	Tisher	Rouse	29 Oct	1969	Yuba	3 Apr	1884	California	Widowed	W-Thomas J Cann, "Mother"
Cann, Philip	P					2002			1911			H-Martha Cann, S-Thomas J. Cann
Cann, Thomas J.	P	Thomas James		Harris	28 Mar	1945	Yuba	24 Jan	1886	England	Married	H-Maud Cann
Cann, Winona F.	P					1985			1918			W-Bert Cann
Carlson, Kim Loring	P		Carlson	Burns	19 Dec	1971	Yuba	28 Feb	1951	Washington	Single	S-James Carlson & Reva Rouse,
Carlson, Rio	P			Burns	26 Aug	1976	Yuba	6 Feb	1946	California	Married	H-Angela Elder Carlson, S-Jim Carlson
Carney, Cora May	P				27 Dec	1987	Yuba	7 Dec	1890	California	Widowed	W-William R. Carney
Carney, Richard William	P				28 Nov	1970	Nevada	19 Sep	1947	California	Single	Grandson of W.R.Carney, S-William R. Carney
Carney, William R.	P	William Raymond		Pitchford	19 Jul	1981	Nevada	16 Nov	1886	Missouri	Married	H-Cora May Carney, F-William Carney, Jr.
Carpenter, William C.					23 Jan	1874		8 Sep	1822			
Carter, Ida Hall	P	Ida Ellen	Jones		6 Sep	1923	Yuba	53 yrs	1869	California	Married	W-W. S. Carter, W-John Clark Hall, D-D. N. Jones
Casey, John J					18 Jan	1916	Yuba	8 Sep	1853	California	Married	H-Abigail P. Casey
Casey, Margaret	P	Margaret J.			27 Jan	1896	Yuba	27 Mar	1871	California	Single	D-Peter Casey
Casey, Mary J.	P	Mary Jane			20 Aug	1882	Yuba	20 Feb	1868		Child	D-Peter & Winifred Casey
Casey, Peter	P				29 Sep	1904	Yuba		1828	Ireland	Married	H-Winifred Casey
Casey, William	P	William Joseph		Howsaw	5 Oct	1917	Yuba	6 Dec	1864	California	Widowed	S-Peter & Winifred Casey
Casey, Winifred	P				2 Nov	1906	Yuba		1836	Ireland	Widow	W-Peter Casey
Chadwick, Jacob F.					20 Dec	1864			1838	Maine		
Chamberlin, W. W.	P	William Wallace			14 Jun	1905	Yuba		1833	Canada	20 Dec 1875	H-Mrs Annie Keith Otis
Clark, Abigale Ann	P					1859			1859		Child	D-L. B. & J. Clark
Clark, Annie		(Clarke)			19 Nov	1877			1827			
Clark, Henry A.				Harworth	24 Mar	1941	Sacramento	1 Aug	1859	California		
Clark, Jane Haworth					3 Jul	1927		16 Feb	1838			
Clark, Levi Blaisdell					15 Jan	1886	Yuba	16 Feb	1821			
Clark, Mary			Clark		3 Nov	1875		26 Jan	1867	California	Child	D-W & M Clark

Smartville Cemeteries
Combined Information Analyses

Name on Headstone	P	Name in Records	Father's Name	Mother's Name	Death	Died	County	Born	Birth	Birth Place	Marriage	Relationships
Clark, Mary Jane	P					1861			1849		Child	D-L. B. & J. Clark
Clark, Sherman Grant	P					1868			1864		Child	D-L. B. & J. Clark
Clark, Wright					18 Dec	1875		12 Apr	1828		2 Apr 1863	F-Mary Clark, H-Mary Clark
Clayton, Charles V.	P	Charles Vernon		Morton	11 Jan	1991	Sutter	18 July	1908	California	Married	H-Esther Clayton
Clayton, Esther	P		Baker	Powell	8 Nov	1991	Nevada	10 Nov	1904	Kentucky		
Clifford, Keith Manville					31 Oct	1905			1832			
Collbran, Arthur Harry	P				15 Jul	1943	Monterey	17 Feb	1876			
Collbran, Arthur, Jr.	P	Arthur Harry, Jr.		Farrish	12 Jun	1982	Marin	24 May	1911			
Collbran, Florence Farish	P					1936			1882			
Collbran, James Farish	P			Bachelder	25 Feb	1993	Marin	6 Jul	1848	Maine		
Collbran, John Stuart	P					1972			1883			
Collbran, Mabel Farish	P	Virginia	Farish	Paddock	6 Aug	1965	San Diego	12 Jan	1884	Colorado		
Collbran, Virginia Bachelder	P		Bachelder			1993			1911			
Collier, John A.		John Arthur		Eggleston	17 Oct	1981	Nevada	25 Sep	1923	Idaho		H-Virginia Collier, S-Clara Reeh,
Collier, Patricia E.		Patricia Ethyl	Shields	Ogden	19 Mar	1981	Sacramento	20 Oct	1926	Montana	Married	W-John A Collier, M-John, Bruce Collier
Colling, Charles Earle	P					1930			1887			
Colling, Infant					9 Nov	1890	Yuba	5 Nov	1890	California	Child	Infant son of Mrs. W. H. Colling
Colling, William Henry	P				12 Sep	1899	Yuba	10 Nov	1854	California	25 Jan 1887	H-Bertha Jeffery Colling, Brother was Charles Coling of Lower CA
Collins, Bertha LeBoeuf	P	Bertha F.			30 May	1995	Nevada	14 Apr	1911	Washington	Widowed	W-Eugene collins
Compton, Brady Farrell				Dougherty	22 Oct	1940	Contra Costa	12 Dec	1872	California	10 Dec 1895	H-Mary Jane Compton, S-Charles & Nellie Compton
Compton, Clarence Albert	P			Byrne	31 Jul	1955	Yuba	14 Sep	1898	California	Divorced	S-Brady & Mary Compton
Compton, Mary Jane	P		Byrne	Havey	22 Apr	1957	Santa Barbara	22 Aug	1876	California	10 Dec 1895	W-Brady Farrell Compton, "Mother"
Conboy, Margaret					2 Jun	1897	Yuba		1851	Ohio	Married	
Congdon, Frances D.	P	Francis Dalbear			20 Jan	1916		18 Jul	1829			
Congdon, Fredrick H.	P				30 Mar	1949			1869			H-Lucretia Congdon (Not the one here with a hdstone)
Congdon, George F.	P				21 Apr	1937	Yuba	10 Oct	1854	Connecticut	Single	S-Frank & Lois Congdon
Congdon, George N.	P				17 Dec	1860		12 Mar	1822			
Congdon, Infant Daughter			Congdon									D-George N. & Amelia A. Congdon
Congdon, Lucretia	P				4 Mar	1878		4 Sep	1798			
Congdon, Lucretia	P				24 Jan	1892	Yuba	26 Dec	1830	Connecticut		D-F & L Congdon
Congdon, Sarah E.	P				1 Jan	1870	Yuba	19 Dec	1857			D-F & L Congdon
Conlin, John H.	P				26 Sep	1912	San Francisco		1866	California		
Conlin, Margaret	P		Mulhare		13 Nov	1908	Yuba	1 Aug	1835	Ireland	Widow	D-Michael Mulhare, W-Thomas Conlin
Conlin, Sarah L.	P				20 Jan	1885		20 Jan	1871		Child	
Conlin, Thomas	P				9 Feb	1898	Yuba	20 Apr	1832	Ireland	Married	H-Margaret Conlin
Conlin, Thomas B.	P	Thomas Benjamin			6 Oct	1887	Yuba	20 Jan	1865			"Ben", S-Thomas Conlin
Conrath, Elizabeth	P		Turnkert		8 Feb	1937	Yuba	1 Jan	1860	Michigan	Widowed	"Mother", W-Lois Conrath
Conrath, Gustav	P					1930			1857			"Brother"
Conrath, Louis	P				2 Sep	1904	Yuba	29 Sep	1854	Wisconsin	Married	"Father", H-Elizabeth Conrath
Cony	P	Mary Connery			23 Apr	1898	Yuba		1822			
Cony	P	Thomas			14 Dec	1898	Yuba		1825		Widowed	Citizenship in August 14, 1855

Page 253

Smartville Cemeteries
Combined Information Analyses

Name on Headstone	P	Name in Records	Father's Name	Mother's Na	Death	Died	County	Born	Birth	Birth Place	Marriage	Relationships
Cooney, Mary	P		Walsh	Dougherty	19 Aug	1945	Yuba	1 May	1868	California	Widowed	W-William Cooney
Cooney, William	P				1 Jul	1928	Yuba	20 Jul	1865	Illinois	Married	H-Mary Cooney
Coopmann, Fredrick F.	P				10 May	1876		14 Nov	1822	Germany		
Cory, Daniel	P				8 Jan	1899	Butte		1829	England	27 Feb 1870	H-Mary E Cory
Cory, Elizabeth	P	Mary Elizabeth	Morrison		20 Nov	1868			1832		27 Feb 1870	W-Daniel Cory
Cory, Wm. H.	P	William H.			10 Oct	1873	Yuba	10 Jul	1853	California		S-Daniel & Elizabeth Cory
Champ, Wilfred R.									1868			S-William Champ
Cramp, Wm.	P	William			1 Aug	1903	Yuba	19 Aug	1826	Maine		F-Wilfred R. Cramp, Co. F. 27 Ill. Inf
Cramsie, Francis James			Harvey		5 Aug	1959	Yuba	30 Dec	1879	California		
Cramsie, John E.	P	John Edward	Harvey		17 Feb	1937	Yuba	7 Nov	1872	California	Married	H-Anna McGienney Cramsie S-William & Elizabeth Cramsie
Cramsie, Joseph		Joseph D.			6 Dec	1911	Napa		1875			
Cramsie, William				Dougherty	18 Mar	1912	Yuba	15 Jun	1837	Ireland	1870	S-John & Margaret Jane Cramsie, H-Elizabeth Harvey Cramsie
Cranshaw, James	P				15 Dec	1877	Yuba		1807	England	19 Feb 1865	H-Catherine E. Cranshaw
Crawford, Charles Jay	P			Cross	20 May	1871		12 Feb	1870			S-James & Catherine E. Cranshaw
Creden, Daniel	P	David			7 Dec	1877	Yuba		1819	Ireland		Native of Maecroom Co.
Crocker, Fredona Damon					7 Aug	1917	Yuba	11 Jul	1830	Canada		
Crosett, J. G.					10 Apr	1878						
Crouch, H. W.					16 Aug	1895						
Crouch, H. W., Jr.						1878						
Curran, John		John J.			1 Aug	1893	Siskiyou	19 Jul	1866	California		
Curran, Mark	P				19 Jul	1873	Yuba		1819	Ireland	Married	Native of Deaney Parish of MnnsReavis Co.
Curran, Mark					5 Feb	1886			1870		Child	
Curran, Mark					4 Nov	1896	Butte		1856	California		Brother of Daniel Curran
Curran, Sarah		Sarah Ann			16 May	1889	Yuba	16 Sep	1863		Single	"Sallie", Sister of Dennis Curran
Curran, Susan					5 Feb	1886	Yuba	2 Nov	1832		Married	
Curry, Artie B.					26 May	1880		11 Jan	1880		Child	
Curry, Bertram Earl					30 Mar	1889		7 Jan	1889		Child	
Curry, Flora J.					28 Jan	1876		20 Mar	1873		Child	
Curry, Wilbur R.					20 May	1876		15 Sep	1875		Child	
Curtis, Thomas		Thomas H.			9 Jul	1890	Yuba	35-40y				Drowned in the Yuba River
Daly, Ann	P				23 May	1906			1838			W-Lawrence Daly
Daly, Cecila	P				18 Apr	1879			1866		Child	
Daly, Infant					28 Dec	1860	Yuba	Apr	1860	California		Infant of Patrick Daly
Daly, James					30 Aug	1860	Yuba		1796	Ireland		
Daly, Jane	P				3 Nov	1886						With Cramsie family plot
Daly, John J.	P				9 Nov	1898	Butte City, MT		1868	California	Single	
Daly, Kate					23 May	1906	San Francisco					
Daly, Lawrence	P				6 Oct	1890	Yuba		1832	ireland		H-Ann Daly
Daly, M.	P				17 Jun	1876			1877		Child	
Daly, Michael					15 Apr	1859	Yuba		1828	Ireland	Single	
Daly, Patrick					3 Jun	1897	Contra Costa		1832	Ireland		

Smartville Cemeteries
Combined Information Analyses

Name on Headstone	P	Name in Records	Father's Name	Mother's Name	Death Died	County	Born Birth	Birth Place	Marriage	Relationships
Daly, T.	P				2 Jul 1879		1874		Child	
Daniel, Elizabeth	P		Otis		9 Aug 1868	Yuba	1773	Scotland		W-W. W. Chamberlain, M-Mrs I B Otis
Davey, Carrie Bell					5 Jul 1879		15 May 1879			
Davey, William H.										
Davies, Elizabeth	P				17 Oct 1889		1809			
Davies, Samuel	P				2 Jun 1867	Yuba	1802	Philadelphia		
Davis, Adeline C.	P		Harris		7 Jun 1978		30 Jun 1883			W-Bert Benjamin Davis
Davis, Bert Benjamin	P			Mollett	25 Nov 1968	Yuba	19 May 1890	Missouri	Married	H-Adeline Harris Davis
Davis, Edward H.	P	Edward Henry		Mason	11 Apr 1957	Alameda	8 May 1881	Indiana	Married	H-May Black Davis
Davis, Julia A.	P	Julia Amelia	Tifft		23 Jan 1909		1 Dec 1870			
Davis, Mae M.	P		Black	Sanford	20 Aug 1962	Alameda	27 Feb 1882	California	Widow	W-Edward Davis, Granddaughter of Benjamin Sanford
Dempsey, Ann Breslin	P	Annie			15 Sep 1899	Yuba	1839	Ireland	Married	W-John Dempsey
Dempsey, Catharine Loretta			Meade		7 Aug 1913	San Francisco	1882	California	Married	W-William J. Dempsey
Dempsey, John	P				2 Aug 1909	Yuba	1831	Ireland	Widowed	H-Ann Dempsey
Dempsey, Joseph William		William J.			Feb 1937	Yuba	1870	California	Widowed	H-Catherine Loretta Dempsey, S-John & Ann Dempsey
Dempsey, Rose					13 Nov 1904	Yuba	13 Dec 1903	California	Child	
Denehey, Mary	P				7 Jun 1872		1844			W-Cornelius Denehey
Denney, Dorothy Ann			Cann		3 Jun 1966	Sacramento	3 Jan 1922	California	Married	W-Albert T. Denney, D-Maud Cann
Dewan, Charles					28 Oct 1881					
Dewan, Ellen					11 Jan 1913	Yuba	1841	Ireland	Widow	
Dewan, James Henry					8 Jul 1918		4 Jan 1876			
Dewan, John		John Dewain			15 Nov 1894	Yuba	2 Jul 1878	California	Single	
Dewan, John					9 Jun 1895	Yuba	9 Sep 1837	Ireland	Married	H-Ellen Dewan
Dewan, William E.		William Edward			29 May 1901	Nevada	1872	California	Single	S-Mrs. Dewan, Proprietress of Dewan Hotel, Smartsville
Dillon, Peter	P				12 Jun 1879		12 Jul 1859	Ireland		
Dittmer, James Henry					15 Sep 1869	Yuba	2 May 1868		Child	S-John & Katie Dittmer
Dittmer, John	P				2 Jul 1871		2 Oct 1831			H-Katie Dittmer, Came from Wheeling, W. Virginia
Dittmer, Mary Elizabeth			Dittmer		25 Sep 1869	Yuba	25 Nov 1862	W. Virginia	Child	D-John & Katie Dittmer
Divver, Charles	P				26 Oct 1881					
Divver, George W.		George M.		Butler	12 Aug 1900	Yuba	1867	California	Single	S-Charles & Mary Divver
Divver, Mary A.	P			Butler	3 Sep 1911	Yuba	5 Aug 1891	California	Single	S-Charles & Mary Divver
Divver, William F.	P	William F.			26 Apr 1890	Yuba	25 Dec 1835	Ireland	Widowed	
Dooley, Louisa A.	P	Louisa Ann		Butler	11 Jan 1910	Yuba	3 Feb 1869	California	Divorced	S-Charles & Mary Divver
Doubt, Elizabeth	P	"Eliza"			11 Nov 1910	Yuba	11 Aug 1878	California	Married	W-James Dooley, D-M Fiene
Doubt, Annie	P	Anne			5 Nov 1895	Yuba	1825	Ireland	Married	W-John Doubt
Dougherty, Catherine					23 Nov 1906	Yuba	1835	Ireland	Married	W-Daniel Dougharty
Dougherty, Daniel	P	Daniel James		Brady	6 May 1916		Nov 1869			
Dougherty, Daniel	P	Daniel John Ambrose			26 Apr 1913	Yuba	27 Mar 1876	California	Single	S-Daniel & Anne Dougherty
Dougherty, Ezekiel C.					6 Jul 1914	Yuba	1825	Ireland	Widowed	H-Annie Dougherty
Dougherty, Grace	P				26 Sep 1882	Yuba	9 Nov 1800	Virginia		Was a Judge
					1931		1867			D-Daniel Dougherty

Page 255

Smartville Cemeteries
Combined Information Analysis

Name on Headstone	P	Name in Records	Father's Name	Mother's Name	Death Died	County	Born	Birth	Birth Place	Marriage	Relationships
Dougherty, John					8 Nov 1887	Yuba		1869			D-Daniel Dougherty
Dougherty, Katie	P				1916						
Dougherty, Sadie		Sadie Josephine			18 Apr 1902	Yuba		1878	California	Single	Granddaughter of Michael Dougherty
Dougherty, Thomas					2 Jul 1896	Yuba		1851	Ireland		
Doyle, Alexander	P	Alexander, Sr.			24 Jul 1877	Yuba		1819	Ireland		H-Elizabeth Doyle
Doyle, Alexander, Jr.					14 Feb 1891	Yuba	10 Nov 1860				W-Alexander Doyle, Sr.
Doyle, Elizabeth J.	P				14 Nov 1886	Yuba		1829	Kentucky		S-James Doyle, Sr.
Doyle, James	P	James Joseph, Jr.			30 Jan 1901	Yuba		1874			W-James Doyle, Sr
Doyle, Mary					2 Jul 1908	Yuba		1868	Ireland	1846	D-E. & A. Doyle
Doyle, Olive R.	P				6 Oct 1875		6 Mar 1873			Child	D-John & Catherine Driscoll
Driscoll	P	Sarah Delilah			18 May 1910		9 Sep 1903			Child	W-John Driscoll
Driscoll, Catherine			Driscoll	Sweeney	3 Sep 1913	Yuba		1838	Ireland	Widowed	H-Catherine Driscoll
Driscoll, John	P	John, Sr.			11 Dec 1907	Yuba	11 Dec 1825		Ireland	Married	
Driscoll, Mary E.	P	Mary Ellen			15 Apr 1891	Yuba	22 Nov 1871			Single	
Driscoll, Theodore	P				1893			1892		Child	
Durham, Lucinda H.					13 Sep 1880		22 Nov 1829				
DuShane, L. June Bishop		Lois June	Bishop		31 Dec 1998		13 Jun 1924				DuShane
Dykes, Grace L.	P		Black		26 Jun 1973	Alameda	13 Apr 1884		California	Married	W-Hugh Thomas Dykes
Dykes, Hugh T.	P	Hugh Thomas		Baker	31 Dec 1954	Alameda	14 Oct 1884		California	Child	H-Grace Black Dykes
Earley, Margaret Ann	P		Earley		9 Mar 1873	Yuba	5 Jun 1868				D-T. & M. Earley
Earley, Timothy Edward					18 Jul 1908	Silver Bow			Ireland	26 Oct 1856	Arrived from New Jersey, H-Mary Early
Early, Mary A.	P	(Earley)	Sullivan		22 Jan 1879	Yuba		1836		26 Oct 1856	W-Timothy Earley, D-P & E. O'Sullivan
Early, Timothy Edward		(Earley)			26 May 1911	Yuba		1874	California	Single	S-Timothy Edward Earley,
Ehmann, George	P				5 Nov 1992	Alameda	8 Feb 1904		California		
Ehmann, Paul	P		Schmidt		10 Jan 1929		27 Aug 1867				H-Rosa Ehmann
Ehmann, Rosa	P				22 Jun 1940	San Mateo	9 Jun 1867				W-Paul Ehmann
Eiden, Mable McCrea		Mable May	Wheaton	Fraser	24 Aug 1968	Nevada	11 Aug 1896		California	28 Jan 1913	W-Albert McCrea, M-Arthur W McCrea
Emery, Mary Rose	P	Mary Violet	Rose		21 Feb 1901	Washoe,NV		1867	California	6 Feb 1895	W-Joseph D. Emery
Eutmon, Alden					23 Mar 1861		18 Sep 1833				
Farish, Adam Thomas	P				1865			1810			
Farish, Etta Paddock	P		Paddock		1919			1858			
Farish, Helen Ruth	P				1903			1892			
Farish, John Bolton	P				1929			1854			
Farish, Mary Wren Prather	P		Prather		1870			1810			
Fenton, Lois	P	Lois Amelia			5 May 1945		8 Oct 1858				
Fiene, Annie Margaret			Doucher		22 Aug 1912	Yuba	12 Jul 1832		Germany		W-Henry William Fiene
Fiene, Henry	P	Henry William			17 Jul 1908	Yuba	25 Mar 1825			Married	H-Annie Fiene, F-Wm. H. Fiene, F-Mrs. Charles Hares
Filcher, Eugenia Hapgood	P				1957			1869			Hapgood Family Plot, W-W.E. Smith, W-Wm. B. Filcher
Pippin, Asa D.	P	Asa David		Singleton	11 Dec 1968	Yuba	18 May 1885		California	8 Nov 1911	H-Emma Annabella Pippin
Pippin, Emma A.	P	Emma Annabelle	Wheaton	Fraser	15 Nov 1981	Nevada	4 May 1893		California	8 Nov 1911	W-Asa David Pippin, D-Allen & Mary Wheaton
Pippin, John A.	P	John Allen		Wheaton	28 Jun 1922	Yuba	5 Oct 1915		California	Child	S-Asa D. & Emma Pippin

Page 256

Smartville Cemeteries
Combined Information Analyses

Name on Headstone	P	Name in Records	Father's Name	Mother's Name	Death	Died	County	Born	Birth	Birth Place	Marriage	Relationships
Fippin, Sidney Wallis	P			Wheaton	18 Sep	1992	Contra Costa	3 Sep	1934	California	Married	H-Betty Fippin, S-Asa David & Emma Annabelle Fippin
Fisher, Grace E.	P				13 Nov	1894	Yuba	9 Apr	1883	California	Child	
Fisher, Rose	P	Rose Evelyn	Fehrown	Funk	2 Jan	1945	Nevada	6 Sep	1856	Kentucky		W-Samuel J. Fisher
Fisher, Samuel	P	Samuel Jackson			15 Dec	1915	Yuba	20 Mar	1839			H-Rose Evelyn Fisher
Fitzhugh, Grace O.	P	Grace Olive		Wilcox	31 Oct	1969	Nevada	18 Sep	1889	New York	Widowed	W-James M. Fitzhugh
Fitzhugh, James M.	P	James Monroe		Mayben	4 Jul	1967	Yuba	5 Jan	1880	California	Married	H-Grace Fitzhugh
Flanagan	P	Michael			20 Feb	1895	Yuba		1838	Ireland	Married	
Flanagan	P	Peter J			11 Mar	1904	Yuba		1845	Ireland	Single	
Flanagan	P	James Henry		Conell	10 Dec	1907	Yuba	20 Jan	1868	New Jersey	Divorced	S-Michael & Ellen Flanagan, "John"
Flanagan	P	Mary			4 Dec	1926			1842			
Flanagan	P	Thomas J.			12 Oct	1935	Nevada	9 Jul	1880	California		S-Michael & Ellen Flanagan
Flint, George W.	P				19 May	1873		7 Sep	1872		Child	
Flynn, Morris	P				27 Aug	1867		15 Aug	1852		Child	
Flynn, Patrick					15 Sep	1893	Orange					S-P. & S. Flynn
Ford, Patrick					23 Feb	1915	Yuba		1844	Ireland	Single	S-Patrick Ford
Foreman, Frank Gilbert	P				30 May	1963	Yuba	18 Jan	1895	California	5 Dec 1916	H-Frieda Foreman
Foreman, Ida		Ida J.	Gassaway		14 May	1912	Yuba	14 Dec	1848		11 Jun 1876	W-Jacob W. Foreman
Foreman, Jacob Washington											11 Jun 1876	H-Ida Foreman
Foreman, Jennie												
Foreman, Violet L.	P	Violet Lillian	Foreman	Busbby	1 Feb	1918	Yuba	2 Aug	1917	California	Single	D-Frank & Blanche Violet Foreman
Fortune, Martin	P				24 Mar	1891	Yuba	74y		ireland		H-Mary Fortune
Fortune, Mary	P				2 Jul	1898						W-Martin Fortune
Foust, Louis Perry					3 Nov	1900	Yuba	4 Oct	1826	Ohio		H-Elizabeth Foust
Foust, Nelson Lewis					18 Nov	1902		29 Nov	1876			
Fraser, Daniel	P	Daniel Fransen			24 Dec	1916	Yuba	18 Nov	1843			H-Elizabeth Fraser
Fraser, Elizabeth	P	Elizabeth Marie	Loomer		7 Sep	1883		23 Aug	1847			W-Daniel Fraser
Fraser, James H.	P			Lommer	2 Dec	1945	Nevada	16 Feb	1872	Canada		S-Daniel & Elizabeth Fraser
Frazier, Benjamin					8 Aug	1902	Yuba	24 Mar	1822	New York		H-Carrie Frazier
Frazier, Carrie	P				1 Sep	1886	Yuba	23 Mar	1817	Virginia		W-Benjamin Frazier
French, Joseph					2 Jan	1947	Yuba	6 Feb	1854	Michigan	Single	"Joe"
Gaffney, Ellen				Callaghan	28 Feb	1958	Yuba	23 May	1867	New Jersey	Single	D-Patrick Callaghen
Galligan, Hugh P.	P	Hugh Peter		Ewing	16 Aug	1931	Yuba	26 Sep	1858	Iowa	Married	H-Mary Elizabeth Galligan, S-Mathew & Susana Galligan
Galligan, Mary E.	P	Mary Elizabeth	McCarty	Flannigan	13 May	1957	Yuba	14 May	1864	California	Widowed	W-Hugh Peter Galligan, D-Andrew & Susan McCarty
Gann, Gladys D.	P		Dupes	Hahn	23 Aug	1994	Yuba	19 Apr	1917	Tennessee	Married	W-Wayne Jack Gann
Gann, Wayne J.	P	Wayne Jack		Shoffiett	6 Nov	2000	Yuba	18 Dec	1916	Tennessee	Widowed	H-Gladys Dupes Gann
Gassaway, Bertha Colling	P	Bertha	Jeffery	Davies	23 Oct	1938	Yuba	25 Jan	1865	California	11 Dec 1904	W-Robert S Gassaway, W-William H Colling 1887
Gassaway, Robert Samuel	P				19 May	1915		20 Mar	1856	California	11 Dec 1904	H-Bertha Gassaway
Geraghty, Alvina			Gunning		13 Apr	1976	Marin	30 May	1887	California	28 Dec 1913	W-John J. Geraghty, D-S. O. & Alvina Gunning
Geraghty, John J.	P	John James			27 Oct	1940	San Francisco	10 Jan	1880	California	28 Dec 1913	H-Alvina Gunning Geraghty
Gilbert, Absalom A.	P	Absalom Andrew		Callaghan	22 Jan	1984	Sutter	5 Jan	1890	California	Widowed	Father of Carl Gilbert and Bernice Rada
Gilbert, Pauline Ella	P		Hemmann		5 Aug	1975	Yuba	17 Aug	1889	Missouri	Married	W-Absalom Gilbert,

Smartville Cemeteries
Combined Information Analyses

Name on Headstone	P	Name in Records	Father's Name	Mother's Name	Death Date	Died	County	Born	Birth	Birth Place	Marriage	Relationships
Gilliam, Dorothy		Dorothy Grace			2 May	1998	Yuba		1940	San Francisco		D-Grace Smith
Gilliam, Wm. Glenn		William Glenn		Price	22 May	1989	Yuba	24 Sep	1913	Kentucky	Married	H-Dorothy G. Vega Gilliam
Gleason, Harold James	P	James Harold		Vineyard	26 Feb	1989	Nevada	28 Feb	1903	California	19 Nov 1933	H-Mildred L. Gleason, S-Elmer P & Lydia L Gleason
Gowne, Ann	P				12 Oct	1866			1836			
Gowne, John C.	P				7 Mar	1867			1834			
Gowne, Maria	P				20 Aug	1865			1843			
Gowne, Thomas H.	P				23 Jan	1865		19 Feb	1863			
Gowne, William	P				14 Dec	1865			1827			
Grace, John Thomas	P				19 Feb	1984	Yuba	25 Oct	1894	Pennsylvania	Single	Pvt U. S. Army WWI, S-John Thomas & Mary Grace
Grace, Rose						1999						
Gratiot, Eliza					2 Nov	1890	Yuba		1824	Canada	Married	
Greever, Aldon Lee	P			Kramer	4 Mar	1998	Yuba	15 Feb	1919	Oklahoma	Married	H-Maxine Minnie Greever
Greever, Hazelle C.	P	Hazelle Claire	Kramer		17 Mar	1974	Sutter	23 Jun	1895	Oklahoma	Married	W-Kyle A. Greever
Greever, Jackie W.	P	Jackie Warren		Maples	22 Jul	1998	Yuba	22 Nov	1943	Oklahoma	11 Jan 1963	H-Dorothy Jean Greever H-Martha Belle Greever
Greever, Kyle A.	P	Kyle Aldon		Fuller	26 Sep	1981	Yuba	21 Aug	1893	Kansas	Widowed	H-Hazelle C. Greever, F-Aldon, Kyle Greever
Greever, Maxine Minnie	P		Maples	Whited	19 Apr	1998	Yuba	11 Jan	1920	Oklahoma	Widowed	W-Aldon Greever, Married 58 years
Gross, Harry						1894						
Gruber, William Henry	P			Cramer	24 Apr	1991	Yuba	7 Aug	1900	Missouri	Divorced	"Wild Bill", S-William Patrick & Antonetta Gruber
Gunning, Baby	P	Infant Son			12 Jan	1893	Yuba	12 Jan	1893	California	Child	S-S. O. & Alvina Gunning
Gunning, Ellen	P								1816	France		
Gunning, Fremon					25 Oct	1886	Yuba					
Gunning, Mary	P											
Gunning, Samuel O.	P				14 Dec	1910	Yuba	14 Jan	1834	Ireland	12 Sep 1886	H-Lavina Lottsen Gunning, Native of Co. Sligo, Yuba Co Recorder
Gunning, Samuel Owen	P			Lotzin	6 May	1982	Nevada	15 Mar	1891	California	Single	U. S. Navy WWI, S-S. O. & Alvina Gunning,
Hacker, Cora E.	P	Cora Elizabeth	Peckham		5 Mar	1962	Sacramento	10 Apr	1880	California	2 Jul 1905	W-William F. Hacker, D-Thomas W. Peckham
Hacker, Darling Emma Pearl	P					1913			1911			In Peckham Fam Plot
Hacker, William F.	P			McMullen	9 Sep	1952	Sacramento	30 Nov	1878	California	2 Jul 1905	H-Cora E. Hacker
Hall, John Clark					12 Jun	1901	Yuba	12 Sep	1869	Colorado		F-L. Otis Hall, Father in Law was D. N. Jones
Hagood, Annie Lilly	P	Elizabeth Ann	Hapgood		24 Aug	1881	Yuba	9 Aug	1865		Child	D-Eugene D. & Elizabeth Hapgood
Hagood, Elizabeth	P				15 Oct	1905	Yolo		1842			W-Eugene D. Hapgood, M-Annie Lilly Hapgood
Hagood, Eugene D.	P	Eugene Delanmore			27 Nov	1913	Alameda	5 Dec	1838	Vermont		H-Elizabeth Hapgood, F-Annie Lilly Hapgood
Hagood, Fannie Marple		Fannie Elizabeth	Marple	Silverthorn	25 Apr	1946	Yuba	28 Mar	1862	California	28 Sep 1894	W-James Mortimer Hapgood D-Wm & Elizabeth Marple
Hagood, James Mortimer				Allen	4 Apr	1959	Nevada	28 Mar	1872	California	28 Sep 1894	H-Fannie Elizabeth Hapgood S-Ed & Elizabeth Hapgood
Harriman, Elsie	P		Platt	Pill	29 Jan	1906	Yuba	30 Jan	1871	New Jersey	Married	W-Edward L. Harriman
Harriman, Susan Ann			Duke		31 Oct	1904	Yuba		1835	Canada	1874	W-Stephen Harriman
Harris, Angela M.		Angela Mary		Hancock	9 Feb	1987	Nevada	31 Oct	1969	California		
Harris, Lee	P	Gerald Lee		Elliott	23 May	1986	El Dorado	8 Dec	1952	California		
Hartley, W. M.					27 Oct	1884	Yuba	6 Apr	1816			H-Mary Hartley, F-Martha Revell
Hartley, Mary A.					3 Feb	1877		2 Nov	1822	England		M-Martha Revell, W-W M Hartley
Havey	P	Amanda Veronica			3 Sep	1889	Yuba	24 Feb	1884		Child	D-John Havey
Havey		Mary			29 Jan	1897	Yuba		1814	Ireland	Married	W-Patrick Havey, "Grandma"

Smartville Cemeteries
Combined Information Analyses

Name on Headstone	p	Name in Records	Father's Name	Mother's Name	Death Died	County	Born	Birth	Birth Place	Marriage	Relationships
Havey		Anna Lucile	Havey	Pryor	21 Dec 1905	Yuba	13 Feb	1896	California		D-John C. & Mary Ann Havey
Havey	P	Margaret			26 Jun 1918			1835			
Havey	P	Louis Patrick			15 Oct 1922		21 Feb	1891		7 Jun 1922	H-Mary Emma Poor Havey, S-John Havey
Havey	P	John C.		Havey	6 Jan 1931	Yuba	16 May	1849	New Jersey	Married	H-Mary Ann Havey
Havey	P	Gertrude					7yrs			Child	
Havey, Mary Emma	P		Poor		18 Nov 1929		16 May	1903			Poor Family Plot
Havey, Pat Jr.	P	Patrick, Jr.			27 Oct 1884	Yuba		1838			"Uncle"
Hawkins, Eldon John				Neetatge	29 Jun 1993	Placer	5 Jan	1921	Nebraska		H-Gloria C Hawkins
Hawkins, Gloria Cecilia			Kiste	Miller	23 Apr 1987	Nevada	8 Feb	1927	Oregon		W-Eldon John Hawkins
Haworth, Abraham	P							1826			
Haworth, Ann	P							1806			
Haworth, Thomas	P				20 Sep 1873	Yuba		1841	England	5 Nov 1875	H-Elizabeth Painten Haworth
Hays, Isaac N.	P				17 Mar 1910	Yuba	5 Dec	1821	Virginia		Col. In the Mexican War-Co H S LA Mil Inf
Heggerty, Mary	P							1832	Ireland		
Heggerty, Maurice	P					1875		1830	Ireland		
Heggerty, William	P					1873		1827	Ireland		
Henderson, Bridgett						1891		1842	Ireland		
Herboth, Elizabeth E.	P		Conrath	Quinkert	16 Jul 1899	Yuba	11 Jul	1890	California	Widow	W-George Herboth, D-Louis & Elizabeth Conrath
Herboth, George	P			Keller	17 Jul 1972	Yuba	11 Nov	1889	California	7 Sep 1912	H-Elizabeth Conrath Herboth, S-John & Elizabeth herboth
Hermann, Carl	P				6 Apr 1964	Yuba	31 Mar	1851		7 Sep 1912	
Hermann, Caroline	P				14 May 1921		23 Apr	1852	Germany	Widow	
Hermann, Charles H.	P				14 Mar 1946	Alameda	22 Sep	1886	Missouri	Single	
Heyne, Bernard W.	P	Bernard Wilford		Wood	1 Nov 1974	Butte	17 May	1905	Oklahoma		
Heyne, Ethel K.	P				17 Jun 1982	Nevada	24 Nov	1917			
Hickeson, Katie J.	P	Catherine Jane	Bushby	Dougherty	29 Aug 2001		31 Dec	1869	California	Married	W-William Hickeson
Hicks, Joseph	P				21 Sep 1917	Yuba		1847	England		Native of Cornwell, England
Hinkle, Archie W.	P				18 Mar 1878			1879			
Hinkle, Noble F.	P	Noble Francis		Franks	28 Feb 1944		1 Jun	1911	Washington	Divorced	Brother to Thelma Quality, S- Maud Curl
Hite, David C.	P	David Calton			13 Feb 1965	Yuba	7 Nov	1907		Child	S-H. L. & I. M. Hite
Hite, Herbert L.	P	Herbert Leslie		Atkins	7 Feb 1909		24 Nov	1878	California		H-I. M. Hite, F-David C. Hite
Hite, Ida May	P	Ida May	Manasco		11 Nov 1960	Alameda	16 Nov	1882	Idaho	Widowed	W-Herbert Leslie Hite, D-Catlon Wilber Manasco
Hogarth, Rose Ann	P			Newbert	20 Oct 1977	Sutter	10 Apr	1826	Ireland		W-M Hogarth, M-Minnie E Newbert
Holbrook, Elsie D.	P			Tucker	22 Jan 1886		19 Mar	1895	California	26 Apr 1918	W-James E. Holbrook, D-Eddington & Albertine Detrick
Holbrook, James E.	P	James Edward			3 Jul 1963	Alameda	14 May	1896	California	26 Apr 1918	H-Elise D. Holbrook, S-Katherine Holbrook
Holbrook, Kathleen Adele	P				11 Jun 1975	Alameda					D-James K. & Mary O'Brien
Holcomb, David					27 Dec 1934						
Holcomb, Thatcher B.					12 Apr 1878			1837		Married	S-Thatcher B. Holmes
Holmes, Thatcher B.					15 Jul 1873	Yuba		1865			
Holmes, Winslow					1 Mar 1866		19 Dec	1858	Ireland	Married	
Horan, Michael	P				18 Sep 1897	Yuba					
Hord, Mary Wright		Mary Belle	Welch	Bast	10 Aug 1937	Yuba	2 Aug	1862	California		W-Charles Hord, In Murphy Fam Plot
House, Anthony											

Page 259

Smartville Cemeteries
Combined Information Analyses

Name on Headstone	P	Name in Records	Father's Name	Mother's Name	Death Died	County	Born	Birth	Birth Place	Marriage	Relationships
Hudson, Laverne J.			Hansen	Jensen	29 Sep 1979	Yuba	15 Jan	1909	Utah	Married	W-Ralph Hudson, M-Richard & Robert Hudson
Hudson, Ralph R.		Ralph Rule		Obert	4 Oct 1993	Yuba	8 Dec	1905	Wyoming	Married	H-Laverne Hudson, F-Richard & Robert Hudson
Hug, Alois			Hug	Truttmann	19 Apr 1955	Yuba	17 Dec	1880	Switzerland	Single	S-Alois & Anna Hug, AKA Hug, Citizen of Switzerland
Hughes, Winifred	P				18 Feb 1874			1839			W-Michael Hughes
Hurling, Charles					22 Feb 1930		4 Jul	1845			
Hurling, George W.					23 May 1894		28 Nov	1872			
Hurling, Lucille					2 Feb 1891		9 Jan	1854			
Hutchinson, David							24 y				
Hyatt, Jacob					13 Sep 1878		23 Feb	1806			
Ingram, Mary			Denault		2 Dec 1877		Nov	1801			
Jackson, M. H.					15 Aug 1882			1813			
Jackson, William H.	P	Malinda H.			4 Sep 1901		3 Jun	1820			
James, William Ray					16 Dec 1974	Yolo	28 Oct	1888	Wisconsin	Widowed	
Jeffery, Ella Clinton			Jeffery	Davis	7 May 1912	Yuba	21 Mar	1863	California	Single	D-George & Mary Jeffery
Jeffery, Ellen H.			Perkins	Magonigal	25 May 1943	Yuba	1 Apr	1877	California	19 Nov 1902	W-William B Jeffery, D-J P & J Magonigal
Jeffery, George		George Archibald			31 Jul 1907	Yuba	21 Dec	1829	Canada	2 Jul 1860	H-Maria Jeffery, F-Clinton,Harry,Ella
Jeffery, Harry Clinton					15 Nov 1890	Yuba	15 Feb	1887		Child	S-G & ML Jeffery
Jeffery, Maria Louise			Davis	Davies	15 Jan 1902	Yuba	29 Dec	1840	Pennsylvania	2 Jul 1860	W-George Jeffery, "Mary"
Jeffery, William		William Barns(Bernard)		Davies	14 Jul 1953	Yuba	27 Aug	1868	California	19 Nov 1902	H-Ellen H Jeffery, H-Maria L Jeffery,S-Geo. & Maria Jeffery
Jewel, Henry C.	P	Henry Clay			28 Mar 1873	Yuba	Jun	1858		Child	S-E. H. & Elizabeth Jewell
Johnson, Carrie R.			Weldon	Morey	31 Dec 1945	Sutter	30 May	1865	California	Widowed	W-William F. Johnson
Johnson, Francis V.	P	Francis Vincent			8 Feb 1925	Alameda	18 Jul	1895	California	25 Jan 1922	H-Cora Elizabeth Crump Johnson, S-W. F. Johnson
Johnson, Wm. Allen	P	William Allen		Murphy	3 Feb 1930	Yuba	3 Feb	1930	California	Child	S-William & Mary Johnson
Johnson, Wm. F.	P	William Franklin			2 Jul 1961	Yuba	14 Sep	1864	New York	Widowed	H-Carrie R. Johnson, S-Charles Johnson
Jones, J. Y.	P				16 Oct 1912		23 May	1912		Child	
Jones, Simon G.					6 Jan 1860			1831	Maine		
Kay, Samuel					10 Jul 1899	Yuba	10 Sep	1829	England		
Keegan, Ann J.					31 Aug 1886		31 Feb	1884			
Keegan, Annie L.	P				7 Jun 1867		2 Sep	1866		Child	D-Jas & Mary J. Keegan
Keegan, Annie Louise	P				1 Jan 1867		13 Oct	1867		Child	D-Jas & Mary J. Keegan
Keegan, Charles M.	P				7 Jun 1873			1873		Child	S-Jas & Mary J. Keegan
Keegan, Infant Child	P				23 Oct 1887	Yuba	23 Feb	1886		Child	Infant-John & Mary Keegan
Keith, Granville C.		Granville Clifford			31 Oct 1905	Sutter		1832			No relatives when when he died
Kelly, James					29 Aug 1909	San Francisco		1848	Ireland	Married	
Kerrigan, Ambrose F.	P				19 Aug 1873		18 Aug	1873			
Kerrigan, Charles M.	P				7 Jun 1873	Yuba		1873			
Kerrigan, Mary	P				30 Jul 1899	Yuba	15 Aug	1833	Ireland		W-Peter Kerrigan
Kerrigan, Mary J. or L.	P				20 May 1866		Jun	1862			D-Peter & Mary Kerrigan
Kerrigan, Patrick	P	Patrick K.			9 Dec 1922	Yuba		1863	California		
Kerrigan, Peter	P				21 Aug 1873	Yuba	21 Nov	1818	Ireland		Native of Co. Sligo
Kerrigan, Thomas J.	P				3 Feb 1899	Yuba		1871	California	Single	S-Peter & Mary Kerrigan

Smartville Cemeteries
Combined Information Analyses

Name on Headstone	P	Name in Records	Father's Name	Mother's Name	Death	Died	County	Born	Birth	Birth Place	Marriage	Relationships
Kershaw, Diane C.	P				29 Oct	1996		14 Nov	1931			W-Theodore G. Kershaw
Kershaw, Henrietta Collbran	P		Collbran	Parish	31 Mar	1986	Santa Clara	22 Feb	1913			
Kershaw, Theodore G.	P							23 Apr	1934			H-Diane C. Kershaw, Dbl headstone - No death date
Kershaw, Theodore Gourdin	P	Theodore Gourdin Sr			12 May	1975	Monterey	23 May	1906	S Carolina		
Kirby, John	P					1934			1882			
Lambert, Sheldon	P				30 Dec			4 yrs			Child	
Lane, Esther Theodora				Arguello	26 Feb	1998		17 Apr	1927	California		W-Larry Lane, M-Linda Lane
Lane, Larry		Lawrence Leroy			15 Aug	1999		4 Sep	1918			H-Esther Lane, F-Linda Lane
Lane, Linda						1993			1948			D-Esther & Larry Lane
Lane, Louisa					7 Jun	1876	Yuba	7 Nov	1873		Child	D-Riley & Louisa Lane
Lavelle, Mary												
Leaky, Michael					15 May	1906	Yuba		1840	Ireland	Single	"Mike"
LeBoeuf, Cecilia Ethelyn	P		LeBoeuf	Burroughs	17 Dec	1967	Yuba	22 Nov	1907	Oregon	Single	D-William & Ethelyn LeBoeuf
LeBourveau, Ernest	P			Page	30 May	1972	Yuba	15 Aug	1877	Canada	Widowed	
Lee	P	Agnes	Lee		13 Dec	1903	Sacramento	5 Aug	1885	California	Single	D-Thomas R. & Mary Ann Lee
Lee	P	Thomas R.			20 Oct	1906	Sutter		1856		Married	
Lee, James R.					27 Jul	1882	Yuba		1882	California	Child	S-Thomas R. & Mary Ann Lee
Leigh, Keana	P				20 Apr	1993						
Leasley, Betty	P	Betty Lee	Greever	Maples	9 Apr	1983	Yuba	12 Dec	1942	Oklahoma	26 Sep 1975	W-Carrol Leasley, D-Aldon Lee & Maxine Greever, "Billie Lee"
Likens, A. L.	P	Alma Lou	Palmer	Conway	3 Oct	1986	Yuba	20 Apr	1913	Texas	Married	W-Hance B Likens
Likens, H. B.	P	Hance Benjamin			10 Nov	1986	Sutter	28 Aug	1912	Texas	Widowed	H-Alma Lou Likens
Likens, R. C.	P	Reginald Collins			12 Sep	1994		16 Nov	1934			
Linehan, Daniel					11 May	1918			1858			
Linehan, Edward Thomas	P				18 Jan	1904	Yuba	6 Jun	1870	California	Single	
Linehan, Ellen	P		Kelly	Madden	8 Jan	1912	Yuba		1837	Ireland	Married	W-Timothy Linehan, D-Peter Kelly
Linehan, John	P	John Henry		Kelly	8 Jun	1922	Yuba	27 Feb	1863	California	Married	H-Anna Linehan, S-Timothy & Ellen Linehan
Linehan, Katherine	P											D-Timothy & Ellen Linehan
Linehan, Mary	P											
Linehan, Peter	P	Peter Francis		Kelly	25 Aug	1953	Yuba	17 Oct	1865	California	Widowed	S-Timothy & Ellen Linehan
Linehan, Timothy	P			Murphy	10 Aug	1924	Yuba	Feb	1824	Ireland	Widowed	H-Ellen Linehan, S-Patrick & Mary Linehan
Linscott, James O.					4 Feb	1869			1833	Maine		
Locmer, J. W.		James Wellington			23 Nov	1884			1820			A Rev.
Locmer, Maggie	P				1 Feb	1881		10 Mar	1852			
Locmer, Sarah I.					24 Sep	1883		23 Feb	1824			
Looney, Catherine	P	Katherine	Haggerty		6 May	1917	Yuba	27 Apr	1827	Ireland	Widowed	
Lyons, Jeremiah	P				27 Sep	1880	Yuba		1828	Ireland		
MacWilliams, Joe	P				27 Mar	1994		25 Apr	1948			
Magonigal				Vineyard		1994						
Magonigal, Alice R.								20 Nov	1912			W-Henry E Magonigal
Magonigal, Clarence B	P			Raehily	16 Jun	1938	Yuba	27 Sep	1909	California	Single	S-T G & Hannah Magonigal
Magonigal, Eleanore Germaine	P		McQuaid	McNamara	8 May	1987	Nevada	5 Apr	1903	California		Native of Buttapouney Co., Cork, "Jerry"

Smartville Cemeteries
Combined Information Analyses

Name on Headstone	P	Name in Records	Father's Name	Mother's Name	Death Died	County	Born Birth	Birth Place	Marriage	Relationships
Magonigal, Ellen J	P			Bowman	7 Jan 1896	Yuba	29 Dec 1865	California	29 Jan 1890	W-W B Magonigal
Magonigal, Emma E.	P	Emma Eliza	Fraser	Loomer	15 Nov 1961	Sutter	3 Apr 1870	Canada	Widowed	W-John Magonigal, D-Daniel Fraser
Magonigal, Frankie	P	Frank Webster		Bushby	14 Nov 1908	Yuba	28 Jan 1907	California		S-T G & Hannah Magonigal
Magonigal, Gerald E.		Gerald Eugene		Vineyard	29 Aug 1994	Yolo	7 May 1944			S-Henry E. & Alice R Magonigal
Magonigal, Hannah	P		Bushby		7 Oct 1970	Placer	17 Jan 1872	California	Widowed	W-TG Magonigal, D-George Bushby
Magonigal, Henry E.		Henry Edwin			15 Jun 1986	Nevada	2 Nov 1905	California	Married	H-Alice R Magonigal
Magonigal, John					24 Sep 1927		1863			H-Emma E Magonigal
Magonigal, John L.		John Lester		Fraser	11 Nov 1980	Nevada	16 Nov 1901	California		S-John & Emma E Magonigal
Magonigal, Lizzie	P	Elizabeth		Magonigal	22 Mar 1926		28 Nov 1854	California	Single	W-S Magonigal
Magonigal, Nancy	P		Boyd		11 Nov 1892	Yuba	15 Apr 1821	Ireland		W-W Magonigal
Magonigal, S.	P	Samuel			19 Jul 1897	Yuba	1831			H-Lizzie Magonigal
Magonigal, T. G.	P	Thomas George		Boyd	8 Jul 1945	Placer	7 Aug 1866	California	Married	H-Hannah Magonigal
Magonigal, Vera Lee			Thedore	Stone	14 Dec 1982	Nevada	29 Nov 1906	Missouri	Widowed	D-John & Emma E Magonigal
Magonigal, William	P	William B.			27 Feb 1891	Yuba	1818	Ireland	Married	H-Nancy Magonigal
Magonigal, William B.					1933					
Magonigal, William Daniel				Fraser	22 Mar 1963	Yuba	13 Sep 1898	California	Married	H-E G Magonigal, S-John & Emma E Magonigal
Manasco, Calton W	P	Calton Wilbur			5 Apr 1915	Nevada	8 Mar 1830	Alabama	Married	H-Amelia Manasco
Manasco, George C.					17 Nov 1943		12 Nov 1884			
Manasco, Veronica A.	P	Veronica Amelia		Schmidt	28 Aug 1922	Yuba	23 Nov 1856	California		
Manford, Dorothy E.	P	Dorothy Elizabeth	Shewood	Grantham	23 Sep 1982	Nevada	17 Nov 1911	Missouri	Married	W-James B. Manford
Manford, James B.	P	James Bunch		Bunch	15 Jan 1985	Yuba	2 Feb 1904	California	Widowed	H-Dorothy E. Manford
Marple, Brant					1876		1860			
Marple, Elizabeth					1860		1836	Pennsylvania		W-William Marple
Marple, Fred				Silverthorn	1929		1872	California		S-William & E Marple
Marple, George		George Washington		Silverthorn	16 Oct 1950	Yuba	22 Feb 1871	California	Never Married	S-William & E Marple
Marple, Harry				Silverthorn	12 Feb 1933	Yuba	26 Apr 1876	California		S-Whitten & E Marple
Marple, Infant		Infant (2 stones)								There are 2 markers, recnts show 5 buried
Marple, John W.					1860		1856			
Marple, Samuel				Silverthorn	2 Aug 1949	Sacramento	25 Jan 1878	California		S-William & E Marple
Marple, Whitten	P	William Whitten			30 Jan 1907	Yuba	1822	Pennsylvania		H-Elizabeth Marple
Martien, Charles					27 Feb 1951	Placer	13 Feb 1862	Pennsylvania	Married	H-Elizabeth Dougherty Martien
Martien, Elizabeth	P	Elizabeth Jane	Dougherty	Brady	8 Apr 1961	Yuba	2 Apr 1871	California	Widowed	W-Charles Martien, D-Daniel & Ann Dougherty
McAllis, Fannie	P				1 Aug 1867	Yuba	1 Jun 1865	California		D-J. & F. McAllis
McCarty, Andrew	P				20 Aug 1909	Yuba	1833	Ireland	Married	
McCarty, James	P				1 Mar 1902	Siskiyou	1873	Yuba	Single	S-Andrew McCarty
McCarty, John T.					4 Feb 1860	Yuba	24 Aug 1828	Indiana		
McCarty, Matthew H.	P	Matthew Henry		Flanagan	9 Feb 1951	Yuba	1 Jan 1869	California	Never Marr	S-Andrew & Susan McCarty
McCarty, Robert E.	P	Robert Emmett		Flanagan	31 Jul 1953	Yuba	30 Sep 1880	California	Married	H-Ella P. McCarty, S-Andrew McCarty
McCarty, Susan	P		Flanagan	Smith	30 Oct 1927	Yuba	Jan 1839	Ireland	Widowed	D-Matthew & Mary Flanagan
McClaskey, John Calvin					31 Jul 1870		9 Sep 1856			
McClure, Alexander Campbell					13 Nov 1901	Yuba	1831	Ireland	Widowed	

Smartville Cemeteries
Combined Information Analyses

Name on Headstone	P	Name in Records	Father's Name	Mother's Name	Death	Died	County	Born	Birth	Birth Place	Marriage	Relationships
McClure, James Alexander				McKindley	24 Jan	1951	Yuba	27 Jul	1878	Ireland	Single	S-Alexander & Jane McClure
McClure, Joseph Patrick				McKindley	4 Dec	1951	Yuba	7 Apr	1886	Massachusetts		S-Alexander & Jane McClure
McConnell, James	P				26 Feb	1867		17 Jul	1865			S-J & Martha McConnell
McConnell, James, M. D.	P				3 Oct	1874		16 Jan	1825			H-Martha McConnell, F-James McConnell
McConnell, John Taylor					2 Nov	1887		15 May	1855			
McConnell, Martha					3 Dec	1918	Marshfield, OR	26 Sep	1837			
McConnell, Mary Jetta					1 Mar	1879		17 Mar	1872		Child	
McCoughlan, Margaret					5 Aug	1903			1818			
McCoughlan, Peter	P				27 Jan	1869			1831	Ireland	Widowed	Native of Co. Loudonderry
McCoughlan, William		McCoughlan, Wm.			4 Jul	1904	San Joaquin		1825	Ireland		
McCrea, Albert Ernest						1931			1885	California	28 Jan 1913	H-Mabel M. Wheaton McCrea Eiden
McCrea, Arthur Wheaton					6 Mar	1923			1913			S-Albert E McCrea & Mable M Eiden
McGanney, Anna	P	Anne Cranisie	McGanney									W-John Edward Cransaie, D-Daniel & Mary McGanney
McGanney, Edward James	P				5 Oct	1918	Argonne, France		1891			U.S.A. 30 Infantry, S-Daniel & Mary McGanney
McGanney, Father	P	Daniel			11 May	1895	Yuba		1828	Ireland	Married	H-Mary McGanney
McGanney, Georgie Gray	P	Georgie Elizabeth	Gray	Burk	5 Feb	1953	Yuba	10 Mar	1879	California	2 Mar 1897	W-F. J. McGanney
McGanney, James F.	P	James Francis			4 Jul	1907	Yuba	14 Mar	1863		Married	S-Daniel & Mary McGanney
McGanney, Mother	P	Mary			26 Jan	1902	Yuba		1834	Ireland		W-Daniel McGanney
McGanney, Ned	P											
McGovern, John					9 Jul	1891	Yuba		1827			
McGovern, Mary Agnes	P			Linehan	9 Apr	1957	San Francisco	28 Aug	1875	California	Widowed	W-Thomas McGovern
McGovern, Mrs. John					23 Feb	1893	Yuba		1833	Ireland		W-John McGovern
McGovern, Thomas James				Flanagan	8 Jun	1926	Yuba	20 Jan	1865	California		S-John & Alice McGovern, H-Mary McGovern
McKeague, Robert	P				8 Oct	1864			1831			
McKeel, John												
McKeel, Thomas												
McKenna, Elizabeth T.			Murphy								29 Mar 1880	W-John McKenna
McKenna, John											29 Mar 1880	H-Elizabeth Murphy McKenna
McKenna, Margaret					24 Nov	1894	Yuba	24 Feb	1828	Ireland		
McLean						1864		6 m				
McNally, Edwin	P										Child	S-Thomas & Mary McNally
McNally, Mary	P				20 May	1900	Yuba		1845	Ireland	Married	W-Thomas McNally
McNally, Thomas	P					1919			1836			H-Mary McNally
McNutt, Alice	P				4 Aug	1896		15 Jan	1821	England		W-J.F McNutt, "Auntie"
McNutt, John F.	P	John Flemming			5 Mar	1899	Yuba	11 Jan	1815	Tennessee		H-A McNutt, Was a Judge
McPhillips, Barney	P				8 Feb	1892			1827	Ireland		
McPhillips, Owen					27 Mar	1861	Yuba		1819	Ireland		
McQuaid, Frank					18 Aug	1911	San Francisco		1862	California	10 Nov 1889	H-Kate McNamara McQuaid
McQuaid, John	P				5 Nov	1913	Yuba	10 Dec	1837	Ireland	Widowed	H-Mary S. McQuaid
McQuaid, John H.	P	John Henry			26 Mar	1932	Yuba	17 Apr	1870	California	15 Oct 1902	S-John & Mary McQuaid, H-Katherine McQuaid
McQuaid, Katherine R.	P	Katherine L.	Rearden	Mulhare		1962			1875		15 Oct 1902	W-John Henry McQuaid

Page 263

Smartville Cemeteries
Combined Information Analyses

Name on Headstone	P	Name in Records	Father's Name	Mother's Nm	Death	Died	County	Born	Birth	Birth Place	Marriage	Marriage	Relationships
McQuaid, Mary	P		McQuaid	Mulhare	18 May	1925	Yuba	8 Apr	1866	California		Single	School teacher, D-John & Mary McQuaid, "Aunt Mary"
McQuaid, Mary Margaret	P			McNamara	21 Aug	1990	San Mateo	31 May	1907	California	10 Nov 1889		W-Frank McQuaid, "Kate"
McQuaid, Mary S. Mulhare	P	Mary S.	Mulhare		20 Jun	1899	Yuba		1836	Ireland	29 Sep 1861		W-John McQuaid, D-Michael Mulhare, Sister to Margaret Conlin
McWilliams, Bridget	P				3 Jan	1900	Yuba		1872	Ireland			Native of Co. Derry
Meade, John	P	John T.		Leahy	17 May	1944	Yuba		1879	California		Single	S-Garrett & Annie Mead, "Jack"
Meade, Leo	P	Leo Garrett		Leahy	15 Nov	1944	Yuba	27 Aug	1883	California		Single	S-Garrett & Annie Mead
Meade, Mary	P												
Meade, William Joseph					30 Nov	1921	Butte	30 Jan	1880	California		Married	S-Garrett & Annie Meade, H-Agnes Weber Meade
Mellarkey, Edward					14 Feb	1909	Yuba	12 May	1814	Ireland		Married	H-Mary Mellarkey
Mellarkey, James		James S.		McLean	1 Apr	1923	Yuba	4 May	1857	California	24 Dec 1879		H-Mary Mellarkey, S-Edward Mellarkey
Mellarkey, John J.	P				25 Jan	1900	Kern	8 Sep	1873				H-E Mellarkey
Mellarkey, Mary					2 Apr	1915	Yuba	27 Nov	1824				W-Edward Mellarkey
Melody, Bridget	P				18 Dec	1869			1840				W-Thomas M. Melody
Meredith, W. J.	P				10 Nov	1878	Yuba		1836				Lived on the Empire Ranch in Smartsville
Michaels, Walter C.	P	William James		Wisotska	19 Jul	1987	Ventura	12 Jan	1916	New York			
Miles, E. Adele	P	Walter Constantine			26 Feb	1931		29 Mar	1916			Widowed	
Miller, George C.		George Charles		Weiterhouse	19 Apr	1987	Solano	9 May	1930	Missouri			
Miller, George W.	P				20 Mar	1865		23 Feb	1862				S-Geo & Jane Miller
Miller, John					24 Feb	1860	Yuba		1837	Ireland			
Miller, John L.					8 Dec	1855			1831				
Miller, John R.	P				26 Mar	1865		17 May	1864			Child	S-Geo & Jane Miller
Miller, Joseph	P				8 Jul	1869		10 May	1866			Child	S-Geo & Jane Miller
Miller, Jane Alice	P				19 Sep	1869		11 Sep	1865			Child	D-Geo & Jane Miller
Miller, May					27 May	1880		4 Sep	1870			Child	
Mills, Homer C.		Homer Clyde		Crabtree	30 Sep	1985	Yuba	3 Jan	1924	Oklahoma	24 Jul 1983		H-Maria Isabel Bettancourt Mills
Mitchell, Catherine					10 Jun	1901	Yuba		1826	Ireland		Married	W-John Mitchell
Mitchell, Elvira Bell	P		Wright	Welch	23 Dec	1914	Yuba	7 Oct	1879	California		Married	W-Thomas Mitchell, Newbert Plot
Mitchell, John					8 Aug	1902	Yuba						H-Elvira Mitchell
Mitchell, John					10 Aug	1902	Yuba						H-Catherine Mitchell
Mohn, Mary H.	P				11 Apr	1857			1841	Pennsylvania		Child	W-Joseph Mohn
Monk, Betsy	P	Elizabeth	Travena		31 Jan	1894		26 Dec	1817				W-James Monk
Monk, Elizabeth	P				11 Jan	1879		8 Oct	1844				W-James Monk
Monk, James	P				12 Apr	1877		2 Mar	1818	England			H-Betsy Monk
Monk, James	P				11 Dec	1895		27 Dec	1849				H-Elizabeth Monk
Moody, Ellen R.	P				9 Jul	1856			1824	Mississippi			W-Tom B. Moody
Mooney, Adelaide			Mooney	Huting	13 Jul	1953	Yuba	19 Nov	1865	California		Never Married	D-Thomas & Mary J. Mooney
Mooney, Arthur	P	Arthur Barton			22 Feb	1915		23 Sep	1874				
Mooney, Clara L.	P	Clara Louise	Mooney	Huting	8 Feb	1927	Yuba	4 Jul	1858	California		Single	D-Thomas & Mary J. Mooney
Mooney, Jessie	P	Jessie Gardner	Mooney	Huting	5 May	1927	Yuba	10 Feb	1870	California		Single	D-Thomas & Mary J. Mooney
Mooney, Lucy J.	P	Lucy Jane	Mooney	Huting	7 Oct	1923	Yuba	13 May	1869	California		Single	D-Thomas & Mary J. Mooney
Mooney, Mary E.	P		Mooney	Huting	6 Jan	1941	Yuba	11 Feb	1860	California		Single	D-Thomas & Mary J. Mooney

Smartville Cemeteries
Combined Information Analyses

Name on Headstone	p	Name in Records	Father's Name	Mother's Name	Death Died	County	Born	Birth	Birth Place	Marriage	Relationships
Mooney, Mary J.	p	Mary Jane	Hiding		7 Dec 1875		1 Sep	1840			W-Thomas Mooney
Mooney, Nellie	p	Nellie McMurry			27 Feb 1896	Yuba	24 Oct	1874	California	Single	A Twin, D-D-Thomas & Mary J. Mooney
Mooney, Thomas	p	Thomas Sr.			6 Apr 1884	Yuba	5 Dec	1823	Ireland		"Father"
Mooney, Thomas Jr.	p		Mooney		15 Jul 1948	Nevada	14 Oct	1872	California		D-Thomas & Mary J. Mooney
Morton, Norman Barclay	p			Hiding	11 Aug 1950	San Mateo	6 Oct	1873			
Mulligan, James	p				30 Aug 1875			1797	Ireland		Native of Roscommon
Mullin, Mary					6 Jan 1879	Yuba		1820	Ireland		
Mullin, Mary Ellen	p				22 Mar 1872						D-J. E. & Julia Mullin
Murch, Frances Marie					12 Mar 1901		2 May	1901		Child	"Infant", In Adkins Family Plot
Murdock, Henry					12 Dec 1876			1807	Canada		
Murdock, James	p		Lavers	Sanford	4 Jul 1942	Yuba	25 Oct	1848	Canada	Widowed	W-James Murdoch
Murphy, Catherine	p		Havey		24 May 1898	Yuba		1852	New Jersey	10 Dec 1873	W-Morris Murphy, "Katie", "Mother"
Murphy, Charles E.					3 Aug 1888	Yuba				Child	S-Morris Murphy
Murphy, Gladys L.	p	Gladys Lillian	Murphy	Wright	2 Jul 1913	Yuba	27 Dec	1911	California	Child	Murphy Fam Plot
Murphy, Ida E.	p	Ida Elise	Wright	Welch	7 Oct 1986	Yuba	3 Apr	1890	California	Widowed	W-John Joseph Murphy, D-Alden & Mary Wright
Murphy, John J.	p	John Joseph			14 Mar 1963	Yuba	14 Mar	1886	Nevada	Married	H-Ida E. Murphy
Murphy, Morris	p				12 Mar 1915	Butte		1847		10 Dec 1873	H-Catherine Murphy, "Father"
Murphy, Rosana					18 Jan 1892	San Luis Obispo					W-Stephen Murphy
Murphy, Stephen					30 Jun 1891	San Luis Obispo	75y				H-Rosana Murphy
Naglee, Joseph C.	p			Reed	6 Nov 1970	Yuba	17 May	1917	New York	Married	H-Elizabeth Bennett Naglee, S-Samuel & Rilla Naglee
Newbert, Ella			Foreman		24 Jul 1990	Yuba	11 Aug	1881	California	25 Dec 1899	W-Horace C. Newbert
Newbert, Horace	p	Horace C.		Jackson	3 Apr 1953	Sacramento	15 Mar	1870	California	25 Dec 1899	H-Ella Foreman Newbert
Newbert, Mary J.	p	Mary Ann	McNally	Brown	3 Mar 1942	Sacramento	27 Jul	1879	California		W-Hood Newbert
Newbert, Minnie E.	p	Millicent E.	Jackson		9 Oct 1902	Yuba	4 Sep	1850	New York		W-Thomas A. Newbert, M-Ada Peardon
Newbert, Nellie W.	p	Nellie Wright	Wright	Wright	2 Dec 1963	Yuba	26 Oct	1887	California	Widowed	W-W. L. Newbert
Newbert, Thomas A.	p	Thomas Andrew			29 Jun 1910	Yuba	17 Sep	1830	Massachusetts		H-Minnie Newbert, F-Ada Peardon
Newbert, William				Jackson	17 Mar 1907		17 Mar	1907			
Newbert, William L.	p				31 Jan 1942	Yuba	10 Nov	1873	California	Married	H-Nellie Wright Newbert
Niles, Anthony Joseph					25 Dec 1985		20 Apr	1914			
Niles, George W.	p				1964		1876	S Dakota	Widowed	H-Theresa Jane Niles
Niles, Theresa J.	p	Theresa Jane	Kerwin		15 Dec 1957	Yuba	21 Jun	1883	Ireland	Married	W-George W Niles, D-Jane Karwin
O'Brien, Agnes M.	p				13 Dec 1954	San Francisco			California	Single	D-James K. & Mary O'Brien
O'Brien, Bridget			O'Brien		13 Jan 1866		6 Apr	1826	Ireland		W-Philip O'Brien, Native of Roscommon
O'Brien, Helen M.	p				30 Nov 1944	San Francisco			California	Single	D-James K. & Mary O'Brien
O'Brien, James	p				29 Aug 1915	Yuba	28 May	1830	Ireland	1860	H-Mary Kirby O'Brien of Massachusetts
O'Brien, James K.	p				23 May 1945	San Francisco		1864	California	Single	S-James K. & Mary O'Brien
O'Brien, Josephine	p	Anna Josephine	O'Brien	Kirby	8 Jun 1935	Yuba	29 May	1864	California	Single	D-James K. & Mary O'Brien
O'Brien, Mary Kirby	p		Kirby		13 Nov 1894	Yuba	15 Apr	1832	Ireland	1860	W-James O'Brien
O'Brien, Philip					6 Jan 1892	Yuba		1812	Ireland	Widowed	H-Mary O'Brien
O'Sullivan, Margaret A.	p	Margaret Ann			25 Apr 1905	Yuba		1836			W-Simon O'Sullivan, D-P & E. O'Sullivan
O'Sullivan, Michael	p				5 Feb 1881		5 Oct	1837			S-P & E. O'Sullivan

Smartville Cemeteries
Combined Information Analyses

Name on Headstone	P	Name in Records	Father's Name	Mother's Name	Death Date	Died	County	Born	Birth	Birth Place	Marriage	Relationships
O'Sullivan, Simon	P				14 Mar	1869			1825			S-P & E. O'Sullivan
Otis, Elizabeth								3 y				Children of Isaac B. Otis
Otis, Eugene W.								8 y				Children of Isaac B. Otis
Otis, Isaac B.	P								1838		5 Mar 1858	H-Anna Otis
Otis, M	P	Margaret			6 Jan	1878			1832			
Otis, Mrs. T. P.	P	Anna	Keith	Daniel	18 May	1867	Yuba		1833	New York		W-T.P. Otis, W-W. W. Chamberlin
Otis, Walter B.					17 Dec	1866			1847			GS-Elizabeth Daniels
Parkison, Chloe Mae	P		Parrish	DePue	5 Jul	1950	Yuba	7 Feb	1888	Iowa	Married	W-John Parkison, M-John, Bert, Bernard, Bob Parkison
Parkison, John	P			Cameron	4 Jan	1981	Yuba	9 May	1884	Iowa		H-Chloe Mae Parkison, F-John, Bert, Bernard, Bob Parkison
Pate, Derrel Austin, Jr.				Audette	10 Mar	1997	Madera	28 Feb	1959	California	Divorced	S-Derrel Pate & Karen Audette
Peardon, Ada E.		Ada Elizabeth		Newbert	21 Sep	1917	Yuba	15 Jul	1887	California		D-T. Newbert
Peardon, Elizabeth				Robbins	5 Oct	1906	Yuba		1849	England		W-John Peardon Sr
Peardon, F. G. R.		Frank G. R.			20 Mar	1973	Santa Clara	23 Sep	1888	California	28 Dec 1929	H-Dollie Peardon, S-John & Elizabeth Peardon, "Brownie"
Peardon, Gladys Elizabeth					28 Dec	1911		25 Mar	1909			
Peardon, John H, Jr.					27 May	1910	Yuba	14 Jun	1872	Pennsylvania	Married	S-John & Elizabeth Peardon
Peardon, John, Sr.	P	John H., Sr.			21 Aug	1907	Yuba	24 Aug	1843	England	Widowed	H-Elizabeth Peardon
Peckham, Miriam E.					26 Jul	1901	Yuba	14 Dec	1822	England		M-Thomas W Peckham
Peckham, Sophronia E.	P	Sophronia Elizabeth	Wallis	Staplin	16 Nov	1957	Sacramento	30 Mar	1861	California	9 Jan 1881	W-Thomas W Peckham, D-J. L. Wallis
Peckham, Thomas W.	P	Thomas William		Elizabeth	11 Apr	1935	Yuba	16 Apr	1855	California	9 Jan 1881	H-Sopronia E Peckham
Pendlebury, Mrs.								65 y				
Perez-Tuasa, Margaret Collbran	P		Collbran						1874			
Perkins, Mrs. M. J.	P	Mary Jane			22 May	1878	Yuba	12 Feb	1816	Indiana		
Petitt, Catherine T.	P								1927			
Petitt, John Joseph	P							5 Mar	1856			
Petitt, Nicholas J.	P				5 Sep	1876			1875		Child	S-Catherine & Nicholas Petitt
Pisani, Bertha E.			Magonigal	Busliby	17 Jan	1959	Yuba	10 Nov	1903	California	Married	W-Leslie Pisani, D-Thomas G. & Hannah Magonigal
Pisani, Leslie Pomp				Ronkey	12 Jun	1959	Yuba	28 Feb	1904	California	Widowed	H-Bertha E Pisani
Poole, Clarence Francis				Shehan	29 Dec	1995	Sutter	19 Jul	1905	California	28 Nov 1926	H-Hazel Mae Poole
Poole, Daisie E.	P	Daisie Elma	Cook		23 Feb	1956		27 Nov	1878			
Poole, Francis	P		Poole		13 Jun	1922	Yuba	20 Feb	1832	England	Married	H-Mary A Poole
Poole, Frank David	P				14 Jul	1948	Sutter	20 Apr	1868	California		H-Daisie Poole, S-Frank & M A Poole
Poole, Hazel Mae			Magonigal	Fraser	17 May	1994	Sutter	29 Dec	1907	California	28 Nov 1926	W-Clarence Francis Poole
Poole, Mary Ann			Keates		10 Jan	1924	San Francisco		1838	England		W- Frank Poole
Poor, Emma	P		Bartlet		14 Jan	1936	Yuba		1872	California		W-Mark Poor
Poor, Mark E.	P				7 Feb	1940	Yuba	11 Jan	1862	California		H-Emma Poor
Poor, Mark Lewellyn	P					1923			1892			
Pott, Darrell A., Jr.					10 Mar	1997		28 Feb	1959			Has his picture in headstone
Powers, Amos Mathew					12 Apr	1919		12 Jul	1832			
Powers, Mrs.												
Presley, A girl	P	Gertie			10 Jun	1877	Yuba					D-A & R R Presley, Parents Married 24 May 1876
Presley, Allen					6 Nov	1895	Marin		1846	Ireland		

Page 266

Smartville Cemeteries
Combined Information Analyses

Name on Headstone	P	Name in Records	Father's Name	Mother's Name	Death Died	County	Born	Birth	Birth Place	Marriage	Relationships
Presley, Mary					23 May	1890	Yuba	21 Aug	1867	California		
Pritchett, Jacob H.					4 Nov	1899	Yuba		1819	New Jersey	27 May 1867	H-Jutin Walsh H-Susanna R. Pritchett
Pritchett, Susanan R.	P				9 Jul	1866			1819	Pennsylvania		W-Jacob Pritchett
Pryor, Mary					28 Nov	1895	Yuba		1835	Ireland	Widowed	W-Michael pryor
Pryor, Mary Ann						1934		21 Oct	1855			
Pryor, Michael					28 Jul	1892	Yuba	24 May	1830	Ireland	Married	
Puff, Edmond J.	P	Edmond Joseph, Sr	Wotherspoon	Hogan	8 Nov	1969	Sutter	18 Sep	1905	Iowa	Married	H-Ethel W. Puff
Puff, Ethel W.	P	Ethel Wotherspoon			3 Dec	1974	Yuba	6 Jul	1907	England	Widowed	W-Edmund J. Puff
Quilty, Patrick J.	P				16 Feb	1964	Nevada	11 Jun	1886	Canada	Married	H-Thelma Quilty
Quilty, Thelma M.	P	Thelma Martha	Hinkle	Finnke	26 Mar	1986	Sacramento	7 Aug	1904	Washington	Widowed	W-Patrick J. Quilty
Quinkert, Clara C.			Campbell		4 Mar	1937	Yuba	5 Feb	1866	Michigan	Married	"Clara", W-John Quinkert
Quinkert, Elizabeth	P				6 Jan	1896	Yuba		1826	Germany	Married	"Mother", W-John Quinkert
Quinkert, John	P				26 Oct	1900	Yuba		1824	Prussia		"Father" H-Elizabeth Quinkert
Quinkert, John	P				1 Sep	1941	Yuba	12 Aug	1857	Michigan	Widowed	H-Clara Quinkert, Deceased
Radworth, Charles Jay					20 May	1871		12 Feb	1871			S-Marcus J. & A. W. Radworth
Ray, Charles Vance	P			Dryden	9 Nov	1958	Nevada	2 Jul	1889	Ohio	Married	H-Lavina Ray
Ray, Lavinia Haggood	P	Lavinia Elizabeth	Haggood	Marple	2 Jan	1996	Nevada	14 Jan	1897	California	Widowed	W-Charles Vance Ray
Reese, Kyle Lawrence	P			Turner	12 Sep	1992	Sacramento	1 Sep	1992	Sacramento	Child	S-Kevin Reese & kristi Turner
Reid, Mary					Jan	1916						
Reid, Samual	P				8 Nov	1881	Yuba		1827	Ireland		H-Mary Reid, Native of County Down, Ireland
Reigh, Alexander						1877						
Revell, Joseph	P				1 Mar	1869		18 Jun	1798			H-Martha Revell
Revell, Martha	P			Hartley	20 Dec	1870		4 Dec	1795			W-Joseph Revel
Rickey, John					27 Sep	1903	Yuba	18 May	1829	Canada		
Rickey, Peter S.					16 May	1879		29 Sep	1804			
Rigby, Albert	P			Campbell	19 Dec	1945	Yuba	27 Sep	1866	New York		H-Ida Rigby
Rigby, Frances M.	P	Frances Mary	Collins	Kelleher	30 May	1994	Yuba	2 Oct	1917	Illinois	Widowed	W-George Rigby
Rigby, George A.	P	George Albert		Dougherty	26 Jan	1992	Yuba	18 Jul	1909	California	Married	H-Frances Rigby, S-Albert & Ida Rigby
Rigby, Ida	P	Ida May	Dougherty	Brady	8 Apr	1951	Yuba	16 May	1871	California	Widowed	W-Albert Rigby, D-Daniel Dougherty
Rigby, Joseph	P	Joseph D.		Dougherty	16 Oct	1962	Sacramento	3 Aug	1906	California		S-Albert & Ida Rigby
Robbins, Nathaniel												
Roberts, C. D., Rev.	P	Charles Dillard			12 Oct	1875	Yuba	24 Apr	1838			
Robinson, Sarah A.	P				28 Nov	1874			1834			W-Captain A. Robinson
Rooney, Eugene F.	P			Saunders	24 Jul	1909	Yuba	16 Jun	1881	California	Single	S-M. & Mary Rooney, Nephew-John & Charles Saunders
Rooney, Mary E.	P				5 Dec	1899	Yuba	13 Jul	1856	New York	Married	
Rose, Ann V.	P	Ann Virginia			8 Jul	1910	San Francisco	28 Sep	1839	Virginia	1857	W-John Rose, D-Ezekial Daugherty
Rose, James E.	P	James Edward			8 Sep	1902	Yuba	6 Mar	1862	California		
Rose, John	P				7 Apr	1899	Yuba	9 Mar	1817	Scotland		H-Virginia Rose
Rose, Oscar J.	P	Oscar John		Daugherty	3 Jan	1964	Yuba	2 Mar	1872	California		S-John & Virginia Rose
Royer, Martin					29 Jun	1865		16 Sep	1855			
Royer, Sarah A.					19 Jan	187		10 Feb	1867		Child	

Smartville Cemeteries
Combined Information Analyses

Name on Headstone	P	Name in Records	Father's Name	Mother's Name	Death Died	County	Born Birth	Birth Place	Marriage	Relationships
Royer, W. J.					12 Jul 1865		13 Aug 1862		Child	
Ryan, Ellen					29 Nov 1883		17 Aug 1880		Married	W-Patrick Ryan
Ryan, Hannah M.	P				7 Aug 1907	Yuba	1847	Ireland	Married	D-Patrick & Johanna Ryan, "Ella"
Ryan, Nellie	P				29 Nov 1882	Yuba	17 Aug 1879		Child	H-Hannah M. Ryan
Ryan, Patrick	P				23 Dec 1910	Yuba	1831	Ireland	Widowed	S-Patrick Ryan
Ryan, Thomas P.	P				29 Nov 1900	Yuba	15 Feb 1878		Single	D-Daniel Sanders
Sanders, Annie B.	P	Annie D.	Sanders	Byrne	1 Feb 1913	Yuba	22 Nov 1853	New Jersey	Single	W-Daniel Sanders
Sanders, Catherine					12 Jan 1917		15 Aug 1822			H-Catherine Sanders
Sanders, Daniel		(Saunders)			7 Jul 1915	Yuba	30 Apr 1825	Ireland	Married	S-Daniel & Catherine Sanders
Sanders, John Henry					9 Oct 1938	Yuba	Oct 1858	California	Single	H-Emily T. Sanford, S-Benjamin & Euphemia Sanford
Sanford, Alfred B.	P	Alfred Bruce			16 Nov 1956	Placer	29 Nov 1869	California	11 May 1901	H-Euphemia Sanford
Sanford, Benjamin	P				24 Jul 1931		25 Apr 1832			W-Alfred B. Sanford
Sanford, Emily T.	P		Murdock	Lavers	21 Jan 1953	Yuba	27 Jun 1870	Canada	11 May 1901	W-Benjamin Sanford
Sanford, Euphemia	P		Wallace	Bruce	1 Mar 1926	Yuba	19 Mar 1830	Canada	Married	W-Wallace J Sanford
Sanford, Eva C.	P		Jones		4 Jul 1915	Yuba	7 Apr 1868		21 Nov 1887	S-Benjamin & Euphemia Sanford
Sanford, N. B.	P	Nathan Benjamin			8 Apr 1943	Sacramento	10 May 1858	California		
Sanford, Sadie E.	P	Sadie Evelyn			4 Dec 1919		23 Sep 1899			
Sanford, Sarah	P				9 May 1897			1866		
Sanford, Thomas	P				2 Mar 1888			1862		
Sanford, Wallace James				Wallace	18 Jan 1956	Modoc	22 Jul 1864	California	21 Nov 1887	H-Eva C Jones Sanford
Sanford, Wallace L.	P	Wallace Lorenzo			19 Aug 1895		5 Jan 1893		Child	
Saunders, Amelia					30 Nov 1904	Yuba	7 May 1823	Illinois		W-John Saunders, Jr.
Saunders, John G., Jr					4 Mar 1911	Nevada	1 Sep 1831			H-Amelia Saunders
Saunders, Karen J.	P	Karen Jean	Heyne	Bach	3 Oct 1987	Sacramento	30 Apr 1941	California	Single	Native of Bavaria, Germany, H-Veronica Schmidt
Schmidt, George, Jr.	P				8 Aug 1900	Yuba	6 Aug 1813	Germany		
Schmidt, Veronica			Merz		24 Jun 1852		23 Sep 1826	Germany	Widowed	D-Joseph Merz, W-George Schmidt
Schmidt, William Henry					9 Oct 1907	Yuba	27 Jan 1900			W-Thomas K Scogland
Scogland, Grace D.					5 Jan 1855		4 Dec 1884			H-Grace D Scogland
Scogland, Thomas K.	P				10 Dec 1972	Sacramento				
Shea, M. J.					27 Jun 1972	Nevada				Native of Co. Cork
Shea, Simon	P				28 Feb 1878		1843	Ireland	Single	"Brother", S-Patrick & Julia Sheehan
Sheehan, Daniel F.	P	Daniel Francis		McAuliffe	11 May 1922	Yuba	22 Feb 1862	Massachusetts	Single	"Sister", School Teacher
Sheehan, Johanna C.	P			McAuliffe	18 Nov 1941	Yuba	21 Nov 1855	Ireland	Widowed	"Mother"
Sheehan, Julia	P				7 Sep 1921	Yuba		1836	Married	H-Myrna Shields
Shields, Lloyd Vernon		Vern			20 Nov 1970	Sutter	29 Apr 1904	Illinois		
Shields, Mary Ellen			Odgen		4 Mar 1957	Sutter	4 Jul 1908	Montana		
Shilling, Charles										
Silva, Erna Jeanne	P						1942			W-Louis J. Silva Sr.
Silva, Louis J. Sr.	P				1991		1942			H-Erna Jeanne Silva
Simpson, John Clayton	P		Thomas B.	Mary H.	11 Mar 1865		29 Mar 1860			S-Thomas B. & Mary H. Simpson

Smartville Cemeteries
Combined Information Analyses

Name on Headstone	P	Name in Records	Father's Name	Mother's Nm	Death	Died	County	Born	Birth	Birth Place	Marriage	Relationships
Sims, Jane Prichard Loomis	P		Loomis	Prichard	20 Feb	1996	San Mateo	22 Apr	1914	San Mateo		
Sims, Louise Bundschu	P		Bundschu	Gundlach	15 Sep	1956	Alameda	27 Jan	1876	California		
Sims, Richard Maury	P					1935			1875			
Sims, Richard Maury, Jr.	P			Bundschu	19 Nov	1985	Marin	20 Sep	1910	California		
Sims, Robert Lee	P				3 Dec	1972	San Francisco	24 May	1912	California		
Singleton, Billie P.	P	Billie Pauline	Greever		25 Jan	1978	Sutter	31 Oct	1913	Kansas	Married	W-Pete T. Singleton, D-Kyle A. Greever
Singleton, Pete	P											
Snathurst, Elizabeth	P				15 Jun	1913	Yuba	20 May	1852	Ireland	Married	W-Geo Snathurst
Smith, Agnes	P					1864			1862			
Smith, Anthony	P				7 Aug	1898	Yuba		1823	Ireland	Married	
Smith, Bessie				Peardon	3 Nov	1905	Yuba	20 Nov	1873	Canada	Married	W-Philip Smith, D-John & Elizabeth Peardon
Smith, Catherine A.	P	Catherine Ann	Terry	Lawhon	10 Feb	1993	Butte	11 Dec	1909	Texas	Married	W-Frank Earl Smith
Smith, Frank Earl	P		Thomas		13 Nov	1981	Yuba	8 Dec	1912	California	Married	H-Catherine A Smith
Smith, Gloria J.	P	Gloria Jane	Cunning	McNabb	3 Feb	1994	San Mateo	23 Mar	1930	California		
Smith, John P.	P				7 May	1897	Yuba		1856		Widower	
Smith, Margaret	P					1910			1827			
Smith, Margaret J.	P					1873			1858			
Smith, Susan T.	P					1886			1870			
Smith, Thomas	P					1870			1865			
Smith, W. E.					30 Jul	1892	Yuba				11 Nov 1891	H-Eugenia Haggood Smith (later Filcher)
Smith, William C.	P					1926			1864			
Snelt, Mrs.						1877						
Spargo, T. J.		Thomas Jeffery			11 Dec	1876	Yuba		1844	England		
Spencer, Carl Allen	P		Wheaton		10 Sep	1935	Yuba	11 Sep	1917	California		S-Arthur & Martha spence
Stanton, William	P				2 Nov	1872			1830			
Stone, Charlie E.	P				24 May	1866		26 Nov	1860		Child	S-W H & P. L. Stone
Stone, Willie A.	P				10 Jan	1865		7 Aug	1858		Child	S-W H & P. L. Stone
Sweeney, Annie		Ann			3 Mar	1908	Yuba		1833	Ireland	Widowed	W-Michael Sweeney
Sweeney, John					20 May	1893	Yuba			Ireland		Native of Tralee, County Kerry
Sweeney, Michael	P				12 Apr	1877	Yuba	4 Mar	1826	Ireland	Married	Native of Easkey Co., Sligo, H-Ann Sweeney
Taylor, Edward Roy	P				16 Aug	1975	Yuba	30 Apr	1942	California		S-Georgina Louise Welch
Taylor, Frank L.	P	Frank Wesley			20 Jan	1916	Nevada	5 Sep	1884	California		With William H Bowman
Teske, Edna					29 May	1888						
Thrush, George W.					5 Aug	1885	Yuba		1820			H-Sarah Arvilla Thrush
Thrush, Sarah Arvilla			Roberts		25 Dec	1887	Yuba	25 Sep	1873			W-George W. Thrush
Tifft, Alice M					13 May	1857		11 May	1856		Child	
Tifft, George W.					23 Jan	1909		1 Dec	1870		Child	
Tifft, Julia A.		Julia Amelia	Tifft		23 Jul	1918	Alameda	22 Sep	1835	California	21 Aug 1889	W-Simon A. Davis
Tifft, Julia A.												W-R W Tifft
Tifft, Ray W.		Ray Wilkins			25 Feb	1893	Yuba	23 Nov	1831	Ohio		H-Julia Tifft
Toland, Elizabeth	P		Bach			1939			1871		24 Feb 1901	W-John Toland, D-John Bach

Page 269

Smartville Cemeteries
Combined Information Analyses

Name on Headstone	P	Name in Records	Father's Name	Mother's Name	Death	Died	County	Born	Birth	Birth Place	Marriage	Relationships
Toland, John	P	John E.		Hunter	9 Aug	1943	Alameda	12 Sep	1872	California	24 Feb 1901	H-Elizabeth Toland, S-Robert Toland
Toland, Joseph					16 Sep	1880		3 Feb	1878			
Toland, Martha					13 Jul	1916	Yuba	22 Aug	1832	Ireland		W-Robert McKeague, W-Robert Toland
Toland, Robert					26 Apr	1905	Yuba		1832	Ireland	1868	
Toland, Robert	P					1938			1903			
Toland, William Joseph					26 Jan	1918	Yuba	18 Dec	1869	California		
Travena, Mrs.		Ann Trevena			12 Jan	1907	Nevada		1823	England		W-Nicholas Travena, Sr.
Travena, Nicholas	P	Nicholas Sr.			24 Jun	1877	Yuba	23 Apr	1820			
Travena, Nicholas, Jr.					3 Dec	1905	Yuba					
Trembly, Kay M		Katherine M.			10 Mar	2000		3 Nov	1923			W-Keith D Trembly
Trembly, Keith D.		Keithen Dale		Kelly	28 Nov	1982	Sacramento	18 Nov	1916	Nebraska		H-Kay M Trembly
Twomey, Rev. Andrew T.	P				8 Mar	1902	Yuba		1866	Ireland		Native of Clonakilty, County Cork, Catholic Father
Vargas, Albert Dee					21 Mar	1987		24 Mar	1917			
Vargas, M. Ginger					13 Mar	1981		2 Aug	1918			Double Hdstone with Sandy Webb
Verdier, Ernest E.	P	Ernest Edmund		Zimmerman	24 May	1997	Alameda	6 Apr	1904	California		
Verdier, Lucille M.	P	Lucille Maxine	Davies	Black	14 Mar	1986	Alameda	21 May	1905	California		
Vineyard, Alice Elizabeth					9 Jul	1891	Yuba	17 Apr	1854			W-William Vineyard
Vineyard, Allen E. T.	P				8 Oct	1881		8 Nov	1875			
Vineyard, Almora G.	P				12 Mar	1877		24 Jan	1849			
Vineyard, John T.					12 Oct	1920	Yuba	16 Jul	1826	Wisconsin	May 1913	H-Sara Eliza Woodroffe Vineyard
Vineyard, Mary E.	P	Mary Emily	Hammonds	Meredith	7 Jul	1908	Yuba	7 Feb	1842	Wisconsin	May 1913	W-J T Vineyard
Vineyard, Sarah E. Woodroffe	P	Sarah Eliza	Woodroffe			1928			1848			W-S. A. Davis, W-J. T. Vineyard
Vinsonhaler, Howard F.				Hutchinson	20 Sep	1993	Yuba	8 Sep	1914	Kansas	Married	U.S.AirForce WWII, H-Mildred L. Marrie
Wah, Yirtg Yim					6 Jan	1909			1834			
Waldron, Henry					8 Nov	1867	Yuba		1823	Vermont		
Walker, John		John C.			2 Jan	1899	Yuba		1859			Brother was E. B. Walker of Sacramento
Wallace, Ellen	P				16 Dec	1891	Yuba		1803	Scotland	Widowed	
Wallace, Michael	P	Michael A.			3 Jan	1921	Yuba		1854	Canada		H-Ellen Wallace, Native of West Gore, Nova Scotia
Walsh, Anna					14 Nov	1912	Yuba		1845	Ireland		W-Michael Walsh
Walsh, Anna Kirby	P					1883			1838			W-G. W. Snelhurst
Walsh, Annie Quick	P	Anna A.	Quick		9 Apr	1923	Yuba		1860	California	Married	W-John E Walsh, See obituary for children
Walsh, Charles	P				13 Jan	1874			1873		Child	
Walsh, Edward	P				9 Jan	1904	Yuba	8 Jan	1864	California	Single	
Walsh, John	P				9 Jun	1902	Yuba		1828	Ireland	Widowed	H-Mary Walsh
Walsh, John E.					20 Apr	1911	Yuba		1824	Ireland		H-Margaret Walsh
Walsh, Maggie	P	Margaret Ellen		Dougherty	29 Apr	1934	Yuba	20 Jul	1864	California	13 Apr 1880	H-Annie Quick Walsh
Walsh, Mary	P		McClosky		14 Oct	1913	Yuba		1851	Louisiana	Married	W-John Walsh
Walsh, Philip	P		Dougherty		15 Feb	1901	Yuba		1829	Ireland	Single	W-John Walsh
Walsh, Richard					30 Nov	1899	Butte City, MT	13 Jun	1866	California		S-John Walsh
Walsh, Richard					6 Dec	1878	Yuba		1846	Ireland		Came from North Hampton, Mass before Smartsville
	P				6 Jan	1879			1847	Ireland		native of Co. Waterford

Smartville Cemeteries
Combined Information Analyses

Name on Headstone	P	Name in Records	Father's Name	Mother's Name	Death Died	Died	County	Born Birth	Birth Place	Marriage	Relationships
Walsh, Richard	P					23 Jan	1890	Los Angeles	1820		H-? O'Brien Walsh
Walsh, Sadie Kane	P	Sadie Matie	Kane		27 May	1984	Pierce, WA	18 Aug	1892 Ireland	Widowed	W-Walter L. Walsh
Walsh, Thomas					1 Jul	1889	Yuba	55 yrs	1834 Ireland		
Walsh, Walter Lucien	P				24 Aug	1975	Santa Clara	27 Jul	1890 California	Married	H-Sadie K Walsh
Ward, Caroline Hazel	P										S-W. W. & Mary E. Ward
Ward, Infant Son					7 Jul	1885	Yuba	24 May	1882	Child	D-W. W. & Amanda Ward
Ward, Mary E.	P								1853		D-W. W. & Mary E. Ward
Ward, Mary E.	P				22 Aug	1875	Yuba			Married	W-W. W. Ward
Warne, J. H.	P	James Henry			13 Feb	1930	Yuba	24 Mar	1851 England	Married	H-Mary J. Warne
Warne, Mary					7 Sep	1909	Nevada		1837		W-J. H. Warne
Weaver, Marie Cora	P				31 May	1978	Nevada	23 Oct	1898 California		
Weaver, Warren W.	P	Warren Wayfield			30 Dec	1978	Nevada	26 Apr	1898 California		
Webb, Alice Toland	P					1987			1910		In Toland Family Plot
Webb, Sandy M.	P	Sandra Marie			15 May	1980		12 May	1963		Deouble Hdstone with M. Ginger Vargas
Webb, Benjamin F.	P				18 Dec	1876		24 Sep	1865		S-Radford E & Serena J Welch
Welch, Chester A.	P	Chester Andrew		Walker	20 Jan	1937	Yuba	14 Apr	1936 California	Child	S-D. N. & Katherine Welch
Welch, Clarence R., Jr.	P	Clarence Ray, Jr.		Kennedy	22 Dec	1967	Sierra	16 Mar	1934 California	Married	H-Georgia L. Welch, S-Clarence Roy & Winifred Helen Welch
Welch, Clarence R., Sr.	P	Clarence Ray, Sr.		Jones	2 Jun	1970	Yuba	13 Jan	1896 California	Widowed	H-Winifred Helen Welch
Welch, Georgina Louise	P		Gnadig	Nathan	28 Aug	1985	Yuba	28 Dec	1924 California	Widowed	M-Edward Roy Taylor
Welch, James	P	James V.		Bast	23 Jan	1923	Yuba	28 Apr	1861 California	14 Mar 1882	H-Lillie E Welch, See obituary for children
Welch, James C.	P	James Chester			24 Jan	1953	Placer	9 Dec	1886		PFC Medical Department - WWI
Welch, James R.	P	James Radford		Kennedy	29 Sep	1974	Yuba	23 Apr	1932 California	25 Aug 1961	H-Sharlene Joyce Welch, PFC U. S. Army
Welch, Joseph B.					29 Nov	1936		4 Nov	1867 California	30 Oct 1898	H-Alice May Reed Welch
Welch, Lillie E.	P	Lillie Elizabeth	Jones	Young	11 May	1938	Yuba	14 Jul	1866 Yuba	14 Mar 1882	W-James Welch
Welch, Radford E.					24 Feb	1893		27 Nov	1826		H-Serena J Welch, F-Benjamin Welch
Welch, Serena Jane				Bast	22 Feb	1893	Yuba	25 Feb	1830 Missouri		W-Radford E Welch, M-Benjamin, James,Joseph,Thomas Welch
Welch, Thomas H.		Thomas Hamilton			29 Sep	1903	Yuba	15 Jan	1847 Missouri	Single	S-Radford E & Serena J Welch
Welch, William Gale					23 Apr	1922	Sutter	24 Aug	1851 Missouri		S-Radford E & Serena J Welch
Welch, Winifred H.	P	Winifred Helen	Kennedy	Groves	25 Jun	1964	Yuba	9 Apr	1914 California	Married	W-Clarence Roy Welch, Sr.
Westerfield, W. P.	P	William P.			Oct	1888		May	1820		
Whalen, Jeremiah, Mrs.					4 Aug	1888			1806	Married	
Wheaton, Allan W.	P	Allen Wallace				1966			1900		
Wheaton, Allen G											F-May Eiden
Wheaton, Evelyn											
Wheaton, Florence Esther					14 Sep	1903		12 Jan	1903	Child	
Wheaton, Lucille											
Wheaton, Robert	P	Robert D.			25 Aug	1929	Sacramento	30 May	1894 California		S-A. G. Wheaton
Whitmore, Alice E.	P	Alice Edna	Montgomery	Freend	25 Nov	1950	Yuba	29 Mar	1909 Iowa	Married	W-Jesse Franklin Whitmore
Whitmore, Jesse F.	P	Jesse Franklin		Davis	2 Oct	1958	Yuba	9 Jun	1891 Texas	Widowed	H-Alice Edna Whitmore
Widener, Alpha Frances					18 Jun	1921	Yuba	7 Jul	1852 Tennessee		W-Russell Widener, M-William Widener
Widener, Russell	P	Russell M.			22 Jun	1899		Jan	1827		

Smartville Cemeteries
Combined Information Analyses

Name on Headstone	P	Name in Records	Father's Name	Mother's Name	Death	Died	County	Born	Birth	Birth Place	Marriage	Relationships
Widener, William	P	William Everett			26 Jan	1908	Yuba		1883	California		
Williams, James O.	P	James Ogden			20 Sep	1879	Yuba	21 Mar	1850	Missouri	29 Oct 1873	H-Lizzie Lowe Williams
Williams, James Paul	P			Vinsonhalen	11 Mar	1976	Yuba	7 Nov	1971	California	Child	S-Joe & Christine Williams
Williams, Joseph	P	Joseph Matthew		Vinsonhalen	11 Mar	1976	Yuba	13 Jun	1968	California	Child	S-Joe & Christine Williams
Williams, Juliana Rebecca	P		Williams	Vinsonhalen	11 Mar	1976	Yuba	28 Jan	1975	California	Child	S-Joe & Christine Williams
Williamson, Linda			Lane		10 Oct	1993		27 Sep	1948			
Williamson, Robert Brush				Walmsley	22 Dec	1966	Yuba	8 May	1913	Nevada		
Wilson, Linda	P								1937			
Wilson, Roy Wayne, III	P				17 Dec	1983		6 Mar	1961			PVT U. S. Army
Wilson, Theadore					20 Dec	1863			1829	Maine		
Winegar, Madge A.	P						2001		1919			
Winegar, Marvin E.	P	Marvin Earl		Wilson	1 Jun	1993	Nevada	21 Jul	1939	California		H-Petra M. M. Winegar
Winegar, Patricia M.	P	Petricia Marie	Rodriquez	Warner	23 Aug	1994	Alameda	29 Feb	1944	California		
Winegar, Petra M. M.	P							11 Jan	1953			W-Marvine E. Winegar
Wirth, Fredrick C.		Fredrick Charles	Fontaine	Bishop	13 May	1968	Contra Costa	30 May	1891	Pennsylvania	Married	H-Mable R Wirth
Wirth, Mable R.		Mable Ruth		Sarsfield	4 Mar	1981	Butte	30 Jul	1895	Massachusetts	Widowed	W-Frederick C Wirth
Wiseman, Sara C.		Sara Constance			24 May	1986	Sacramento	28 Oct	1983	Arizona		"Poncharelli"
Woehler, Otto A., Jr	P				12 Feb	1922		12 Jul	1896			S-Otto A. Woehler, Sr.
Woehler, Otto A., Sr	P				22 May	1924		1 Jan	1859			F-Otto A. Woehler, Jr.
Woodroffe, Bruce D.							1935		1884			
Woodroffe, Earl E.					17 Jun	1962	Yuba	23 Aug	1893	California		
Woodroffe, Janie E. Cooley	P	Janie E.	Gilbert	Cooley	24 Sep	1972		10 Sep	1881			W-Sherman E. Woodroffe
Woodroffe, John S.	P				26 Jun	1903	Nevada		1837			F-Annie Woodruffe
Woodroffe, Pearl	P				10 Apr	1886			1886		Child	
Woodroffe, Sherman E.	P			Loomer	31 Jul	1923	Yuba	16 Apr	1880	California	Married	H-Janie Woodroffe, S-J T Vineyard, Obituary for Children
Woods, W. H.					1 Nov	1882			1835			
Wright, Adam	P	Adam Winfield			25 Jan	1921	Yuba	2 Oct	1844			In the Murphy Fam Plot
Wright, Mabel Bennett	P						2001		1903			
Yankey, Roberta Beatty Bobbe	P	Roberta	Beatty	Manasco	24 Jun	1995	Solano	29 Jul	1932	California	Married	W-John S. Yankey, III, "Bobbie"
Young, James					13 Dec	1903	Yuba	Jul	1836	Illinois		B-Mrs. D. N. Jones
Young, John	P	John J			27 Jun	1908	Yuba		1841	New York	Widowed	
Zssisloft, Christopher Scott	P				2 Jan	1993	Sacramento	10 Apr	1971	New Jersey	Married	H-Dianna Zssisloft

About the Author

RENIE RICCOBUANO started doing research in this and other areas fifteen years ago when she volunteered as a librarian at the Family History Center for the Church of Jesus Christ of Latter-day Saints. Since that time she has become Director for that institute, and has, for the past ten years, conducted yearly seminars that are widely attended. In addition to the seminar classes, she also teaches introductory classes that focus on research techniques in her community and other related subjects.

www.ingramcontent.com/pod-product-compliance
Lightning Source LLC
Chambersburg PA
CBHW070726160426
43192CB00009B/1328